Pam'

Rec
197

CW01192822

Willie Harrigan

Laboratory Methods in Food and Dairy Microbiology

Laboratory Methods in Food and Dairy Microbiology

W. F. Harrigan

Department of Food Science,
University of Reading, England

and

Margaret E. McCance
Girvan, Ayrshire, Scotland

Revised Edition prepared by W. F. Harrigan

1976
Academic Press
London New York San Francisco
A Subsidiary of Harcourt Brace Jovanovich, Publishers

ACADEMIC PRESS INC. (LONDON) LTD
24/28 Oval Road
London NW1

United States Edition published by
ACADEMIC PRESS INC.
111 Fifth Avenue
New York, New York 10003

Revised Edition of "Laboratory Methods in Microbiology", 1966

Copyright © 1976 by
ACADEMIC PRESS INC. (LONDON) LTD.

All Rights Reserved

No part of this book may be reproduced in any form by photostat, microfilm, or any other means, without written permission from the publishers

Library of Congress Catalog Card Number: 76 1083
ISBN: 0 12 326040 X

PRINTED IN GREAT BRITAIN BY
UNWIN BROTHERS LIMITED, OLD WOKING, SURREY

PREFACE to "Laboratory Methods in Microbiology"

This laboratory manual is based on the experience of the authors over several years in devising and organizing practical classes in microbiology to meet the requirements of students following courses in microbiology at the West of Scotland Agricultural College. The primary object of the manual is to provide a laboratory handbook for use by students following food science, dairying, agriculture and allied courses to degree and diploma level, in addition to being of value to students reading microbiology or general bacteriology. It is hoped that laboratory workers in the food manufacturing and dairying industries will find the book useful in the microbiological aspects of quality control and production development.

Part I is concerned with basic methods in microbiology and would normally form the basis of a first year course. Abbreviated recipes and formulations for a number of typical media and reagents are included where appropriate, so that the principles involved are more readily apparent.

Part II consists of an extension of these basic methods into microbiology as applied in the food manufacturing, dairying and allied industries. In this part, the methods in current use are given in addition to, or in place of, the "classical" or conventional techniques. For example, in Part I, quarter-strength Ringer's solution is mentioned as the standard diluent for counting procedures, but in Part II it is recognized that this has been superseded by reinforced clostridial medium as the diluent for anaerobic counts, that diluents for halophilic counts need to contain 15 per cent sodium chloride and that dilute peptone water may be preferred to quarter-strength Ringer's solution for some types of aerobic count.

The appendices are devoted mainly to procedures for the indentification of those micro-organisms most commonly encountered in the industries mentioned, and to recipes for the preparation of media, reagents and stains. It is often of value in investigations into factory processes and in quality control and production development to identify the organisms isolated. Such identification frequently is not carried out, as the use is required of time-consuming dichotomous keys or of diagnostic tables involving a large number of tests. Both keys and tables are here combined to permit rapid identification of the majority of probable isolates (to generic or, occasionally, specific level) using a minimum number of tests. We hope that this may encourage laboratory workers to attempt more frequent identification of their isolates, and believe that this would be sufficient justification for the loss of accuracy involved in the simplification of the diagnostic keys. After determining the genus to which an isolate belongs, reference can then be made to works such as "Identification Methods for Microbiologists", edited by B. M. Gibbs and F. A. Skinner (Academic Press, London, 1966), or other books as indicated in the text, for identification to specific level.

We should like to express our thanks to Dr. S. Baines, the head of the Department of Bacteriology, The West of Scotland Agricultural College, for his co-operation and for the facilities he has provided. We should also like to acknowledge the assistance and helpful comments of our colleagues in the bacteriology department and in other departments of the College. We are grateful to Dr. R. W. A. Park and Dr. A. J. Holding for their help with the diagnostic key on Gram-negative bacteria and for allowing us access to the manuscript of their paper prior to its publication. The probability tables in Appendix 3 are reproduced with the permission of the Controller of Her Majesty's Stationery Office.

Auchincruive
July 1966

W. F. HARRIGAN
MARGARET E. MCCANCE

PREFACE

The preceding edition fulfilled our hopes in proving to be helpful to food microbiologists working in quality control laboratories in the food industry, in addition to being used as a laboratory handbook for students reading food science and allied subjects. This Revised Edition has been extensively reorganised and up-dated. The sections on counting methods, interpretation of counts, techniques for selective isolation and the identification schemes for bacteria and fungi have been expanded and considerably modified. It is hoped that these changes will make this book a valuable aid for all those engaged in microbiology in the food industry.

Since *Laboratory Methods in Microbiology* appeared several works have been published of great significance to food microbiologists. These include the two volumes on methodology published by the International Commission on Microbiological Specifications for Foods (of the International Association of Microbiological Societies); the 8th edition of "Bergey's Manual of Determinative Bacteriology"; the *Technical Series* of the *Society for Applied Bacteriology*; and the 8-volume "Methods in Microbiology" (edited by J. R. Norris and D. W. Ribbons).

In preparing this edition, once again it has been written with a view to being used for daily bench-top reference. The book is intended to provide adequate information for the assessment of the microbiological quality of foodstuffs or of equipment in the majority of quality control situations. For those occasions when further information be required on particular topics detailed references have been provided, especially to the works named above.

I should like to express my thanks to Margaret McCance for a number of contributions and suggestions, especially in the section on the examination of food processing plant. Dr. R. W. A. Park read the manuscript and suggested very many valuable amendments throughout the work, for which I am most grateful. Nevertheless I alone am responsible for all errors and omissions. I am indebted to my wife, Rita, for her tremendous encouragement and assistance during all stages in the preparation of the manuscript, and for her help in checking typescripts and proofs, and in the preparation of the index.

The diagram on page 153 is reproduced with the permission of Dr. R. W. S. Harvey, Mr. T. H. Price and the Cambridge University Press.

Reading
July 1976

W. F. HARRIGAN

CONTENTS

Part II
TECHNIQUES IN APPLIED MICROBIOLOGY

Part III

SCHEMES FOR THE IDENTIFICATION OF MICRO-ORGANISMS

Appendix 1

RECIPES FOR STAINS, REAGENTS AND MEDIA

Appendix 2

PROBABILITY TABLES FOR THE ESTIMATION OF MICROBIAL NUMBERS BY THE MULTIPLE TUBE TECHNIQUE

Appendix 3

Appendix 4

PART I

BASIC METHODS

SAFETY PRECAUTIONS IN THE MICROBIOLOGICAL LABORATORY

The laboratory worker is exposed to many hazards. In the laboratory there are chemicals which are toxic, flammable, explosive, corrosive or carcinogenic, and on occasions there are dangers from the use of high voltages, ultra-violet and other radiation. These hazards are described in "Hazards in the Chemical Laboratory" (edited by G. D. Muir, published by The Royal Institute of Chemistry) and it is recommended that a copy of this book or its equivalent should be held in a prominent place in every laboratory. In addition to these hazards, microbiologists are also exposed to hazards from the micro-organisms with which they are working.

It should always be assumed that the micro-organisms with which you are working are capable of causing disease—the assumption will often be true. Great care should therefore be taken in handling cultures, slides, and all material that may contain, or have been in contact with, living micro-organisms. Remember that the main routes of entry of infection to the body are: by inhalation, by ingestion, through cuts and abrasions, and by infecting the eyes (the eye can serve as the portal of entry for infections which do not produce a local pathology). A few types of organisms (e.g. *Brucella*) can enter through the unbroken skin.

1. Laboratory coats must be worn. (For microbiological work, a surgical gown is preferable to the traditional front-opening laboratory coat.) Ideally the laboratory coat should be removed before leaving the laboratory area, and other protective clothing used when entering a processing area for the purpose of taking samples.

   When highly infectious organisms are likely to have been encountered disposable surgical gloves should be worn, the laboratory coat should be autoclaved before being sent for laundering, and the gloves decontaminated by autoclaving.
2. Do not eat, drink or smoke in the laboratory.
3. Labels should be of the self-adhesive type to avoid the temptation of moistening gummed labels with the tongue.
4. Accidents such as spilled cultures, cuts and abrasions should be reported or written into a book kept for the purpose. Existing cuts and abrasions should be adequately covered with a waterproof dressing; if they occur in the laboratory make sure that suitable first aid treatment is received. A spilled culture should be flooded with a suitable disinfectant solution (e.g. an iodophor), and this left for 15–30 minutes before clearing up.

   It must be emphasised that the absence of breakage does not imply that no danger exists since the dropping of a culture in a plastic Petri dish, for example, can result in the release of a microbial aerosol into the atmosphere. The microbial aerosol is dangerous because it can reach a highly susceptible target—the lung—undetected, and in the lung can produce maximum effect in low dosage. When a culture is dropped on the floor you should therefore not bend down to clear it up immediately, but treat the area with a bactericidal solution and leave time for a reduction in the concentration of any aerosol which may have been generated.
5. Remember that aerosols can also be generated using pipettes, wire loops, and even during the removal of a screw cap or a rubber bung from a culture.
6. Sporulating cultures of fungi offer hazards of respiratory infection or allergic reaction even in the absence of aerosol generation and should be handled slowly and without sudden movement in a draught-free atmosphere. (Pathogenic fungi are best handled in suitable inoculation chambers—see item 12.)
7. The use of a wire loop requires considerable skill if risk of contamination of the air and the working area is to be avoided. Inoculating needles and loops must be sterilised before and AFTER use, by heating in a Bunsen flame until red hot along the entire length of the wire. Spattering of material from the wire should be avoided by very gradual introduction into the Bunsen flame if hooded burners or safety loop sterilisers are not being used.

8. Test tube cultures should always be kept in test tube racks. Never lay the test tubes on the bench top.
9. A plate count on a food sample prepared using an "innocuous" medium such as nutrient agar or milk agar cannot be regarded as harmless merely because the food was suitable for consumption. Many pathogens such as *Staphylococcus* and *Salmonella* can grow on such media. The micro-organisms originally present in the food in small numbers are now in the plate culture in billions, and can cause infection by inhalation.
10. The cotton wool plug in a pipette is there to prevent the contamination of the liquid being pipetted, and not to prevent infection of the user. Cultures should never be pipetted by mouth, but with rubber teats or bulbs used in conjunction with the pipette, and this precaution should be taken when pipetting samples or dilutions of samples which may contain dangerous pathogens or toxins. A pipette controller which can be used for this purpose and which is autoclavable is the "Volac" (manufactured by John Poulten Ltd.).
11. Rapid and forceful ejection of the contents of a blow-out pipette can produce an aerosol. In general slow and unhurried movements are to be preferred in microbiological work, but with minimum delays between operations.
12. Used pipettes must be placed in pipette jars containing disinfectant solution. Microscope slides and coverslips must also be discarded into jars of disinfectant solution, cover slips first being separated from the slides.
13. Homogenisers and blenders must not be used in conjunction with bacterial *cultures* without adequate precautions against the spread of air-borne contamination. Their use to obtain dilutions of food samples will rarely present a hazard because of the low microbial concentrations encountered.
14. Positive pressure inoculation chambers are used for sterility testing. They MUST NOT be used for working with microbial cultures. (Dangerous pathogens are handled in either completely closed inoculation chambers or in *negative* pressure inoculation hoods whose air outlets incorporate suitable filters to remove organisms and to render the exhaust air safe.)
15. On completion of work in the laboratory, the work area should be cleared. Contaminated items and cultures no longer required should not be left on the bench but placed on trolleys designated for items requiring decontamination by autoclaving, etc. The work area should finally be swabbed down with an appropriate disinfectant solution (e.g. an iodophor).
16. The hands should be thoroughly washed before leaving the laboratory.
17. There should be at least one staff member on duty at any given time who

has passed a recognised course in first aid and who is trained to provide both respiratory resuscitation by mouth-to-mouth and Silvester methods and external heart compression in the event of cardiac arrest. Every laboratory worker should know the identity and location of "first aiders". It is good practice for such people to be identifiable by a red cross worn on their laboratory coats.

18. The location of first aid cabinets, eye irrigation bottles and fire extinguishers should be known by all laboratory workers. A list of the specific hazards presented by the chemicals and the micro-organisms in the laboratory should be determined and made available, so that in the event of any accident full information can be provided to the doctor or hospital.

(See also Darlow, 1969; Shapton and Board, 1972.)

LABORATORY REPORTS

All work carried out in the laboratory should be recorded fully in a laboratory report book. If you do not record work in the laboratory report book at the time it is performed, very full notes should be taken in a laboratory records book. It cannot be emphasised too strongly that the notes made in a laboratory records book should be as full as possible. Often, during the course of an experiment, observations are made which are not required in a report bearing on a particular aspect of the work. It may be that, at a much later date, these apparently superfluous observations will prove extremely useful. This is particularly the case when a research project is being carried out, since a rather narrow line must frequently be followed, with the necessity at the time to ignore many observations and results which suggest follow-up experiments.

It is always a good policy not to destroy laboratory records books. In the quality control laboratory this will enable reports to be substantiated to a certain extent should the need arise. In this connection, it is perhaps advisable to use bound books, not loose-leaf folders, and to date all entries.

Reports of experiments and tests performed should be written up in a standard form. For example:

1. Title of the experiment or test performed and the date.
2. (*a*) The object of the experiment or test.
   (*b*) Summary of results and conclusions.
3. An outline of the methods used and, if stock cultures were employed, the names and brief descriptions of the organisms. When the methods depart from those detailed in the manual of methods customarily used, they must be described in full.
4. A description of the results obtained. This is often best given as a written report with tables or graphs used to clarify the results, but in some cases tables with any necessary footnotes may be more suitably employed to give the results obtained.
5. A statement of the conclusions which can be drawn from the results.

Laboratory reports on quality control work which are being issued to factory managers should be drafted in a form which will provide easy access to the substance of the report and to any recommendations. Thus it is useful to write a summary of the findings, and recommendations for action, towards the beginning of the report rather than at the end. The laboratory methods used will ordinarily not be described at all, but full information should be given of the sources of the samples and the reasons for choosing them. Results are best given as grades (e.g. A, B, C, D), with the actual counts being retained in the laboratory records books. The preferred system for recording counts is discussed on pages 29, 36 and in relevant sections of Part II.

In research reports references should be given in a manner similar to that used in this book. If the original paper has not been seen, then the reference of the source of the citation should be given.

PROCEDURE FOR THE USE OF A MICROSCOPE WITH AN OIL-IMMERSION OBJECTIVE

1. First ensure that objectives, eye-piece, condenser and mirror are clean, removing any dust with a soft brush and lens tissue.

2. Align the microscope and artificial light source. If an external light source is used place the lamp about 15 cm from the mirror.

3. Adjust the draw tube to the correct length (usually this will be 160 mm when a microscope with a rotating nose-piece is used).

4. With the ×10 objective in position and the eye-piece removed, adjust the plane side of the mirror to send the light centrally up the microscope draw tube.

5. Replace the eye-piece and place the object slide on the microscope stage. Focus the ×10 objective on the object using the coarse adjustment.

6. Focus the light source on the object by placing a pencil close against the light source and racking the condenser up and down to get a sharp image of the pencil in the same field as the focused object. (This is termed "critical illumination", the image of the light source being in the same plane as the focused specimen.)

7. Using ×10 and/or ×40 objectives select a suitable field for subsequent viewing with the high power (×90 or ×100) oil-immersion objective.

8. (*a*) Rack up the objective (this may be unnecessary if parfocal objectives are fitted) and rotate the oil-immersion objective into position.

(*b*) Place one drop of immersion oil on the microscope slide.

(*c*) Using the coarse adjustment, gently and very slowly lower the ×90 (or ×100) oil-immersion objective until the oil layer is flattened without the front of the lens touching the slide.

(*d*) Whilst viewing the object through the microscope, rack very slowly upwards with the coarse adjustment until the specimen comes into view. The restriction in the use of the coarse adjustment to upward travel only, while looking down the microscope, lessens the danger of slide breakage and consequent damage to the objective lens. Focus sharply with the fine adjustment.

9. Remove the eye-piece again and adjust the mirror to ensure that the back lens of the objective is symmetrically filled with light. To avoid glare, close the iris diaphragm until the back lens of the objective is about $\frac{3}{4}$ full of light.

10. Replace the eye-piece and refocus with the fine adjustment if necessary.

After use, all immersion oil MUST be wiped off the objective lens and elsewhere with lens tissue. Failure to remove the immersion oil may allow the oil to penetrate the lens mount and cause a severe loss of definition.

Procedures for setting up phase contrast microscopes and dark field microscopes are described, *inter alios*, by Quesnel (1971).

EXAMINATION OF CULTURES FOR MOTILITY BY "HANGING DROP" PREPARATIONS

Cultures for examination should be broth cultures 18–24 hours old. Alternatively, a small amount of culture from an 18–24 hour agar slope can be emulsified gently in a drop of broth or normal saline, taking care that the emulsion is not too dense. By the following procedure, a drop of culture is suspended from a coverslip over the depression in a hollow-ground slide.

First place a little immersion oil round the edge of the depression in the slide. Then with a wire loop transfer a small loopful of the culture to a clean dry coverslip laid on the bench. Do not spread the drop.

Invert the cavity slide over the coverslip so that the drop is in the centre of the cavity and press the slide down gently but firmly so that the oil seals the coverslip in position. Invert the slide quickly and smoothly and the drop of culture should now be in the form of a hanging drop. The preparation should be examined without delay and as quickly as possible.

When examining hanging drop preparations the substage condenser of the microscope should be racked down slightly from its normal position and the iris diaphragm should be partly closed. Not only does excessive illumination render the unstained organisms invisible, but also the heating effect may cause them to lose their motility.

First, use the low power objective to focus on the edge of the drop, moving the slide until the edge of the drop appears across the centre of the field. This should be easily recognised since minute droplets of condensed water can usually be seen on the other side of the line which represents the edge of the hanging drop. Then place the high power ($\times 40$) *dry* objective in position and refocus the edge of the drop. The bacteria should now be easily seen, particularly towards the edge of the drop where the reduction in the depth of liquid assists in keeping the organisms within the depth of field. The oil-immersion objective should *not* be used since the focusing movements of the objective would be mechanically transmitted to the coverslip causing streaming in the culture liquid which would seriously hinder observation and may even be misinterpreted by an untrained observer as motility of the organisms.

It is necessary to distinguish between Brownian movement (a continuous agitation of very small particles suspended in a fluid, which is caused by unbalanced impacts with molecules of the surrounding fluid) or drift in one direction caused by the slide being slightly tilted, and true motility.

STAINING METHODS

A. Preparation of Smears for Staining

Smears of bacteria from cultures on solid media are made upon clean glass slides as follows.

1. The slide may be divided up into sections, one section for each smear, using a wax pencil or a diamond.
2. With a wire loop, place a small drop of water on each section of the slide.
3. Sterilise the wire by holding it vertically in the Bunsen flame until it is heated to redness along its entire length.
4. Allow the loop to cool.
5. Holding the loop like a pen, pick off a little of the bacterial growth. Try to avoid transferring any agar medium.
6. Transfer to a water drop on the slide and, using the flat of the loop, emulsify the growth, finally smearing it evenly over the area of the slide allocated to it. Aim at obtaining *thin*, evenly spread smears which are almost too thin to be seen when dry. The individual bacteria will then be well spaced for examination under the microscope. If the suspension does not spread evenly over the slide, but collects in small droplets, the slide is greasy and should be discarded.
7. Sterilise the loop.
8. Prepare the other smears in the same manner.
9. Leave the slide to dry in the air, then heat-fix by passing the slide twice through the Bunsen flame. This coagulates the cell contents and causes the bacteria to adhere firmly to the slide.

Smears of bacteria from cultures in liquid media are made in a similar manner except that dilution with water is not required. The preparations may be less clean since some ingredients of the medium will remain and be stained.

B. Simple Stains

Place the heat-fixed smear on a staining rack over the sink and flood with any of the staining solutions given below. After the time indicated, rinse the slide gently in water and blot dry with clean blotting paper or filter paper. Examine as described on page 9.

(*a*) *Crystal violet:* stain for 1 min.

(*b*) *Loeffler's methylene blue:* stain for at least 5 min as this stain is weaker in action than crystal violet.

(*c*) *Carbol fuchsin* (*dilute*): stain for 30 sec only.

C. Gram's Staining Method

This is a differential double-staining method which forms the basis of most examinations and the preliminary identification of bacteria. In this method, bacteria are first stained with crystal violet and are then treated with iodine solution. The bacterial smears are next treated with ethanol or acetone, which entirely removes the violet stain from Gram-negative bacteria, but not from Gram-positive bacteria. Dilute carbol fuchsin or safranin may be used as a counterstain, to stain the Gram-negative bacteria.

This method divides bacteria into two classes:

(*a*) *Gram-positive*. These do not decolorise with ethanol, thus appearing *purple*. Examples of Gram-positive bacteria are *Staphylococcus*, *Bacillus*.

(*b*) *Gram-negative*. Gram-negative bacteria appear *pink*. Examples of Gram-negative bacteria are *Pseudomonas*, *Escherichia*.

Note that the test should be carried out on young cultures (18–24 h), or cultures of various ages including young cultures should be examined, since some bacteria change in their Gram reaction as well as in their morphology as the cultures age.

There are in existence many modifications of Gram's staining method (see for example Silverton and Anderson, 1961). The reagents and times employed depend upon the nature of the specimens most commonly being examined in a particular laboratory, but user preference also plays a large part in the choice of the modification. The reagents used in the method described below are chosen for having a greater latitude than many others towards deviations from the recommended staining times, and they are thus particularly suitable for class and routine use. Nevertheless, every effort should be made to adhere to the times recommended, particularly for the decolorisation stage.

Procedure

1. Prepare a heat-fixed smear from an 18–24 h culture in the usual way.
2. Stain with crystal violet solution for 1–2 min.
3. Rinse rapidly with water, add Gram's iodine solution and leave for 1 min.
4. Pour off the iodine, blot dry and wash the slide with 95 per cent ethanol (or industrial methylated spirits) until no more violet stain runs from the slide (only 5–15 sec in the case of well-prepared thin smears).
5. Rinse under the tap and stain with dilute carbol fuchsin solution for 20 sec.
6. Wash the slide well and blot dry.

D. Ziehl-Neelsen Method for Staining Acid-fast Bacteria

Members of the genera *Mycobacterium* and *Nocardia* can be differentiated from many other organisms by this staining technique.

The Ziehl-Neelsen method consists firstly of staining the organisms with a hot, concentrated dye. Once stained, the cells resist decolorisation with acid; they are thus "acid-fast". Decolorisation is effected with suitably strong acid and the smear is then counterstained with methylene blue solution. Acid-fast bacteria stain red, other bacteria and the background stain blue. This method (with 20 per cent sulphuric acid) is used clinically for the detection of *M. tuberculosis* in body tissues and fluids (e.g. lungs, liver, sputum, urine). When 1–5 per cent sulphuric acid (or hydrochloric acid) is used instead of 20 per cent sulphuric acid, the technique is useful for identifying saprophytic mycobacteria, nocardiae and certain coryneform organisms.

Procedure

1. Cover the slide with strong Ziehl-Neelsen's carbol fuchsin and heat the underside of the slide with a lighted alcohol-soaked swab. Stop heating when the slide steams. Keep the slide hot and replenish the stain if necessary, taking care not to allow the smear to become dry. Heat for 5 min, not allowing the staining solution to boil.
2. Wash well.
3. Decolorise with acid-alcohol or with 1, 5 or 20 per cent sulphuric acid. The excess stain is removed as a brownish solution, and the smear will become brown. Rinse in water, when the film will appear pink once more. Apply more acid and repeat the rinsing several times until the film appears faintly pink upon washing.
4. Wash well.
5. Counterstain with Loeffler's methylene blue for 5 min.
6. Wash well and carefully remove the stain deposits from the back of the slide with filter paper. Blot dry and examine.

E. Staining of Bacterial Flagella

The diameter of a bacterial flagellum is below the limit of resolution of the light microscope, and for this reason flagella are normally not visible. By the use of special staining techniques, an appropriate stain can be made to build up around each flagellum, thus increasing its apparent diameter. This enables the flagellum to be visualised under the light microscope, and the arrangement of the flagella on the bacterial cell to be determined. Specifically, the flagella stain is of use in differentiating members of the Pseudomonadaceae, which have polar flagella, from members of the Enterobacteriaceae, which have peritrichous flagella (when motile).

Good results in flagella staining are obtained only with difficulty, since the extremely delicate flagella easily become detached from the bacterium. In addition, most of the staining methods result in the production of a background precipitate which, on occasions, renders observation of the flagella difficult.

The flagella stains most commonly used fall into two categories: firstly, those in which a silver salt is used to deposit silver on the flagella (Rhodes, 1958) and, secondly, those in which basic fuchsin is deposited on the flagella (see Leifson 1951, 1958; Leifson and Hugh, 1953).

The silver deposition method described below is a modified Fontana method (Rhodes, 1958).

Procedure

Only slides which are completely clean (and therefore grease-free) should be used. This is best achieved by firstly cleaning in chromic acid (see page 315), followed by rinsing in distilled water and finally storing in clean 95 per cent ethanol in a screw-capped jar until required. When a slide jar becomes empty, the ethanol should be discarded, fresh ethanol being used for each batch.

The bacteria should be grown on agar slopes at 3–5° below the optimum growth temperature, the slopes having first been moistened with one or two drops of sterile distilled water.

1. Remove a microscope slide from the ethanol and flame, the slide in a Bunsen burner for about 10 sec. Place on a staining rack to cool, and then draw a halfway division with a wax pencil (one end of the slide can be used for holding, using forceps).

2. With a pipette or Pasteur pipette add 2 ml of sterile distilled water to a young, actively growing slope culture (usually about 18 h old) and gently suspend the growth by careful agitation and rotation of the test-tube. It is advisable not to use a wire loop to suspend the growth if it can be avoided. Transfer to a *clean* test-tube, check for motility with a hanging drop preparation, and dilute the suspension with distilled water until only slightly turbid. Place in an incubator at 20–30°C for 30 min and then remove a large loopful of the suspension to one end of the cool microscope slide. Tilt the slide until the drop runs to the central pencil line. Dry in air at room temperature. Do *not* heat-fix the film.

3. Cover with mordant for 5 min.

4. Rinse gently but thoroughly with distilled water to remove all traces of mordant.

5. Cover with hot Fontana silver solution, and stain for 5 min, renewing the stain once a minute. (The Fontana silver solution should be heated over a boiling water bath.)

6. Wash with water, allow to dry in air, and examine.

Alternative procedure

If Leifson's flagella stain is used substitute steps 3–6 above by the following:

3. Cover the smear with 1 ml of Leifson's flagella stain and allow to act until a very fine rust-coloured precipitate has formed (about 10 min).
4. Rinse gently but thoroughly with distilled water.
5. Counterstain for 5–10 min with 1 per cent methylene blue.
6. Wash with water, allow to dry in air and examine. Without a counterstain the cells and their flagella stain pinkish-red. When the counterstain is used, the cells stain blue, and the flagella are red.

Note. Since flagella are often detached rather easily, a culture showing large numbers of peritrichously flagellate cells indicates that the strain is peritrichously flagellate, but an apparent polar arrangement of flagella does not *necessarily* contra-indicate a peritrichous organism.

F. The Demonstration of Bacterial Capsules

The capsules of bacteria examined in blood or animal tissues can often be seen as unstained haloes separating the bacteria from the stained background even when ordinary staining procedures (e.g. Gram's method) are used. In order to visualise the capsules of bacteria which are being studied in pure culture, special techniques are usually necessary. Leifson's flagella stain (Leifson 1951, 1958) may be used for bacterial capsules, the capsules staining red, and the bacteria blue when methylene blue is used as a counterstain. A simpler method of demonstrating capsules is by the use of a wet Indian ink film.

Production of a detectable capsule is often dependent on the composition of the medium and the age of the culture (Wilkinson, 1958). Most capsules are of carbohydrate and are best detected when grown to stationary phase in the presence of an excess of a utilisable sugar (Duguid and Wilkinson, 1961).

The Wet Indian ink Film

Procedure

1. Place a loopful of Indian ink on a *very clean* microscope slide.
2. Mix into the Indian ink a little of the bacterial culture or suspension.
3. Place a coverslip on the mixture, avoiding air bubbles, and press firmly with blotting paper until the film of liquid is very thin.
4. Examine with the high power dry objective or the oil-immersion objective. The capsule will be seen as a clear area around the bacterium.

N.B. A film of the ink alone should always be prepared and examined as a control, since Indian ink may occasionally become contaminated with capsulate bacteria (Cruickshank, 1965).

G. Negative Stain

This is a very simple and effective method for demonstrating the external shape of bacteria in a smear preparation. The bacteria are surrounded by a thin film of black dye and appear as white objects upon a grey background.

Procedure

1. Prepare a very thin smear in the usual way, using a clean, grease-free slide.
2. At one end of the slide place one drop of nigrosin solution (2 per cent).
3. Take another microscope slide, lay one end on the first slide at an angle of 30° touching the drop of nigrosin, and use it to push the nigrosin across the surface of the first slide. The smear will thus be covered with a thin, even film of dye.
4. Allow the dye to dry and examine the preparation under the oil-immersion objective.

This can be used in conjunction with a simple staining technique in order to demonstrate the presence of capsules. In this case, the smear should be stained with dilute carbol fuchsin, washed and dried, before treating with nigrosin.

H. The Staining of Bacterial Spores

Bacteria in the genera *Bacillus* and *Clostridium* produce endospores, which are highly resistant to high temperature, lack of moisture, and toxic chemicals. The endospores are also resistant to bacteriological stains and, in a smear stained by Gram's method, they can be seen as colourless areas in the vegetative organisms which stain Gram-positive. However, once stained, the spores tend to resist decolorisation.

Bartholomew and Mittwer's Spore Staining Method

Procedure (Bartholomew and Mittwer, 1950)

1. Prepare a smear in the usual way, but heat-fix very thoroughly by passing through a Bunsen flame 20 times.
2. Stain for 10 min with a saturated aqueous solution of malachite green.
3. Wash gently with cold water for 10 sec.
4. Counterstain with a 0·25 per cent solution of safranin for 15 sec.
5. Wash with water and blot dry.
6. Examine under the oil-immersion objective.

N.B. The times in steps (2) and (4) may need modifying when staining some species.

CULTIVATION OF MICRO-ORGANISMS

The characteristics of bacteria used in the study and identification of bacteria depend upon the behaviour of populations or cultures rather than of individual organisms, so that materials and methods for achieving growth and multiplication of the individual organisms are a basic requirement in bacteriology. Nutrient materials provided in a form suitable for growth are known as culture media. Details of some commonly used media are given on pages 52–58.

Culture media may be distributed in various ways in test-tubes, flasks, or screw-capped bottles depending on the method of inoculation to be used. Test-tubes or flasks are stoppered with closely fitting plugs of non-absorbent cotton-wool, metal caps, plastic caps, or specially designed rubber bungs. An advantage of screw-capped containers is that evaporation is prevented and the medium therefore does not dry out on storage.

A. Types of Culture

(*a*) *Liquid batch cultures.* Cultures in liquid media, in which no fresh nutrient is provided during growth.

(*b*) *Agar slope* (*or slant*) *cultures.* Test-tubes or small bottles containing about 5 ml of solid medium, e.g. nutrient agar, dissolved and allowed to cool in a sloping position. The inoculum is either spread over the surface of the medium or applied in a thin streak using a wire loop.

(*c*) *Stab cultures.* Tubes or bottles containing an agar or gelatin medium are allowed to solidify in the upright position. The medium is inoculated by plunging a long straight wire, charged with inoculum, vertically into the centre of the tube.

(*d*) *Semi-solid cultures.* Cultures grown in a medium containing sufficient agar (0·02–0·3 per cent) to increase the viscosity of the medium, but insufficient to solidify the medium completely (see also p. 60).

(*e*) *Shake cultures.* Test-tubes or bottles containing a solid medium are used. The medium is dissolved, cooled to 45°C, inoculated, mixed well by rotating the tube between the hands and allowed to solidify in the upright position.

(*f*) *Plate cultures.*

1. *Streak plates.* Streak plates are used when well isolated colonies are required either for studying colonial form or in the separation of mixed cultures. In either case the solid medium is allowed to form a thin layer in a Petri-dish, and the surface of the medium is covered with inoculum by a suitable streaking technique. Details of this procedure are given on page 23.

2. *Pour plates.* The inoculum is added to the tube of molten medium (at about 45°C) as for shake tubes and mixed well before this is poured into the

plate. When isolated colonies are required by this method, as in the study of colonial form, it is advisable to inoculate and pour several tubes in a dilution series, i.e. inoculating a second tube with some contents from the first before pouring, and similarly inoculating a third tube by transfer of inoculum from the second tube. When a quantitative technique is required, as in the estimation of viable numbers, the inoculum is placed in the Petri-dish and not in the tube of molten medium (for precise details, see page 25).

B. Incubation of Cultures

In studies for identification, cultures should be incubated at the optimum temperature of the organism concerned except in the case of certain special media (e.g. gelatin stabs are usually incubated at 22°C). Cultures to be examined for motility or flagellar type should be incubated at a temperature 3–5°C below that which is optimum for growth. In studies of spoilage potential the incubation temperature is determined by the predicted or recommended storage conditions for the food in question. Plate cultures should normally be incubated in the inverted position to prevent condensed moisture from falling onto the bacterial growth. Incubator doors should not be left open longer than necessary.

C. Method of Inoculation: Aseptic Technique

In the laboratory the study of micro-organisms is based usually on pure cultures, i.e. cultures derived from the growth of a single organism and consisting therefore of one strain of organism only. It is essential to avoid contamination of the cultures and to use an aseptic technique during inoculation and laboratory procedures.

Transfer of micro-organisms is normally carried out with a platinum or nichrome wire (used either as a straight wire or as a wire loop) inserted in a wire-holder. The internal diameter of the loop may be standardised (e.g. 4 mm) by turning the wire round a standard gauge. Straight wires and loops are sterilised before use by holding vertically in the Bunsen flame and heating momentarily to redness. A few seconds should then elapse before using the wire to allow it to cool, thus avoiding killing the organisms on contact.

Cotton-wool plugs removed from test-tubes prior to subculture should be held at the projecting surface only, the inner surface must be protected from contamination (e.g. never laid on the laboratory bench) and may be lightly flamed before replacement. The mouth of a tube or flask should be flamed momentarily after removing, and again immediately before replacing, the plug. Once the cotton-wool plug is removed, the exposed tubes are liable to atmospheric contamination. This may be minimised by holding the tubes in an inclined position in the vicinity of the Bunsen flame, and by not delaying the replacement of the plug. The wires used in subculturing are sterilised following transfer of culture by heating gradually to intense heat in the Bunsen flame. Gradual heating is important in order to avoid spattering of material.

D. Maintenance of Pure Cultures in the Laboratory

(Society of American Bacteriologists, 1957; Lapage, Shelton and Mitchell 1970; Lapage *et al.*, 1970)

The most suitable method for the maintenance of a pure culture depends on the characteristics of the particular organisms and the method used must be chosen accordingly.

(*a*) *Agar slope cultures.* Many organisms can be maintained on the surface of agar slopes (e.g. nutrient agar, malt extract agar), the choice of medium depending on the organism. Slopes may be in test-tubes or screw-capped bottles, the latter being preferable as there is less risk of drying. Cultures once grown may be stored in the dark at room temperature, or in the refrigerator, but in either case must be subcultured at intervals of between 1 month to 2 years, depending on the species.

(*b*) Cultures of lactic-acid bacteria will not grow well on the surface of solid media incubated aerobically, and are more suitably maintained in a liquid medium (e.g. yeast glucose chalk litmus milk or Robertson's cooked meat medium), being subcultured at intervals of 2–4 months. Media containing milk or added sugars should also contain chalk to buffer against the development of too low a pH. After bacteria have died in the coagulated milk layer in yeast glucose chalk litmus milk, viable organisms may still remain associated with the chalk sediment.

(*c*) Cultures of anaerobic bacteria will not grow on the surface of solid media incubated aerobically and are best kept in a medium providing reducing conditions (e.g. Robertson's cooked meat medium), and subcultured at intervals of up to 1 year.

(*d*) *Preservation under oil.* Cultures are first grown on agar slopes as in (*a*) above or in stab cultures, and then completely covered with sterile liquid paraffin or mineral oil. (To sterilise liquid paraffin, dispense in flasks in shallow layers and sterilise in the hot air oven at 160°C for 1–2 h). Cultures maintained in this way will generally remain viable for several years without subculturing.

(*e*) *Freeze-dried cultures.* In this process the cultures are freeze-dried or lyophilised and stored in sealed glass ampoules under vacuum. Ampoules may be stored at room temperature or in the refrigerator and cultures can be expected to remain viable over several years. This method is particularly useful for the storage of culture collections and for despatch of cultures. (See also Lapage *et al.*, 1970; Kusay, 1972.)

(*f*) *Desiccated serum suspensions*. Laboratories not possessing freeze-drying apparatus can obtain good results with the procedure described by Alton and Jones (1963). Suspend a loopful of growth from a 24–48 h culture in 2 ml of sterile serum, and place one drop into each of a number of sterile, plugged 50 × 6 mm tubes. Dry in an evacuated desiccator over phosphoric oxide for 7 days at 5°C, re-evacuating once after 24 h. Place each tube (containing a thoroughly dry suspension) in a larger soda glass test-tube, evacuate and seal in a Bunsen flame. Store refrigerated or in a deep freezer. Such cultures may remain viable for years.

E. Methods for Bulk Cultivation

Liquid media are usually employed for bulk cultivation of cultures (e.g. when a metabolite or cell component is being obtained either for analysis or for utilisation). Problems associated with batch cultures increase with bulk. For example, there are problems in the sterilisation of the containers of media, aeration and temperature control (Calam, 1969).

Continuous cultivation is a procedure whereby nutrients are supplied continuously or recurrently and a proportion of the culture is concomitantly removed. The methods, which are of much use both for theoretical studies and in some industrial processes, are discussed in Volume 2 of "Methods in Microbiology" (Norris and Ribbons, 1970). A variety of "off-the peg" continuous cultivation units can be obtained from many manufacturers including A. Gallenkamp & Co. and LKB Produkter AB.

THE DESCRIPTION OF THE MORPHOLOGICAL AND CULTURAL CHARACTERISTICS OF AN ORGANISM

The following are general characteristics which may be recorded for cultures grown on any media including non-selective and non-diagnostic media. In addition, when biochemical test media have been employed, the specific biochemical reaction (if any) should be noted (see pages 66 ff., 241 ff).

For the characterisation and identification of bacterial cultures, the Society of American Bacteriologists suggests the use of a standard form of descriptive chart which contains a series of descriptive terms to be underlined as appropriate and a number of blanks to be filled in against the various characteristics and tests (see Society of American Bacteriologists, 1957).

A. Morphological Characters

1. Gram reaction.
2. Shape, size and arrangement of organisms.
3. Motility.
4. Presence of endospores, capsules, flagella (detected by appropriate stains).
5. Reaction of Ziehl-Neelsen and any other special stains.

A drawing of typical organisms and their arrangement should be made. The size may also be recorded.

B. Cultural Characters

(Wilson and Miles, 1975)

Surface colonies on solid media

1. *Shape.*

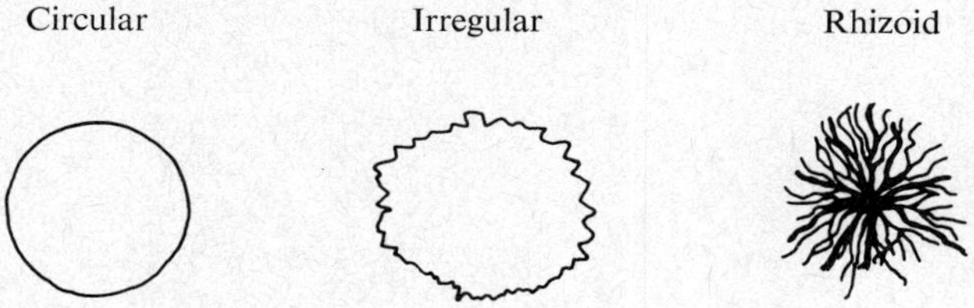

2. *Size.* Record diameter in mm. Punctiform (pin-point): less than 1 mm in diameter.
3. *Chromogenesis.* Colour of pigment, soluble or insoluble in medium.
4. *Opacity.* Transparent, translucent, opaque.

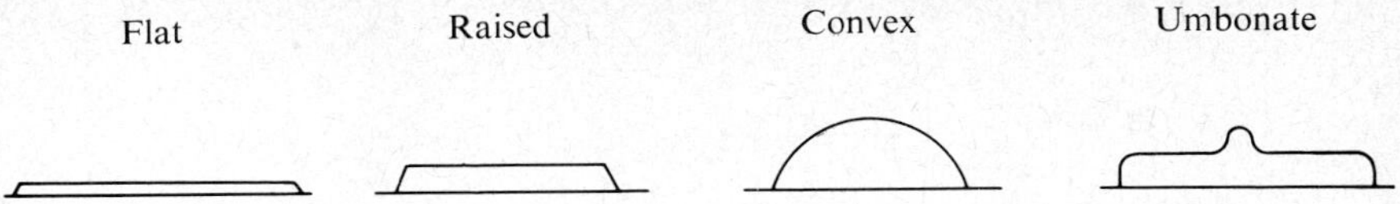

5. *Elevation.*
6. *Surface.* Smooth, rough, dull, glistening.
7. *Edge.*

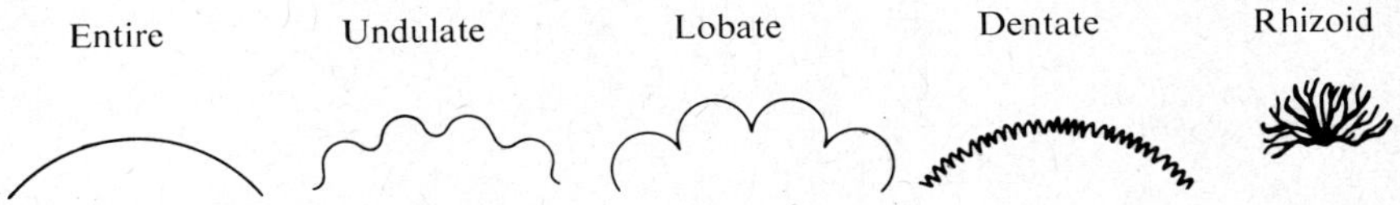

A drawing of a typical colony, plan and elevation should be made.

8. *Consistency.* (tested by touching with a sterile wire loop). Butyrous, viscid, granular.

9. *Emulsifiability.* Easy or difficult in water; forms uniformly turbid suspension; forms granular suspension; does not emulsify.

10. *Odour.* Present, absent, identification of the odour if possible.

Broth culture

Amount of growth. None, scanty, moderate, profuse.
Surface growth. Present or absent; formation of a ring; pellicle which disintegrates or not on shaking.
Turbidity. Uniform, flocculent or absent.
Deposit. Amount; granular, flocculent, viscid, disintegrates or not on shaking.

Shake culture

Growth. On surface; in depth of tube (record depth to which growth occurs, bearing in mind that insufficient mixing of the inoculum may have stopped the organisms reaching the bottom); position of optimal growth.

PLATE CULTURES

A. Preparation of Plates for Streaking

Where a large surface area of medium is necessary, as in the separation of organisms from mixtures, the agar medium is allowed to solidify as a thin layer in a Petri-dish. 10–15 ml of the melted sterile medium are poured into a sterile Petri-dish, care being taken to avoid contamination. The outside of the test-tube or bottle should be wiped dry before pouring to avoid water droplets falling into the dish. The mouth of the flask or tube containing the medium should be flamed after removal of the cotton-wool plug, and the lid of the Petri-dish raised only enough to allow easy access.

When the plates have been poured and the medium allowed to solidify, they must be dried, since the presence of moisture on the surface of the medium would interfere with the production of discrete colonies. The plates are dried in an incubator at 37°C for from 20 min to 1 h. The lid of the Petri-dish is first laid in the incubator and the part of the dish containing the medium is inverted and placed with one edge resting on the lid.

This method of drying helps to avoid contamination from dust. Alternatively, the plates may be dried in the incubator for one or more days until the condensed water droplets present on the lids have evaporated, the lids not being removed and the plates being stacked inverted in the usual way. This will also help to reveal any plates contaminated during preparation.

B. Streak Plates: The Separation of Mixed Cultures

The term "plating" is generally applied to the inoculation of a medium in Petri-dishes, usually by successive strokes with a wire loop.

A sterile wire loop is charged with the bacterial mixture, or the material containing the bacteria to be isolated. Several methods of streaking the plate can be used, two of these being as follows.

Method A

1. The inoculum is spread evenly over a small area towards the edge of the plate.

2. The wire loop is sterilised and used to make two strokes from this small area. One of the strokes is made only to the centre of the plate.

3. The wire loop is again sterilised and a series of strokes made at right angles to the first two strokes.

4. After sterilising the loop once more, a series of strokes should be made at right angles to the previous series, the strokes being made towards the "pool". Care should be taken that the loop does not touch the "pool" at this stage.

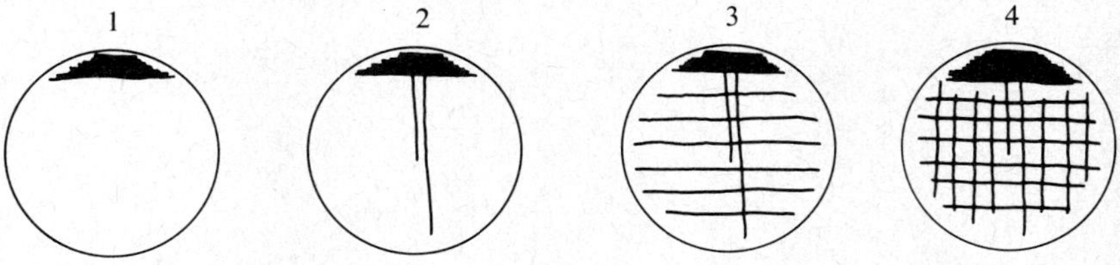

On incubation of plates inoculated in this manner, discrete colonies should be obtained. A pure culture can be grown by selecting a well isolated colony to inoculate a broth culture. After incubation, the broth culture is restreaked as above.

Method B

A similar procedure is used, but streaking in the directions indicated. Note that the loop should be sterilised between each of the steps illustrated.

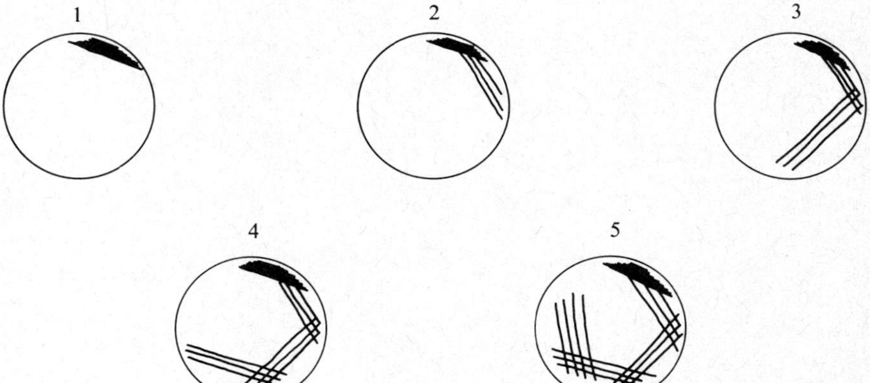

DETERMINATION OF THE NUMBER OF VIABLE ORGANISMS IN A SAMPLE

A. Colony Count Methods

1. Pour-plate method

This enables the number of living organisms or clumps of organisms (i.e. colony-forming units) in a sample to be counted, subject to the appropriate medium and incubation conditions being used. Solid materials which are water soluble or give fine suspensions in water (e.g., soil, flour, dried milk, sugar) can be examined by shaking a known weight of the sample in sterile diluent and proceeding as below. Some solid materials, including some foodstuffs (e.g. cheese, meat) need to be macerated in sterile diluent in order to prepare the suspension. Commonly used diluents include quarter-strength Ringer's solution ($\frac{1}{4}$-R) and peptone water diluent.

A measured amount of the suspension, or of a known dilution of the suspension, is mixed with molten agar medium in a Petri-dish. After setting, the plates are incubated and then the number of colonies is counted. Counts should be made on plates which contain fewer than 300 colonies. Whenever possible, duplicate or triplicate sets of plates should be incubated at each temperature.

The procedure described below is a standard method for liquid samples. The precise details of procedure may vary for different substances or products (see Part II).

(*a*) *Mixing the sample.* It is important that the sample be thoroughly mixed before proceeding further. In the case of milk (or other liquid samples), if the sample bottle is only partly filled, the sample should be mixed by shaking the bottle 25 times with an excursion of 30 cm. If the sample bottle is full it should be inverted 25 times with a rapid rotary motion to mix the contents thoroughly, about a quarter of the contents should then be poured away and the sample shaken 25 times as described above.

(*b*) *Preparing the dilutions*

1. Holding a sterile 1 ml blow-out pipette vertically, introduce the pipette tip not more than 3 cm below the surface of the sample and suck up and down 10 times to the 1 ml mark. Withdraw 1 ml of the sample, touching the tip of the pipette against the neck of the bottle to remove the excess of liquid adhering to the outside of the pipette. Transfer the pipette to the first tube of the dilution series with the tip touching the side of the tube 2–3 cm above the level of the diluent. The pipette must not contact the diluting fluid. Blow out the contents of the pipette, allow three seconds to elapse, then blow out the

remaining drops. Discard this pipette and label the first dilution tube 1/10, or 10^{-1}.

2. Using a fresh sterile pipette, mix the contents of the first dilution tube by sucking up and down to the 1 ml mark 10 times (or by rotating the tube between the hands). The tip of the pipette should be not more than 2–3 cm below the surface of the diluent. Then withdraw 1 ml of this first dilution and transfer to a second tube of sterile diluent expelling the contents of the pipette as described above. Discard this pipette and label the second dilution tube 1/100 or 10^{-2}.

3. Taking a fresh sterile pipette and a further dilution tube, prepare in the same way a 1 in 1000, or 10^{-3} dilution.

4. Further dilutions of 10^{-4}, 10^{-5}, etc., can be made similarly as required, depending on the probable bacterial content of the sample. It is useful to assume that most bacteria have a volume greater than $1 \mu m^3$ so that a solid bacterial mass would be unlikely to give a count of more than 10^{12} per ml. Spoilt meat could give a colony count of about 10^9 per cm^2 of surface. River water could give a colony count of 10^4–10^6 per ml.

(c) Preparing the plates

1. Using a fresh sterile pipette, mix the contents of the final dilution tube, e.g. 1/1000, by sucking up and down ten times. Withdraw 1 ml of the dilution, touching the tip of the pipette against the side of the tube to remove excess adhering to the outside, and transfer the contents to a sterile Petri-dish. Allow three seconds to elapse, then touch the tip of the pipette against the dish away from the previous inoculum, and gently blow out the remaining drops.

2. The same pipette can be used to transfer 1 ml from the 1/100 dilution to a sterile Petri-dish, but before taking the sample, raise and lower the 1/100 dilution in the pipette three times in order to rinse the sides of the pipette and also to give the dilution a final mixing.

3. In the same way, the same pipette may be used to transfer into Petri-dishes 1 ml of the 1/10 dilution and then of the original sample.

4. An alternative method which is somewhat quicker is to inoculate the plates at the same time as the dilutions are made. Thus, as soon as the particular dilution has been mixed by sucking up and down ten times, transfer 1 ml to a sterile Petri-dish before carrying over 1 ml to the next tube of sterile diluent.

(d) Pouring the plates

1. To each plate, add 10 ml of molten agar medium at 45°C and immediately mix the medium and inoculum by a combination of to-and-fro shaking and circular movements lasting 5–10 sec. The exact procedure consists of

movements five times to-and-fro, five times clockwise, five times to-and-fro at right angles to the first and five times anticlockwise. This procedure ensures complete dispersal of the sample. Take care not to get the agar on the lid of the Petri-dish.

2. Allow the plates to set, then invert and incubate at the appropriate temperature. It is essential to label the plates adequately. It is worth giving some thought to labelling. Use codes chosen with care to avoid ambiguity. Remember that more writing takes time. Consistent use of different colours reduces the need for extensive detail. To facilitate rapid equilibration of temperature, the plates should not be stacked in piles of more than six.

N.B. At all times, take full precautions to prevent contamination. The pipettes should be rapidly passed through the Bunsen flame before use. The mouths of the sample bottle and test-tubes should be flamed after removing stopper or plug and *again* before replacing stopper or plug. The lids of the Petri-dishes should be raised only sufficiently to allow for easy access. Before pouring in the molten agar, dry the outside of the tubes or bottles containing the medium, and flame the necks of the containers after removing the stoppers (and again before replacing the stoppers if the containers held multiples of 10 ml).

Not more than 15–30 min should elapse between the dilution of the sample and its admixture with the medium, because suspension in the diluent may be lethal or may encourage growth or separation of clumps.

(*e*) *Counting the plates*

Select a dilution which yields fewer than 300 colonies per plate, since with colonial concentrations exceeding this the count will usually be depressed to an unknown degree by overcrowding and microbial antagonism. If only a single plate at each dilution has been prepared the dilution chosen for counting should also provide more than 30 colonies on the plate, as the statistical error involved in counting fewer than 30 colonies becomes overwhelmingly great. When replicate plates have been prepared at each dilution, the arithmetic mean of the colony counts at the chosen dilution is used to calculate the microbial concentration in the original sample. The significance of the count can be assessed by determining the 95 per cent confidence limits (that is, the range in which the true count should lie in 95 cases out of 100). This range is given by

$$\frac{(n\bar{x}) \pm 1{\cdot}96\sqrt{(n\bar{x})}}{n}$$

where $(n\bar{x})$ is the total number of colonies counted on all the plates at the chosen dilution, and n is the number of plates at that dilution. ($\bar{x}$ is the arithmetic mean of the colony counts at the chosen dilution.)

The fewer the colonies counted, the wider will be the 95 per cent confidence limits. Maximum practical precision is obtained by counting 600–1000 colonies (Meynell and Meynell, 1970).

In daily laboratory records counts should initially be recorded as a colony count at the given dilution, e.g. "182, 200, and 260 colonies at 10^{-3}". This information is then used to calculate the microbial concentration in the original sample by multiplying the arithmetic mean by the dilution factor; in our example this would be 210,000 per gram or per ml. Note that the final estimate should only be given to 2 significant figures in the report. In more demanding investigations both this estimate and the 95 per cent confidence limits should be quoted in the report thus: "210,000 per gram; 95% C.L. 197,000 to 231,000 per gram". Unfortunately in some laboratories it is the practice to record the counts directly as the microbial population in the original sample, often to the usual 2 significant figures – if replicate plates are prepared the arithmetic mean is worked out in the head or on a piece of scrap paper, and the multiplication by the dilution factor is similarly performed. It is obvious that this means that data are irretrievably lost at the time of counting. The inaccuracies inherent in counting only 1, 2 or 3 colonies at a given dilution are completely masked, and a count of 1 colony at 10^{-4} dilution on one occasion will wrongly appear to be the exact equivalent of a count of 300 colonies on triplicate plates at the 10^{-2} dilution on another occasion. It is therefore advisable to record full data in the laboratory records book.

Other frequent omissions are the details concerning the medium used and the temperature and time of incubation. These details are of great importance since they obviously affect the count obtained, sometimes profoundly. It can be argued that in any one laboratory the techniques are standardised so that a count performed, for example, on sliced roast pork will be a count on plate count agar at 37°C. This may be true, but alterations and modifications in the methodology should be very well documented. If at some future date a retrospective survey of trends is required then the precise points will be known at which the standardized method was modified – such modifications may have a profound effect upon the apparent counts. Consider how easy it would be to deal with the following questions: "When did we replace the plate count agar prepared from basic ingredients in the laboratory by Oxoid dehydrated Plate Count Agar?" "When was the temperature of incubation altered from 37°C to 35°C?"

When the colonies are being counted a hand lens or bench magnifier should be used with a good source of illumination (preferably oblique or lateral illumination), the plate being placed against a dark background. Ideal conditions are provided by various commercially available plate counting chambers, but it is quite easy to construct at a very low cost plate illuminators which perform just as well, using matt black Formica and a small tungsten strip light, shaded from view, at each side of the plate.

All colonies should be counted, including "pinpoint" colonies. Spreading colonies (e.g. *Bacillus* colonies spreading over the agar surface) may cause trouble because of the suppression of the growth of other organisms as well as causing possible masking of other small colonies, so plates containing spreading colonies should preferably not be used for counting but if there be no choice but to use them then the presence of spreading colonies should be noted in the record.

Occasionally compact clusters of very small colonies occur at the agar/glass or agar/plastic interface on the bottom of the dish. It is customary to count each cluster as one colony as it is usually assumed that they derive from individual organisms contained within a clump (e.g. *Staphylococcus* or *Micrococcus*) or chain (e.g. *Streptococcus*) which have spread at an early stage but which in the depths of the medium would have acted as a single viable unit and resulted in only one colony.

(*f*) *Recording the results*

The following is an example of the type of entry recommended for laboratory records books.

Date	Sample	Type of Count	Dilutions examined (duplicate)	Incubation
16/2/76	Cooked ham	General viable count (Plate count agar)	10^{-2}; 10^{-3}; 10^{-4}	3 days at 30°C

Diluent: 0·1% peptone
Blended by Colworth Stomacher

Numbers of colonies found

> 300 > 300; 150, 162; 12, 17.

Mean count per g of ham: 156×10^3

Report as: $1{\cdot}6 \times 10^5/g$ (95% confidence limits $\frac{312 \pm 1{\cdot}86\sqrt{312}}{2}$)

Grade: C

2. Roll-tube method

This is a modification of the pour-plate method in which inoculated nutrient agar is solidified as a thin, even film on the inside surface of a test-tube or bottle. The Astell Roll Tube Apparatus (Astell-Hearson) which uses a specially designed roll-tube with a rubber seal, is commonly employed in Great Britain 4·5 ml of medium (containing 2 per cent agar) is measured into each bottle which is plugged with a rubber seal and sterilised by autoclaving. The sterilised, filled bottles can be stored at room temperature for long periods before use without risk of contamination or drying out. The principle advantages of this modification are in savings of incubator space and media. Amongst the disadvantages are that colonies cannot readily be sampled from the bottles, and that spreading colonies are very common with certain types of sample.

The agar medium is melted before use by partially immersing the bottles in boiling water after first slightly loosening the stoppers, and the bottles are then transferred to a thermostatically controlled water bath to cool the medium to 45°C.

0·5 ml amounts of appropriate tenfold dilutions are transferred to roll-tubes which are then placed on a special spinner. The spinner holds seven bottles on spinner heads driven by an electric motor, and as the bottles rotate, jets of water play on the surface of each bottle causing rapid setting of the agar as a thin, even film.

The bottles are incubated with the stoppers downwards so that any condensate drains from the surface of the agar, helping to prevent spreading of the bacterial growth. After incubation, the colonies that have developed in each bottle are counted using a colony illuminator which incorporates a magnifying lens. Care should be taken to allow for inclusion of 0·5 instead of 1 ml of diluted sample, when making calculations.

The roll tube technique may be used for cultivation of very strict anaerobes by employing a stream of oxygen-free gas during inoculation (Hungate, 1969).

3. Use of agar droplets

Considerable savings in plates and media can be obtained by performing colony counts on small droplets of agar. Such a system has been described by Sharpe and Kilsby (Sharpe, 1973; Sharpe and Kilsby, 1971) using apparatus now obtainable from A. J. Seward. This apparatus ("Droplette" Model BA 6013) consists of an automatic pipetting device which accepts sterile disposable polypropylene pipettes (or glass Pasteur pipettes), and a viewing system for counting in which the agar droplet acts as its own optical condenser producing a ×10 image of the droplet on a viewing screen.

In use a number of bottles or tubes containing 9 ml amounts of agar medium are held on the bench in a water bath at 45–50°C. One ml of the sample or the first dilution of it, is pipetted with a 1 ml pipette into the first bottle and mixed. The pipetting device is used to take up automatically 1·5 ml of this dilution and by the use of two separately operated solenoids it can be made to deliver a row of five 0·1 ml drops of the inoculated agar into a sterile petri dish, followed by 1 ml into the next bottle of agar to provide the next decimal dilution. A standard Petri dish can hold three rows of five droplets and thus one dish allows the preparation of quintuplicate counts from three successive dilutions.

The principal disadvantages are that the system does not allow good differentiation of colony types when used with diagnostic media (e.g. brilliant green agar) and smaller colonies at the edges of the droplet can be difficult to see on the viewing screen.

4. Miles and Misra surface colony count (Miles and Misra, 1938)

The method consists of placing drops (0·02 ml) of serial dilutions on the surface of poured agar plates and counting the colonies which develop upon incubation of the plates.

This method is useful when the bacteria are best grown in surface culture (e.g. when the presence of obligate aerobes is suspected) or when an opaque medium is employed. Surface counts are also indicated where the microbial population in a sample is likely to include bacteria which are killed by the brief exposure to 45–50°C which occurs when using melted agar for the pour-plate techniques (Mossel, 1964).

1. Pour agar plates and dry the surface of the medium for 24 h at 37°C with the lids closed, followed by 2 h at 37°C with the lids and bases separated as described on page 23. This enables the medium to absorb the water of the inoculum quickly.
2. 0·02 ml drops of tenfold dilutions are dropped onto the surface of the agar from a height of 2·5 cm, using a calibrated dropping pipette. For six dilutions, six plates are used for each count, one drop from each dilution being placed on each plate. Allow the drop to be absorbed before inverting and incubating the plates.
3. After incubation, counts are made of the drops showing colonies without confluence, the maximum number obtainable depending on the size of the colony and ranging from 20 to 100. The total count of the six drops of the appropriate dilutions is divided by six and multiplied by 50 to convert the count to colonies per 1 ml of the *dilution used.*

A less precise method of counting involves the recording of growth or no growth, and the use of probability tables to give a MPN (Harris and Somers, 1968). This is particularly useful for counting groups defined by physiological characteristics detectable on solid media (e.g. counting of starch-hydrolysing bacteria).

Preparation of pipettes

Pull and cut pasteur pipettes of a sufficiently fine bore to enable the capillary to be inserted through a standard wire gauge of diameter 0·91 mm. Insert the capillary of a pipette in the gauge and with a glass-cutting knife score the capillary at the point of insertion. It should then be possible to snap the capillary level with the gauge in such a way as to get a square-cut end (a bending and slightly pulling movement will be found to give the best result).

Pipettes prepared in this manner will give drops of about 0·02 ml when the drops are delivered at a rate of one per second, with the pipette held vertically. Nevertheless, it is recommended that pipettes should be calibrated for the liquids being handled since solutions, suspensions and emulsions differ in their effect upon the actual volume delivered. For example, Wilson (1922) found that, when used to deliver dilutions of bacterial culture, pipettes gave drops of more or less constant volume irrespective of the dilution used. He suggested that this was probably because the increase in density of the emulsion counterbalanced the increase in viscosity, causing the drop volume to remain more or less constant. On the other hand, when using calibrated pipettes to deliver dilutions of milk, the volume of the drop depended upon the dilution (Wilson, 1935). The following table, prepared by the late Dr. T. Richards (unpublished work) indicates the correction factor required for the pipettes prepared as described above when used with varying dilutions of milk:

Liquid being delivered	Volume of 1 drop	No. of drops per ml
Tap water	0·0209	48
Milk 10^{-3}	0·0193	52
Milk 10^{-2}	0·0181	55
Milk 10^{-1}	0·0164	61
Sep. milk	0·0153	65

Stainless steel dropping pipettes with interchangeable glass barrels suitable for this method are obtainable from Astell-Hearson.

B. Membrane Filtration

The membrane filter consists of a thin, very porous disc composed of cellulose acetate or mixed cellulose esters. Membrane filters obtainable from Oxoid Ltd., are available in a single grade only, having a pore size within the range 0·5–1·0 μm in diameter. Membrane filters manufactured by the Millipore Filter Corporation, are available in a range of pore-size grades ranging from 10 nm to 8 μm or more in pore diameter. Millipore filters recommended for bacteriological work are those of grade HA with a pore size of 0·43–0·47 μm. Similarly, membrane fiters manufactured by Gelman Instrument Co. and Sartorious-Membranfilter GmbH, are also available with a range of graded pore sizes. The Millipore Filter Corporation manufacture a very wide range of apparatus for use with their membrane filters.

The membrane will allow large volumes of water or aqueous solutions to pass through rapidly when under positive or negative pressure, but the very small pore size prevents the passage of any bacteria present. The bacteria, which remain on the surface of the membrane, can be cultivated by placing the membrane on an absorbent pad saturated with a liquid medium. The capillary pores in the membrane draw up the nutrient liquid and supply each bacterium with nutrient. The bacteria will give rise to individual colonies after incubation. The media used are specially designed for the membrane filter method and contain ingredients in concentrations different from those in the standard media. Such media can be made up in the laboratory, or can be obtained in dehydrated or ready-prepared form from, for example, Difco Laboratories Inc. and Baltimore Biological Laboratories.

The membrane filter colony count has the advantage that, provided the sample will pass through the filter readily, small numbers of organisms can be detected in large amounts of sample. Membrane filter counts thus may achieve the sensitivity of the multiple tube count whilst retaining the accuracy of the colony count method.

Alternatively, a microscopic examination can be carried out if the bacteria are stained and the membrane filter then rendered transparent by treating it with a liquid such as immersion oil or cottonseed oil. Suitable stains are Loeffler's methylene blue and crystal violet. (See page 42 for details.)

Membrane filters can be used for the routine examination of water, air, sugar solutions and some beverages. Milk and dairy products can be examined if fat globules are first broken down by the use of an appropriate wetting agent such as Triton X–100 (made by Rohm and Haas). They have also been found of use in the testing of disinfectants, separation of phage from bacteria, and the sterilisation of fluids including bacteriological media and gases. The use of selective media such as specially formulated MacConkey's broth, and non-selective media such as tryptone soya broth enables a viable count and coliform count to be carried out on water samples, for example. For further details of these specific techniques, refer to the relevant *Application Data Manual* published by the Millipore Filter Corporation (see also Mulvany, 1969). The general procedure is as follows.

(*a*) *Sterilisation of equipment*. The membrane filters should be placed between sheets of filter paper, wrapped in paper and autoclaved for 15 min, at 121°C. At the start of every day, the filter holder should be loosely assembled, wrapped in paper and autoclaved for 15 min at 121°C. Between samples, the funnel top can be sterilised by swabbing with alcohol and flaming.

Incubating tins or dishes, each containing an absorbent pad, should be wrapped and sterilised by autoclaving or by dry heat.

(*b*) *Preparation of incubating dishes.* To the sterile absorbent pad in each incubating dish, add aseptically sufficient sterile medium to achieve saturation of the pad without an excess of liquid being visible.

(*c*) *Filtration of the sample and incubation of the membrane.* Attach the sterile filter holder to a filter flask which is connected to a suction pump. Using sterile forceps place a sterile membrane filter, grid side up, on the platform of the filter unit, after first removing the funnel top (taking care not to contaminate the interior of the funnel). Replace the funnel and lock into place.

Pour the liquid sample into the funnel, and draw the sample through the membrane filter by applying suction. After all the sample has passed through the filter, rinse the funnel with sterile quarter-strength Ringer's solution. Reduce the suction and after removing the funnel top aseptically transfer the membrane filter with sterile forceps to an incubating dish containing an absorbent pad previously saturated with an appropriate liquid medium. The filter should be placed on the absorbent pad with a rolling action in order to avoid trapping air bubbles between the filter and the pad. After replacing the lid of the incubating dish, incubate with the lid uppermost. The period of incubation normally required is somewhat less than for poured plates.

After incubation, the numbers of colonies are counted and the viable count calculated per ml or g of sample. Sometimes, there may be insufficient contrast between the colour of the colonies and the colour of the membrane and in this case a staining procedure (staining either the membrane or the colonies) may assist accurate counting.

(*d*) *Malachite green staining of the membrane.* After incubation, the membrane should be flooded with a 0·01 per cent aqueous solution of malachite green for 3–10 sec, and then the excess stain poured off into a jar containing disinfectant (Fifield and Hoff, 1957). In this method, the membrane is stained, while the colonies remain unstained.

(*e*) *Methylene blue staining of bacterial colonies.* Saturate a fresh pad with a 0·01 per cent aqueous solution of methylene blue and transfer the incubated membrane bearing the colonies to this pad for 5 min. Then transfer the membrane to a pad saturated with water. The methylene blue tends to be removed from the membrane more rapidly than from the colonies and the colonies can be counted as soon as there is sufficient contrast with the membrane.

C. Multiple Tube Count

The multiple tube count (or Most Probable Number count) provides an estimate of the number of living organisms in a sample which are capable of multiplying in a given liquid medium. A liquid medium is chosen that will support the growth of the bacteria under investigation. Sometimes the count is based on the production of a certain reaction in the medium (e.g. coliform counts).

Tubes containing the liquid medium are inoculated with, for example, 1 ml quantities of serial dilutions of the material or suspension of material being examined (see page 25). After incubation, the highest dilution (i.e. the lowest concentration) giving growth is noted and this enables an *estimate* of bacterial numbers in the original sample to be made.

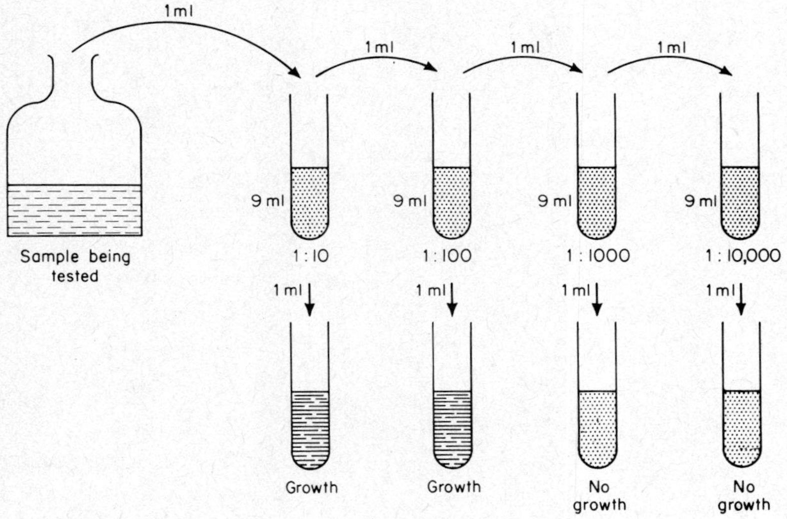

In the example illustrated, growth has occurred in tubes inoculated with 1 ml quantities of dilutions of 1 in 10 and 1 in 100, but there has been no growth at the level 1 in 1000. Therefore, there are probably at least 100 but not as many as 1000 bacteria per ml of the original sample.

Unfortunately, this method has a very large sampling error. In order to reduce this error, several tubes are inoculated at each dilution and the approximate numbers of living organisms in the original suspension are estimated by reference to probability tables (see pages 381 ff.).

Commonly 5 tubes are inoculated at each of a number of 10-fold dilutions. The MPN table for this arrangement is given in Table 3 of Appendix 2 (page 386).

After deriving the MPN from the appropriate probability tables the 95 per cent confidence limits should be indicated. Cochran (1950) suggested that an approximation is given by $\frac{\text{MPN}}{x}$ to MPN.x, where the value of x depends on both the dilution factor and the number of tubes inoculated at each dilution. For 5 tubes at each of a number of 10-fold dilutions, $x = 3 \cdot 30$; for 3 tubes at each of a number of 10-fold dilutions, $x = 4 \cdot 68$; for 2 tubes at each of a number of 10-fold dilutions, $x = 6 \cdot 61$.

Recording the results. The following is an example of the type of entry recommended for laboratory records books.

Date	Sample	Type of Count	Dilutions examined (×5)	Incubation
16/2/76	Cooked ham	Coliform count	$10^0, 10^{-1}, 10^{-2}, 10^{-3}$	2 days at 30°C
Diluent: 0·1% peptone		(MTC: MacConkey		
Blended: by Colworth Stomacher		broth) positives:	5 5 1 0	
	MPN per g. of ham: 35 (95% confidence limits $\frac{35}{3 \cdot 3}$ to $35 \times 3 \cdot 3$)			
	Grade: C			

The multiple tube method of counting is particularly useful (and may be the only method available) when an estimate is required of numbers present only in very low concentrations, e.g. less than 1 per ml, when the large amount of material to be examined makes the plate count method impracticable. A comparatively large volume of sample will be required and the presence or absence of micro-organisms in the various quantities tested (e.g. 5 × 100 ml, 5 × 10 ml, 5 × 1 ml) may be determined. In such sampling regimes it is convenient to add the sample or its dilution to an equal volume of double-strength medium. It should be noted, however, that if the sample to be tested is a liquid capable of passing through a membrane filter, a more accurate count may be obtained by use of the membrane filtration procedure described previously.

The multiple tube count procedure may be used in conjunction with selective and diagnostic media to determine the numbers of particular groups of organisms, for example the numbers of coli-aerogenes and faecal streptococci in food samples.

D. Dye Reduction Methods

These methods depend on the ability of micro-organisms to alter the oxidation-reduction potential of a medium. They are in consequence a measure of the activity of micro-organisms in the test system rather than of the numbers in the sample. Suitable indicator dyes include methylene blue and resazurin. The length of time taken to reduce the dye depends upon the mass and activity of bacteria present in the sample, the greater the number present the shorter the time required for reduction. However, many other factors are important, including the nature of the sample, the medium used, and the types of organisms present. The organisms must be capable of metabolism and growth in the medium to which the dye is added, and if the sample itself is incapable of supporting growth the dilution liquid should be a nutrient liquid. For reproducible and interpretable results the test system including the sample must be of a sufficiently constant chemical composition to have an invariable effect on the micro-organisms present.

The test procedure adopted is dependent on the type of sample being examined, and since it must result in standardised conditions a dye reduction test method can only be applied to those foods or food constituents which have a reasonably constant and predictable composition. Tests may be based on determining either the time taken for decolorisation or the amount of reduction obtained after a set time. The test methods allow interpretation relative to arbitrary grades only. In Part II there are examples of standard test methods as applied to milk (pages 171–175), ice-cream (pages 194–195), cream (pages 190, 192). Test procedures have been described elsewhere for frozen foods (Straka and Stokes, 1957) and sugar syrups (Kissinger, 1969).

The following outline experimental procedure is suggested as a simple way of assessing the feasibility of adopting a dye reduction test method with any given food or food constituent.

Model system for study of the dye reduction test

A series of tubes of growth media, each containing a different redox indicator dye which changes colour at a known redox potential, is inoculated and incubated.

As each becomes decolorised the bacterial population is counted.

(*a*) *Using pure cultures.* In this experiment the food or an appropriate dilution of it should be sterilised (probably by autoclaving) so that the effects of pure cultures can be studied.

Into each sterile 150 × 16 mm test-tube with airtight closure place:

4 ml of sterile food sample or appropriate (usually 10^{-1}) dilution*
4 ml of sterile distilled water or tryptone soya broth*
0·1 ml of 0·005% (w/v) solution of redox indicator‡
1 ml of a 24-hour bacterial culture, or appropriate dilution†.

* If the sterilised sample is of a food of reasonably constant composition (as a microbiological medium) distilled water can be used as the diluent. If the food is of variable composition or is unlikely to support the growth of the micro-organisms, tryptone soya broth should be used.

† Suitable pure cultures used may include *Streptococcus lactis*, *Str. faecalis*, *Escherichia coli*, *Bacillus cereus* or other species particularly relevant to the food.

The culture dilution is chosen so that the incubation time necessary for the

After preparing the tubes, mix well and then immediately seal the contents from the atmosphere by adding 1 ml of sterile liquid paraffin to each.

Place the tubes simultaneously in a water bath (e.g. at 37°C) and examine at frequent intervals (15 min intervals in the case of a 4 h test duration). Observe the colour changes which occur. A redox indicator will start to change colour at $E_h = E_0' + 0{\cdot}05$ v and will cease to change colour when $E_h = E_0' - 0{\cdot}05$ v. In each case, when the colour change is complete, estimate the bacterial population (this estimate will be performed on a tube 15 min after ($E_0' - 0{\cdot}05$ v)), microscopically or by a colony count. Graphs of E_h: time, population: time, E_h: population can be plotted.

N.B. A given culture in a given system will reach an equilibrium E_h which may be above the E_0' of some of the indicators used.

(*b*) *Using the naturally contaminating microflora.* A similar set of experiments can be performed in which the food sample is not sterilised in the laboratory, but provides its natural microflora. This can lead to a preliminary test regime based on a single redox indicator (e.g. methylene blue or resazurin).

In designing a standard test method to be used with a particular food the following need to be defined:

a. The *objective* of the dye reduction test. For example, although it is a dynamic metabolic test, the methylene blue test is often used on ice-cream samples purportedly to assess hygienic quality. The test measures the ability of micro-organisms present in the ice-cream to grow in a dilution of the ice-cream when kept at a temperature of 20–37°C. Yet ice-cream will be kept frozen until consumption, and many micro-organisms which could be present, especially certain pathogens, would not affect the test result. In these circumstances it is difficult to see how

colour changes is acceptable (e.g. up to 4 h for a half-day experiment, up to 7 h for a full day experiment, or 17–30 h for an overnight experiment).

‡ At least 8 tubes should be set up; each with one of a range of redox indicators, and one tube without indicator to serve as a colour control.

A very wide range of redox indicators is available (e.g. from B.D.H. Ltd.). The E_0 of a redox indicator is batch dependent but the following list shows a typical test range.

	E_0
Phenol blue	+0·224
Phenolindo-2,6-dichlorophenol	+0·217
Lauth's violet	+0·063
Methylene blue	+0·011
Janus green	−0·225
Phenosafranine	−0·252
Neutral red	−0·325

the methylene blue test assesses either the keeping quality of the ice-cream, or its potential health hazard.

b. The *constituent species* of the expected microflora, and their relative effects on redox potential.
c. The *incubation temperature* to be chosen having due regard to a. and b.
d. The *levels of microbial population* in the sample to be detected as separate grades. Can these be related to a set of reduction times needed for decolorisation to occur at the test temperature?

DETERMINATION OF THE TOTAL NUMBER OF ORGANISMS IN A SAMPLE

A. The Breed's Smear Method for Direct Microscopic Counts

This is a method in which the number of bacteria or yeasts in a sample (e.g. broth culture, cell suspension, milk) may be determined by direct microscopic examination. Since the staining procedure does not differentiate between living and dead organisms, the bacterial count obtained is known as a "total count". Any suitable staining technique may be used (e.g. simple stain or Gram's method). In the case of samples of milk and other foods with a high fat content it is necessary first to defat (e.g. with xylene). Newman's stain conveniently combines both defatting and staining processes although not allowing the determination of the Gram's staining reaction. The Breed's smear method is also used for counting the animal cells – predominantly leucocytes – which are found in much larger numbers in mastitis milk (see page 179) than in milk from healthy animals and so can be diagnostic of mastitis.

It is essential in this procedure that the microscope is first calibrated so that the area of the microscopic field is known. A known volume (0·01 ml) of the sample or an appropriate dilution is then spread over a known area (1 cm^2), on a glass slide. An alternative procedure often used which gives the same sample concentration on the slide is to draw two straight lines across the width of the slide to enclose a length of 2 cm. Since the standard microscope slide is 2·5 cm wide, this marks off an area of 5 cm^2, over which 0·05 ml of sample is spread. The sample is allowed to dry, then stained and examined microscopically. The average number of bacteria or cells per field is determined and it is then possible to calculate the number of bacteria or cells per ml of original sample.

Determination of the area of the microscopic field

1. Set up the microscope and, using the $\times 10$ eye-piece and low power objective, adjust the stage micrometer so that the graduated scale (1 mm divided into 100 units of 10μm each) is in the centre of the field.

2. Place a drop of immersion oil on the stage micrometer and focus with the oil-immersion objective. Determine the diameter of the microscopic field in μm, using the micrometer scale, adjusting the tube length of the microscope slightly if necessary. In subsequent observations this same tube length must be maintained.

3. Calculate the area of the microscopic field in mm for the oil-immersion lens and $\times 10$ eye-piece using the formula

$$\pi r^2 = \text{Area of field}$$

where r = radius of field. Knowing the area of the microscopic field, it is then possible to determine the microscope factor (MF) by calculation:

(*a*) MF = Number of fields in 1 cm^2 (100 mm^2)

$$= \frac{100}{\text{Area of field in mm}^2}$$

(*b*) Average number n of organisms or cells per field =

$$\frac{\text{Number counted}}{\text{Number of fields counted}}$$

(*c*) MF $\times$ n = Number of organisms or cells present in 1 cm^2.

(*d*) Since 0·01 ml of sample was spread over 1 cm^2 (or 0·05 ml over 5 cm^2),

MF $\times$ n $\times$ 100 = Number of cells or organisms present in 1 ml of sample.

For most microscopes as used above, one organism per field represents approximately 500,000 organisms per ml of sample.

Preparation of smear

1. Thoroughly mix the sample to be examined. Prepare dilutions as necessary – to produce smears with no more than 20 organisms per field. (If pure cultures are being examined a barely detectable turbidity will be given by a suspension containing about 10^6 cells per ml.)
2. Deliver 0·01 ml of sample onto a clean glass slide and spread over an area of 1 cm^2 using either a guide card or a marked slide. The sample may be delivered either (*a*) by means of a capillary pipette calibrated to deliver 0·01 ml or (*b*) by using a wire loop. A closed loop of 4 mm internal diameter will hold a drop of approximately 0·01 ml of sample, if withdrawn with the plane of the loop perpendicular to the surface.
3. Dry the smear immediately by placing on a warm level surface or in an incubator at 55°C. Drying should be complete within 5 min so as to prevent possible bacterial multiplication.
4. Stain by an appropriate method.
5. Examine, using the oil-immersion objective.

Counting

1. Count each single bacterium as one. Chains and clumps of bacteria are also counted as one. Bacteria further removed from the clump than the

longest dimension of the constituent bacteria should be counted as separate. In this way the "direct count" is also a "clump count" and bears a closer relationship to the probable plate count, since each clump or chain would probably give rise to one colony only.

2. The precision of the count depends on the number of bacteria counted and, as in the plate count technique, maximum practical precision would be obtained by counting some 600 items (bacteria in this case, colonies in the case of the plate count). However, unless an automated system is being employed, operator fatigue will occur sooner when performing microscopic counts so that it is recommended that a number of microscope fields are scanned and the numbers of bacteria observed in each field summed, until the total number of bacteria observed is 150 or more. Counting such a number will still provide acceptable precision (see Meynell and Meynell, 1970). Note the number of fields scanned. The final field to be examined which achieves the target should be counted in its entirety. Due to the way in which the drop dries on the slide during the preparation of the smear, the micro-organisms tend to become concentrated towards the centre of the smear (Hanks and James, 1940). In order to make some allowance for this non-random distribution of organisms on the smear, fields should not be sampled at random but should be sampled and counted during edge-to-edge traverses of the smear, successive traverses to be taken at right angles.

3. Determine the average number of organisms per field and hence calculate the number per ml or per g of sample.

B. Direct Microscopic Counts by Membrane Filtration

See page 34 for the general procedure for filtration of the sample. After filtration, turn off the vacuum pump but do not remove the membrane filter from the filtration apparatus. Pour sufficient staining solution (e.g. Loeffler's methylene blue) into the funnel to completely cover the membrane filter. Allow to stain for the usual time (5 min in the case of Loeffler's methylene blue) and then start the vacuum pump once more to remove the staining solution. Pass distilled water through the membrane filter until either the filter or the filtrate is colourless. Then remove the membrane filter and allow to dry in air. When completely dry, saturate the filter with immersion oil by floating the filter on a small amount of immersion oil placed in a Petri-dish. This procedure renders the membrane filter transparent. Remove the filter and mount on one or more microscope slides (in the case of the usual 4·7 cm or 5 cm filters it will be necessary first to cut the filter in half), after which it is ready for microscopic examination with the oil-immersion objective. If the microscope factor is known, the number of bacteria in the sample can be calculated from the effective area of the membrane filter and the amount of sample passed through the filter.

N.B. Since staining solutions and washing water are required to pass through the membrane filter during the staining procedure it is necessary to filter the reagents before use first by passage through a fine filter paper, then by membrane filtration.

C. Turbidimetric Methods

These techniques depend on the micro-organisms in a suspension blocking a light beam by scattering or absorption, causing the suspension to appear turbid. The greater the concentration of organisms, the less the light can penetrate the suspension and the more light is scattered.

It is obvious, therefore, that turbidimetric methods can only be used for estimating concentrations of micro-organisms which are suspended in liquids (e.g. distilled water, saline, nutrient broth) which have a low innate turbidity. In addition the light-blocking power of an organism will depend upon its size, shape and its transparency (Powell, 1963). Therefore a turbidity measurement can be correlated with microbial population only when a pure culture is examined and when the growth conditions of the culture have been standardised to ensure reproducibility of size and shape of the organisms in the population.

The main applications of turbidimetric methods are:

(*a*) the standardisation of suspensions of pure cultures of bacteria to be used in laboratory experiments (e.g. for evaluation of disinfectants or of preservation processes). Such standardisation ensures that the bacterial concentration is within the desired range, but it will still be necessary to determine the actual viable count of the suspension used (i.e. the count at t_0).

(*b*) the assay of vitamins and other growth factors, and of antibiotics and other growth inhibitors.

(*c*) the determination of the effects of environment (e.g. temperature) on growth.

Photoelectric colorimeters and spectrophotometers can be used for measuring the amount of light lost from a reference light beam as the result of passing it through a turbid suspension, but they are relatively insensitive to low bacterial concentrations. Another approach is to measure directly the amount of light scattered using a nephelometer, this instrument being much more sensitive to low turbidities than are most colorimeters and spectrophotometers.

1. Brown's Opacity Tubes

This is a simple method of determining the approximate number of bacteria in a suspension by comparing the turbidity (opacity) of the suspension with the graded turbidities of a series of ten standard tubes.

1. Transfer the bacterial suspension, or a suitable known dilution to a tube of dimensions similar to the opacity tubes, to give a column 5–7 cm high.
2. Lay the tube on a clearly printed page in a good light and compare the opacity tubes one by one with the sample tube.
3. Select the opacity tube which most nearly matches the suspension.
4. Calculate the approximate number of bacteria per ml by multiplying the number on the opacity tube and the figure shown in the table for the appropriate cytological type.

Organism	Count (millions per ml)
Clostridium perfringens	300
Corynebacterium	750
Escherichia coli	750
Pseudomonas aeruginosa	700
Serratia marcescens (broth)	550
(agar)	1400
Streptococcus faecalis and *Strep. lactis*	600

If no figure is given for a particular species, a reasonable approximation is obtained by using a figure given for a morphologically similar organism.

N.B. The table given is a shortened form of a table formerly issued by Wellcome Reagents Ltd. Similar standard turbidity tubes are obtainable from Difco.

For more accurate determinations, the tubes can be calibrated for the particular organism and growth conditions being studied by reference to some other counting method.

2. Nephelometry

When using a nephelometer such as the "EEL" manufactured by Corning-EEL Ltd., two reference standards are required: a turbidity standard to set the sensitivity of the galvanometer and a blank (distilled water or uninoculated medium) to set zero. A ground perspex or ground glass turbidity standard can be purchased for use with this instrument.

The limit of sensitivity of the EEL nephelometer is about 10^5 organisms per ml (the actual figure being dependent on species).

One advantage of the EEL is that it can be used with 150 × 16 mm test-tubes, so that cultures do not need to be transferred to cuvettes. However, the test-tube forms part of the optical system in the nephelometer, the hemispherical base of the tube acting as a condenser lens for the light beam. Consequently the test-tubes need to be standard not only in their diameter, wall thickness and glass colour but also in their base configuration. Specially matched nephelometer tubes can be purchased but these are very expensive. The authors prefer to order a number of boxes of "Pyrex" test-tubes (James A. Jobling's type 1626–12 is suitable), and using a standard turbid suspension check the effect of the tubes on the turbidity reading. "Pyrex" test-tubes have been found to be very uniform within a batch, so that it is a simple matter to get a large number of optically matched tubes. These should then be kept exclusively for nephelometer use.

Use of EEL nephelometer

1. Connect the nephelometer head with cover removed to the galvanometer (Unigalvo) input socket.
2. Turn the sensitivity control fully anti-clockwise.
3. Unclamp the galvanometer movement.
4. Switch on the electricity supply.
5. Bring the reading to zero, using the "set zero" control.
6. Insert the Perspex turbidity standard into the nephelometer, and put the metal cover in position.
7. Select the blank aperture (no filter) on the nephelometer. (With some media it may be advantageous to select a colour filter showing minimum absorbence.)
8. Adjust the galvanometer sensitivity control to produce a reading of 100 on the linear scale.
9. Exchange the standard for a test tube containing uninoculated medium (check that the tube is clean). Adjust the "set zero" control to give a zero reading.
10. Repeat steps 6–9 until stability is achieved.
11. Now insert a sample tube (checking that the tube is clean), replace the cover, and note the scale reading, which indicates the turbidity of the sample relative to the standard.
12. Repeat step 11 with all other sample tubes.

N.B. (*a*) The galvanometer sensitivity alters with time so it is advisable to check blank and standard readings from time to time, and re-adjust if necessary.

(*b*) It may be found that the test tube closure affects the turbidity reading, depending on the amount of light reflected by the closure back down the tube. If this is the case standardise your procedure, preferably by taking all readings with the tube closures removed. In growth response experiments requiring continuing incubation the authors use black rubber seals (e.g. Astell seals) as they do not seriously affect the readings.

(*c*) The accuracy of the readings is affected by the optical cleanliness and standard dimensions of the tubes. Avoid scratching or abrading the test tubes, and ensure that the tubes, at the end of the experiment, do not enter the general washing-up but are kept separately.

(*d*) It is necessary to use sufficient liquid in the tubes to ensure that the meniscus is above the photo-electric cells.

(*e*) Since response is not completely linear a reference curve can be made using a series of dilutions of a dense suspension.

STATISTICAL METHODS FOR THE SELECTION AND EXAMINATION OF MICROBIAL COLONIES

When samples of material containing mixed populations, e.g. milk, soil, water or foods, are plated onto non-selective media for an estimate of numbers, it is frequently also required that an estimate should be made of the relative numbers of the different kinds of micro-organisms which develop on the plates. It is then possible to calculate the percentage distribution of the various organisms present in the original sample. Representative samples of colonies cannot be secured by uncontrolled picking from Petri-dishes and various methods have been devised to ensure that a truly random sample of colonies is obtained. In this exercise it should be appreciated that a minor but still possibly important component may be overlooked completely. For example amongst a population of 10^9 pseudomonads there may be 10^7 *Aeromonas* but on average 100 colonies would have to be examined to reveal one *Aeromonas*. The particular method selected will depend on the numbers of colonies required for examination and on the numbers of colonies present on the isolation plates. Some of the methods in use are as follows.

A. Select every colony on a plate. This is a suitable method provided the numbers present are roughly equivalent to those required for examination.

B. Select every colony occurring in either a single sector, or in opposite sectors of a plate. Various methods for marking off sectors are available.

1. The sectors, e.g. quadrants, may be drawn with felt-tipped pen on the surface of the plate.

2. The plate may be superimposed on a Harrison's disc, and the appropriate sectors or colonies marked. Harrison's disc was devised by Harrison (1938) during a study of the numbers and types of bacteria occurring in cheese. It provides a convenient method of obtaining a representative sample when only a few colonies can be studied.

The disc is constructed by first drawing a circle the size of a standard Petri-dish base, then drawing 2 diameters at right angles to each other (i), and 2 further diameters at angles of 9° to the first pair (ii). The areas between the first diameters and those at 9° form four opposite sectors and constitute $\frac{1}{10}$ of the total area. An ideal plate should contain about 30 colonies in this area, but in practice this number might be too great, so that this area is further divided into four.

Three concentric circles with radii equal to 86·5 per cent, 70·7 per cent and 50 per cent of the radius of the disc are drawn with their centres at that of the disc (iii). This gives four equal areas in each of the 4 sectors defined by the original diameters.

Drawing a Harrison's disc

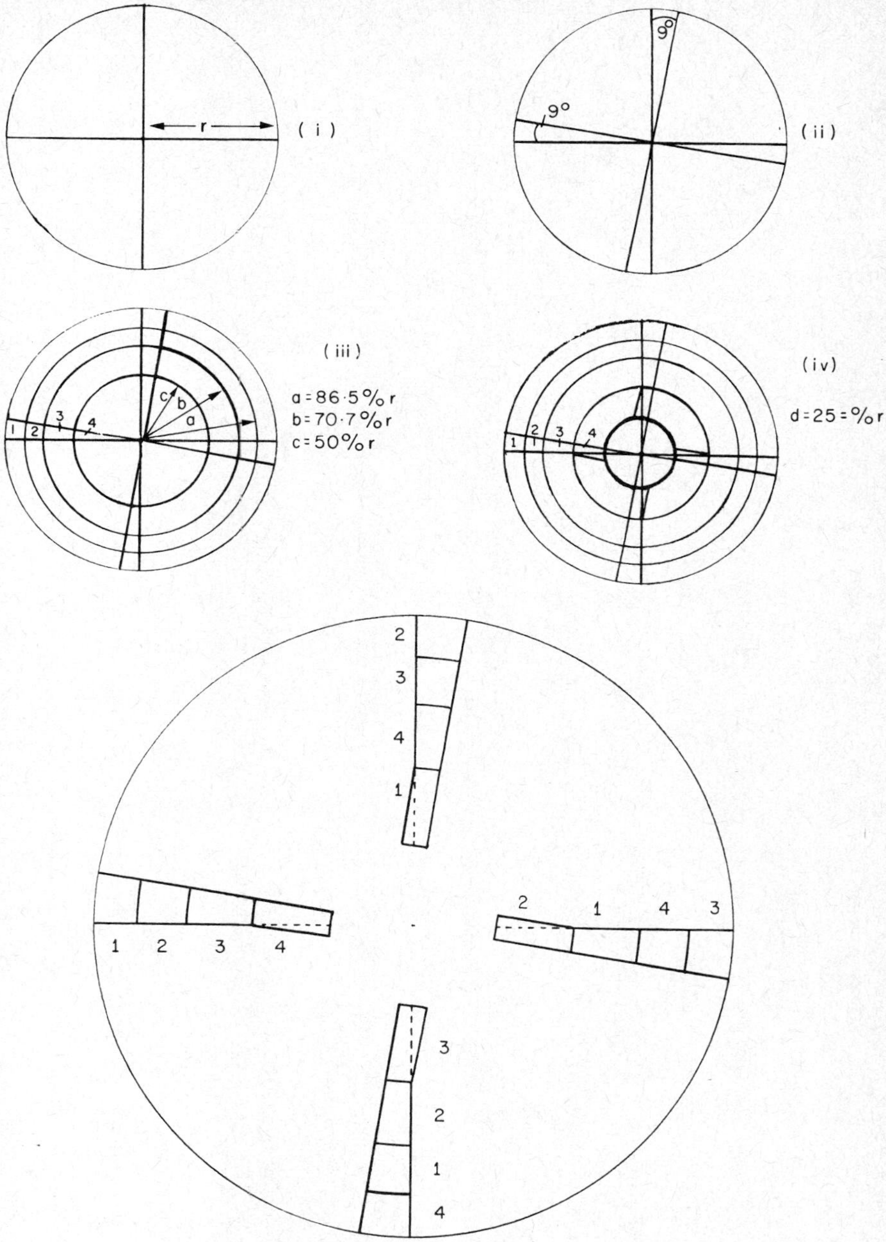

(v) The finished disc showing sampling areas only (this may be used as a template).

To avoid the necessity for sampling in the sharply pointed area at the centre of the disc a further concentric circle is drawn, having a radius of 25 per cent of the radius of the disc (iv). The sharply pointed portion at the centre can then be replaced by an equal area added alongside the rest of the fourth sector which then becomes roughly rectangular (v).

The four smaller areas thus produced in each sector are numbered 1 to 4, but in a changing sequence for each sector, thus: 1, 2, 3, 4; 2, 3, 4, 1; 3, 4, 1, 2; and 4, 1, 2, 3.

The Petri-dish to be examined is placed concentrically on the disc. First, all colonies occurring in the areas marked 1 are selected for examination. If this does not provide a sufficient number, all colonies occurring in the areas marked 2 are selected. The selection can be continued if necessary into areas 3 and 4, until the required number is obtained. Once a series of numbered sectors has been started, selection must continue until all colonies in this series have been marked off. All colonies lying over lines should be ignored.

C. If the plates have been surface inoculated all the colonies can be transferred by a replicator device. The simplest type of replicator is a sterile velveteen pad as described by Lederberg and Lederberg (1952). A suitable cylinder (e.g. of alloy) of a diameter slightly smaller than a Petri-dish is used. Over one end of the cylinder is stretched a piece of sterile velveteen (when autoclaving, reduce the pressure rapidly to ensure drying of the velveteen). This can be pressed lightly onto the surface of the medium bearing the colonies, and then pressed successively on plates of appropriate sterile media and finally on to a control medium to check successful transfer. Usually transfers to 9 plates + a control can be made. In this way every colony can be examined simultaneously for a number of physiological characteristics. If the velveteen is to be re-used it should be discarded after use into a container of water. It is then decontaminated by autoclaving, washed, rinsed, dried and brushed, before sterilising by autoclaving.

COMPOSITION OF CULTURE MEDIA

A. Introduction

A *culture medium* is any nutrient liquid or solid which can be used in the laboratory for the growth of micro-organisms. Such a medium may resemble the natural substrate on which the micro-organisms usually grow (e.g. blood serum for animal pathogens, milk for milk micro-organisms, soil extract for soil micro-organisms). Whatever the medium it must include all the necessary requirements for growth which vary according to the organism it is desired to grow but will include:

1. water;
2. nitrogen-containing compounds, e.g. proteins, amino acids, nitrogen-containing inorganic salts;
3. energy source, e.g. carbohydrate, protein;
4. accessory growth factors.

The nutritional requirements of bacteria range from the simple inorganic requirements of autotrophs to the many vitamins and growth factors required by some of the more fastidious bacteria (including pathogens and the lactic acid bacteria). Therefore, it is not possible to formulate a medium capable of supporting the growth of all micro-organisms. However, the commonly used *empirical media*, e.g. nutrient broth and nutrient agar, are capable of supporting the growth of many bacteria. Furthermore, such a medium as nutrient agar can be used as a *basal medium* to which is added, for example, blood to 5–10 per cent, serum, or milk, to provide the complex growth factors needed by the more fastidious bacteria. Lactic-acid bacteria require B-group vitamins which can be provided by the addition of yeast extract.

A nutrient medium can be made selective (see page 55) or biochemically diagnostic (see pages 56 and 66 onwards) by the addition of suitable compounds.

The media described on pages 52 to 58 are given to demonstrate the basic principles of media making. For a more comprehensive description of the design and formulation of media see Bridson and Brecker (1970).

B. Dehydrated Media

At one time it was necessary for microbiological laboratories to prepare not only the culture media, but also their constituents. It is now possible to obtain both the media constituents and complete culture media in dehydrated form. The dehydrated medium is dissolved in water to the stipulated concentration, and sterilised. Manufacturers' recommendations for methods of preparation should be carefully noted: in particular dehydrated media containing agar should be adequately soaked (15–30 min) with occasional agitation before any heat is applied. The medium will be of the correct composition and pH.

Most media are available in dehydrated form, but a few media cannot satisfactorily be prepared in this way. Dehydrated culture media are available from Oxoid Ltd., Difco Laboratories Inc., and Baltimore Biological Laboratories.

C. The Determination of the pH of Culture Media

1. pH indicator papers

Place a drop of medium onto the indicator paper and compare the resulting colour with the given colour chart. This method can only give an approximate idea of the pH and should always be checked by a more accurate method as in (2) or (3) below. Excellent indicator strips are manufactured by E. Merck (obtainable from B.D.H. Chemicals).

2. Comparator method with added indicator

The Lovibond comparator (manufactured by Tintometer Sales Ltd.) is convenient for this purpose and provides space for two tubes of standard size. A known amount of medium is added to each tube (5 ml or 10 ml) and to one tube only is added a standard amount of a suitable indicator (i.e. the final pH required in the medium falls within the pH range of the indicator). The tubes are placed in the comparator and the appropriate indicator colour disc placed in position. The disc is rotated until a match in colour is obtained. The pH of the medium will then be given by the disc reading. Some indicators used in pH adjustment of culture media include:

Indicator	Range of pH	Colour change
Phenol red	6·8–8·4	Yellow to purple-pink
Bromthymol blue	6·0–7·6	Yellow to blue
Bromcresol purple	5·2–6·8	Yellow to violet
Methyl red	4·4–6·0	Red to yellow

3. The pH Meter

This is the most accurate method of measuring the pH of culture media. Before use the pH meter must first be standardised against standard buffer solutions of known pH (usually pH 4 and pH 7). The electrodes of the pH meter are rinsed with distilled water from a wash bottle and then immersed in the culture medium contained in a beaker. The pH of the culture medium may then be read on the pH meter scale. Detailed instructions are provided by the manufacturers for the operation of each particular pH meter.

4. To adjust the pH of a culture medium

(*a*) *Using the comparator method*

(i) Determine the amount of 0·1 N NaOH (or 0·1 N HCl) needed to adjust 10 ml of medium to the required pH by using a graduated pipette slowly to add the base or acid to 10 ml of medium containing a standard amount of suitable indicator and mixing the contents by inverting the tube, until a colour change indicating the required pH is obtained.

(ii) Calculate the amount of N NaOH (or N HCl) required to adjust the pH of 1 litre of culture medium.

For example, 0·6 ml of 0·1 N NaOH were required to adjust the pH of 10 ml of medium to 7·2, so 6 ml of N NaOH are required to adjust the pH of 1 litre of medium to 7·2.

(iii) Add the required amount of N NaOH to the bulk medium as indicated by the calculation. Mix well. Check that the pH is satisfactory and readjust if necessary.

(*b*) *Using a pH meter*

The pH of a batch of culture medium is most easily adjusted using a pH meter by employing mechanical agitation.

The culture medium in a suitable container (e.g. a large beaker) should be continuously agitated, preferably by using a magnetic stirrer with a polythene- or PTFE-coated magnet. The pH meter is first standardised against standard buffer solutions of known pH. The glass and reference electrodes and resistance thermometer (if there is one) are then placed in the medium, and N HCl or N NaOH (as appropriate) added *very slowly* to the medium on the far side of the beaker from the electrodes until the required pH is reached and maintained. It is advisable to use one of the combined glass/reference electrodes which are partially enclosed in a Pyrex sheath, as these are more robust than the usual separate electrode assemblies. A spear electrode is most suitable for use with agar-containing media as it is easily cleaned. It should be remembered that the temperature compensation of the pH meter allows for the effect of temperature on the electrode system: in some diluents and media there is a considerable shift of pH with change of temperature.

D. Examples of Non-selective Culture Media

1. Nutrient broth

An empirical medium of general use for the cultivation of most bacteria (with the exception of lactic-acid bacteria and some fastidious pathogens).

Ingredients

Peptone	10 g
Meat extract (Lab-Lemco)	10 g
Sodium chloride	5 g
Distilled water	1000 ml

Preparation

1. Weigh out ingredients and heat in steamer until dissolved.
2. When cool, adjust pH to 7·6, with N NaOH. (Approximately 5–10 ml N NaOH are usually required.)
3. Autoclave at 121°C for 15 min, to precipitate phosphates.
4. Filter.
5. Adjust pH with N HCl to pH 7·2.
6. Distribute in bottles (screw caps should be left slightly loose except where bijou (5 ml) bottles are employed) or in test-tubes as required.
7. Autoclave at 121°C for 20 min. Tighten any screw caps on cooling.

2. Nutrient agar

A medium based on nutrient broth but solidified with agar, and of general use as indicated above.

Ingredients

Nutrient broth (pH 7·2)	1000 ml
Agar	15 g

Preparation

1. Weigh out the agar and dissolve in nutrient broth in the autoclave at 121°C for 20 min.
2. Check that the pH is 7·2, and adjust if necessary.
3. Prepare a filter in a Buchner-type filter funnel by first soaking two large (46 cm × 57 cm) sheets of Whatman No. 1 filter paper in water, mashing to a pulp, bringing the mixture to the boil in a large beaker and then pouring into the filter funnel while applying suction. This will result in a layer of hot paper pulp being evenly distributed over the base of the filter funnel and also will pre-warm the funnel.
4. Immediately place into position the filter flask to be used for the collection of the medium, and filter the nutrient agar while still hot.
5. Distribute in bottles (any screw caps should be left slightly loose except where bijou (5 ml) bottles are used) or test-tubes as required.
6. Autoclave at 121°C for 20 min. Tighten any screw caps on cooling.

3. Litmus milk

This is a natural medium, used to indicate the effect of pure cultures of bacteria on milk constituents, and to detect milk spoilage organisms in rinses, etc.

Ingredients

1000 ml of skim-milk, reconstituted from skim-milk powder.

Alternatively, good quality separated milk may be used if available.

Litmus solution.

Preparation

1. Add sufficient litmus solution (*c.* 1 per cent of a 4 per cent aqueous solution) to give a pale mauve colour.
2. Distribute in test-tubes or bottles as required.
3. Sterilise in a steamer at 100°C for 30 min on each of three successive days (intermittent sterilisation) (see also page 346). *Check sterility by pre-incubation prior to use*, preferably at two incubation temperatures and including that to be used in the investigations.

E. Separation of Mixed Cultures—Enrichment Procedures, Elective and Selective Media

A mixed culture can generally be separated into its constituent organisms (provided that the individual strains are present in approximately equal numbers) by using a streak plate method as described on page 23. Isolated colonies can then be picked off and subcultured as required. Some of the problems in the use of selective media to detect organisms (especially pathogens) in foods are discussed further on pages 139 ff.

Enrichment procedures

When, however, the required organism is outnumbered or is accompanied by many unwanted species, as in soil, milk or water, it may be necessary to use an enrichment technique to increase the numbers of the required organism. The methods adopted should be chosen to take advantage of any known physiological characters of the particular organism, e.g. optimum temperature or pH, nutritional requirements, tolerance of added inhibitors.

In some cases, pretreatment of the material itself is appropriate, e.g. heat-treatment of soil suspensions or milk for the isolation of sporing or other heat-resistant organisms, followed by plating on solid media. Advantage may also be taken of the differing optimum temperatures of micro-organisms in a mixed population, so that incubation of material at the optimum temperature of the required organism should increase its numbers in relation to other organisms present. Since, however, enrichment takes place in liquid media, the particular enrichment procedure adopted is unlikely to yield a pure culture, and the final isolation procedure still involves isolation of separate colonies on solid media.

Elective media

In other cases, particularly in the isolation of soil micro-organisms, a medium which satisfies the minimum nutritional requirements of the organisms concerned—known as an elective medium (van Niel, 1955)—may be very useful, particularly if the required organisms have unusual nutritional characteristics. For example, "wild" yeasts can be isolated using lysine agar, on which organisms cannot grow unless they can use lysine as a sole source of nitrogen (Morris and Eddy, 1957).

Selective media

Another important method of separating mixed cultures is to make use of a selective medium. This is a basic medium which may support growth of many types but which has been modified to include one or more inhibitory agents, thereby restricting the growth of organisms not required. The choice of a selective medium must therefore be appropriate for the isolation of the particular organism concerned with reference to the nature of the samples being examined. Inhibitory substances used in the preparation of selective media include dyes, antibiotics, bile salts and various inhibitors affecting the metabolism or enzyme systems of particular species.

One form of a selective medium is that in which the pH of the medium has been modified so that it is suitable for the growth of only acid-tolerant or alkali-tolerant species. For example, yeasts, moulds and lactobacilli are acid tolerant organisms and can grow on media at pH 4–5, whereas less acid-tolerant organisms are unable to grow.

(*a*) *Selective media for Gram-negative bacteria.* Crystal violet used in a medium at a final concentration of 1 : 500,000 inhibits the growth of many Gram-positive bacteria (though not streptococci) while permitting that of Gram-negative bacteria (Holding, 1960). Penicillin incorporated at a final concentration of 5–50 units/ml inhibits many Gram-positive organisms, and the medium thus becomes selective for Gram-negative bacteria.

MacConkey agar is a selective medium used for the detection and isolation of coliforms. In this case the bile salts act as the selective agent, intestinal and coliform organisms being inhibited to a lesser extent than other organisms.

(*b*) *Selective media for Gram-positive bacteria.* Potassium tellurite, thallium acetate and sodium azide added to media to give a final concentration of 1 : 2000–1 : 10,000 have been found to inhibit the growth of Gram-negative bacteria, and these substances are therefore frequently used in selective media for Gram-positive bacteria. Glucose azide broth, for example, is used in the detection of faecal streptococci in water supplies (see Part II, page 229). Similarly, thallium acetate in a glucose agar has been useful for the isolation of lactic streptococci from sour milk, and potassium tellurite for the isolation of *Corynebacterium*.

Differential media

A differential medium is one in which certain species produce characteristic colonies which can easily be recognized. For example, haemolytic and non-haemolytic species can be distinguished by the examination of colonies formed on blood agar, a non-selective medium. In many cases, however, a medium may be both selective and differential. For example, lactose-fermenting coliforms produce red colonies on MacConkey's agar (as a result of acid production affecting the neutral red indicator), while non-lactose-fermenting intestinal organisms such as *Salmonella* spp. produce colourless colonies.

F. Examples of Selective Culture Media

MacConkey's broth (bile salt broth)

This is a combined selective and differential medium used in the enrichment and detection of coliforms (i.e. the lactose-fermenting Enterobacteriaceae) in, for example, milk and water.

Ingredients

Peptone	20 g
Bile salts (Oxoid L55)*	5 g
Sodium chloride	5 g
Lactose	10 g
Bromcresol purple (a 1 per cent ethanolic solution)	1 ml
Distilled water	1000 ml

N.B. Recipes for milk testing may specify the addition of 2·5 ml of a 1·6 per cent solution of bromcresol purple in place of the amount indicated above.

Preparation

1. Add all the ingredients except lactose and heat in a steamer for 1–2 hours.
2. Add the lactose and dissolve by steaming for a further 15 min.
3. Cool and then filter.
4. Adjust the reaction to pH 7·4 with N NaOH.
5. Add 1 ml of 1 per cent ethanolic solution of bromcresol purple (or 2·5 ml of a 1·6 per cent solution according to the specification and standard procedure being followed).
6. Distribute in 5 ml amounts in 150 × 16 mm test-tubes provided with Durham tubes.

* Different proprietary preparations of bile salts may not have equivalent activity at the same concentration.

7. Sterilise in a steamer at 100°C for 30 min on each of three successive days, or in the autoclave at 121°C for 15 min.

MacConkey's agar (bile salt neutral red lactose agar)

This of general use in the detection and isolation of members of the Enterobacteriaceae.

Ingredients

Peptone	20 g
Bile salts (Oxoid L55)*	5 g
Sodium chloride	5 g
Lactose	10 g
Neutral red (a 1 per cent aqueous solution)	7 ml
Agar	20 g
Distilled water	1000 ml

Preparation

1. Dissolve the peptone, bile salts and sodium chloride in the water by steaming.
2. Cool and adjust the pH to 7·4–7·6 with N NaOH.
3. Add the agar and dissolve in the autoclave at 121°C for 15 min.
4. Prepare a filter in a Buchner type funnel (see p. 53) and warm with hot water.
5. Filter the medium while hot.
6. Check the pH and adjust to 7·4 if necessary.
7. Add the lactose and 7 ml of neutral red solution.
8. Dissolve in the steamer.
9. Distribute in bottles and test-tubes as required.
10. Sterilise in the autoclave at 115°C for 15 min.

On this medium after 24 h incubation, colonies of lactose-fermenting organisms are pink or red, while those of non-lactose-fermenters are colourless. Typical appearances are as follows: *Escherichia coli* – red, non-mucoid; *Enterobacter aerogenes* – pink, mucoid; *Salmonella paratyphi* – colourless.

Modifications of MacConkey's agar

The basic MacConkey's agar medium given above may be modified to give improved selectivity for, and definition of, coliform colonies. In the medium prepared in dehydrated form by Difco as Bacto MacConkey Agar and by Oxoid as MacConkey Agar No. 3 the modification consists in (*a*) the addition of crystal violet at 0·001 g/litre (1 ppm), and

(*b*) by including a more refined and more effective bile salt* at lower concentration.

Violet red bile agar. In this medium (also available in dehydrated form) the MacConkey's agar has been modified by

(*a*) lowering the concentration of peptone, bile salt (a more refined and more effective preparation* is used);
(*b*) the addition of crystal violet to 0·002 g/litre (2 ppm); and
(*c*) the addition of yeast extract.

In the pour plate technique, many workers prefer to cover this medium when poured and set with a further 5 ml of melted medium at 50°C, so as to restrict surface colony formation, this being particularly advisable if the medium is being used for the quantitative estimation of coliforms. Submerged coli-aerogenes colonies appear dark red and are usually 1–2 mm in diameter.

* See footnote page 56.

METHODS OF ANAEROBIC CULTURE

In order to grow obligate anaerobes, cultures must usually be kept in an oxygen-free environment. This can be achieved by the use of a natural medium containing reducing substances (e.g. Robertson's cooked meat medium) or by the modification of other media through the addition of various reducing substances (e.g. glucose, sodium thioglycollate, ascorbic acid, cysteine). Media should not be stored for long periods. Dissolved oxygen should be removed from a medium by heating, and atmospheric oxygen subsequently excluded by sealing, by solidification of an agar medium in deep layers, or by incubating the cultures in an oxygen-free atmosphere. Usually a combination of chemical and physical methods is used to achieve anaerobiosis in a culture (for a further discussion of methods see Willis, 1969).

Some very strict anaerobes, particularly members of the asporogenous family, Bacteroidaceae, require complete protection from oxygen if their death is to be avoided. If such organisms are being studied it is desirable to adopt special techniques for the handling of samples and for the preparation of cultures and even of culture media (Hungate, 1969). Because of their extreme oxygen sensitivity, however, these organisms, which are common intestinal inhabitants, apparently die rapidly in most foodstuffs so that they have not yet been studied extensively by food microbiologists.

A. Robertson's Cooked Meat Medium

The medium consists of minced meat (usually bullock's heart) in a nutrient broth, which is distributed in 25 ml screw-capped McCartney bottles. The minced meat contains reducing substances which produce and help to maintain anaerobic conditions in the medium. This medium is useful for the culture of anaerobes. It indicates proteolytic activity (e.g. of *Clostridium sporogenes*) by blackening and disintegration of the meat, and saccharolytic activity (e.g. of *Cl. perfringens*) by reddening of the meat particles. It is also useful for the preservation of stock cultures of aerobes, microaerophiles and anaerobes.

Although this medium is available in dehydrated form it has been our experience that such preparations are not as reliable for the cultivation of anaerobes as cooked meat medium prepared in the laboratory.

Preparation (Lepper and Martin, 1929)

Mince 500 g of fresh, fat-free bullock's heart and simmer for 20 min in 500 ml of boiling 0·05 N sodium hydroxide. After cooking, adjust the pH to 7·4. Strain off the liquid and dry the meat by spreading on filter paper.

Distribute the meat in 25 ml McCartney bottles to a depth of 2 cm. Add 10 ml of peptone water or nutrient broth (see page 52). Sterilise by autoclaving for 20 min at 121°C with the screw caps slightly loose. After autoclaving, tighten the caps. Immediately before use boil for 10 min to remove dissolved oxygen and cool rapidly.

B. Shake Cultures

These are test-tubes or bottles containing a solid medium. The melted medium is inoculated and then allowed to solidify. The simplest medium to use is nutrient agar with the addition of 1 per cent glucose.

1. Liquefy the medium and maintain at 100°C for 10 min to drive off any dissolved oxygen. Cool to 45°C and use immediately.
2. Add the inoculum and mix the contents of the tube well by rotation between the hands, avoiding the introduction of air bubbles.
3. Solidify the agar by placing the tube in cold water.
4. Incubate at the required temperature.

Since oxygen will diffuse only very slowly through the solid medium, the conditions in the lower parts of the tube will be suitable for the growth of anaerobes. This method is also suitable for the growth of micro-aerophiles and facultative anaerobes.

C. Semi-solid Media

Semi-solid media contain 0·02–0·3 per cent agar, which is sufficient to prevent convection currents and therefore helps to prevent the diffusion of oxygen into the medium.

One of the most suitable media is prepared from a nutrient medium (e.g. nutrient broth) with the addition of 0·5 per cent glucose (a reducing compound as well as being an energy source), 0·1 per cent sodium thioglycollate, and 0·02–0·2 per cent agar to prevent convection currents, dispensed in McCartney bottles or in deep layers in test-tubes. As an indicator of oxidation-reduction potential 0·0002 per cent methylene blue (0·2 ml of a 1 per cent aqueous solution of methylene blue per litre of medium) can be added.

The additional substances should be added to the nutrient broth, mixed well and steamed until dissolved. The medium is then distributed in 12-ml amounts in 150 × 16 mm test-tubes, which are plugged and sterilised by autoclaving at 121°C for 15 min. The methylene blue acts as an indicator of redox potential, being blue in the oxidised and colourless in the reduced condition. Should the medium show signs of turning greenish-blue after storage, it should be reheated before use to remove the oxygen, 10 min in a boiling water bath usually proving sufficient.

Reinforced clostridial medium (RCM)

This is a semi-solid medium developed by Hirsch and Grinsted (1954) which, although it does not allow the growth of all obligate anaerobes, is very useful for the enumeration of anaerobes in, for example, food samples. RCM contains glucose and cysteine as reducing compounds and, when distributed into 25 ml screw-capped McCartney bottles, provides anaerobic conditions without the need for a paraffin or other seal. It is particularly recommended that RCM be used also as the *diluent* in the determination of viable counts of anaerobes.

D. Vaseline, Paraffin Wax or Agar Seals

By the use of sterile sealing compounds, liquid media can be maintained anaerobically. The tubes containing the liquid medium should be heated in boiling water for 10 min to drive off the dissolved oxygen and sterile melted Vaseline or paraffin wax added to each tube to seal the medium from the air. The tubes are inoculated by means of a capillary pipette after melting the Vaseline. (Alternatively, the tubes can be inoculated immediately àfter heating and cooling, and *then* the seal added.)

This method can be used to demonstrate the presence of *Clostridium perfringens* in water or milk by the "stormy-clot" reaction, in which the medium is sterile milk rendered anaerobic and the surface sealed as above.

Vaspar is a useful sealing compound prepared by melting together approximately equal amounts of petroleum jelly (Vaseline) and paraffin wax. Sterile liquid paraffin can also be used, but it will not allow the detection of gas production.

E. The Anaerobic Jar

Petri-dishes or tubes containing any medium to be incubated in an oxygen-free atmosphere for the isolation and growth of anaerobes, can be enclosed in an anaerobic jar. Such a jar also enables viable counts of anaerobes to be performed more readily than by other methods. Counts of the less exacting anaerobes, e.g. *Cl. perfringens*, can often be carried out without the necessity for using anaerobic jars by employing deeply poured plates or shake tubes using suitable media (see page 229 of this manual).

The air in the anaerobic jar may be replaced by oxygen-free nitrogen. Alternatively the anaerobic jar makes use of the catalytic non-explosive reaction between hydrogen and oxygen to remove free oxygen from the atmosphere inside the jar, leaving the cultures in an oxygen-free atmosphere composed largely of hydrogen. There are two types of jar available, one using a cold catalyst to bring about the reaction, the other—known as McIntosh and Fildes' jar—using a hot catalyst. The catalyst in the McIntosh and

Fildes' jar is heated by the passage of an electric current, the heating coil being surrounded by wire gauze to prevent the occurrence of an explosive reaction between the hydrogen and oxygen. In order to minimize further the risk of explosion, it is advisable to remove most of the air from the jar with a vacuum pump before passing in hydrogen. The area in which the jars are being filled with hydrogen should have suitable safety screening. Suitable safety screens can be obtained from many laboratory furnishers and suppliers (e.g. Fisher Scientific Co., A. Gallenkamp and Co.). Windows and partitions of ordinary glass in existing laboratories can be rendered safe by the application of the self-adhesive polyester film "Safetykling" (Ozalid Co. Ltd.).

While the cold catalytic jar carries no risk of explosion, it has the slight disadvantage of needing occasional renewal of the catalyst since the catalyst gradually becomes "poisoned". Nevertheless, the advantages of the cold catalytic jar—its safety, and the fact that the reaction will continue between hydrogen and oxygen as long as both gases are present, rather than be terminated when an electric current is removed—outweigh this one slight disadvantage. The following procedures therefore describe the use of a cold catalytic jar (such as the B.T.L. Anaerobic jar, manufactured by Baird & Tatlock (London) Ltd.), but the use of a McIntosh and Fildes' jar is broadly similar, and for further details, the reader is referred to Cruickshank (1965).

1. Procedure using external hydrogen supply

Place the cultures inside the jar. If Petri-dishes are included, separate the two halves of each dish slightly with a narrow strip of filter paper to prevent sealing by condensed water.

In addition to the cultures, also include a redox indicator. If the B.T.L. Cold Catalytic Anaerobic jar is used, indicator capsules are provided which attach to a side arm of the outside of the jar. This indicator is blue in the presence of oxygen but becomes colourless under anaerobic conditions. Should a different type of jar be used, the indicator can consist of a tube of the thioglycollate medium described on page 60. Alternatively, an alkaline glucose solution containing methylene blue can be prepared as described by Cruickshank (1965) by mixing in a plugged tube equal quantities of 0·006 N sodium hydroxide, a 0·015 per cent aqueous solution of methylene blue, and a 6 per cent solution of glucose. This requires boiling until it becomes colourless and then it is immediately placed inside the anaerobic jar.

Next, clamp the lid in position. If the jar and lid are not fitted with a rubber or silicone-rubber gasket, the rim of the jar should be first smeared with high-vacuum silicone grease.

Connect one of the two taps in the lid of the jar to a filter pump fitted with a vacuum gauge. The hydrogen supply must be connected to the jar via a variable reducing valve with pressure gauge and fine adjustment which allow the hydrogen to be delivered at a pressure of 14-kN/m^2 or less. If a wash-bottle is included in the line between the reducing valve and the anaerobic jar, the passage of the hydrogen may be observed. An alternative to using a hydrogen cylinder is Kipp's apparatus.

First apply suction with the filter pump until the jar is evacuated to a pressure of *c*. 5 cm of mercury and then close the tap. Next, adjust the flow of hydrogen through the wash-bottle to approximately one bubble/sec and connect to the second tap of the anaerobic jar, open the tap and allow the hydrogen to pass slowly into the jar for 5 min after the pressure inside the jar reaches atmospheric pressure. Close the tap, remove the hydrogen supply and vacuum line and incubate at the desired temperature. The indicator should decolorise after a few hours, and remain colourless until the jar is opened at the end of the incubation period. Failure of the indicator to decolorise may be due to a leak through side arm, lid gasket or taps or to poisoning of the catalyst.

Usually, it will be found after incubation that there is a slight negative pressure inside the jar which should be released by opening one of the taps.

2. The GasPak System

This system (manufactured by Baltimore Biological Laboratories) provides a hydrogen supply within the anaerobic jar by the use of a disposable foil envelope. The envelope in one section contains sodium borohydride, which generates hydrogen when exposed to water, and in a second section a mixture of citric acid and sodium bicarbonate which generate carbon dioxide when water is added. The GasPak system has been described by Brewer and Allgeier (1966).

The hydrogen is reacted with the oxygen present in the air within the jar by using a cold catalyst. However, the jar is not first evacuated, and because of the high oxygen content of the jar the catalyst can become very hot, so that it is important to use a double layer of wire gauze to hold the catalyst. Therefore, the GasPak system should not be used in the B.T.L. jar unless the catalyst container is suitably modified. The GasPak polycarbonate jar offers the distinct advantage over other types of jar of automatic venting. However, a methylene blue indicator must be used to monitor catalyst activity since the lack of gas outlets prevents the use of a manometer (and additionally prevents the use of gas cylinders).

When anaerobic counts are to be performed infrequently, the GasPak system is convenient as although the system is more expensive it takes up very little space, does not require special precautions against fire hazards, and cylinder hire charges are avoided.

It is unfortunate that the manufacturers do not offer hydrogen-only envelopes. Although many anaerobes are stimulated by carbon dioxide concentrations up to 5 or even 10 per cent, this is not an invariable rule, and Futter (1967) found that the presence of 5 per cent carbon dioxide suppressed the recovery rate for *Clostridium perfringens* spores to 53 per cent compared with the corresponding carbon dioxide-free anaerobic atmosphere.

3. Replacement by Nitrogen

After placing the cultures in an anaerobic jar (which possesses gas outlets) and clamping down the lid, apply suction with a filter pump until the jar is evacuated to a pressure of 4 cm of mercury and then close the tap.

Fill the jar with oxygen-free nitrogen to atmospheric pressure. Repeat evacuation and refilling twice more.

The level of residual oxygen (i.e. derived from the original air in the jar) after three cycles of evacuation and replacement by nitrogen will be approximately 0·003 per cent. Futter (1967) reported that commercially available "oxygen-free nitrogen" may contain 0·3 per cent v/v oxygen, but this can be removed by passing the gas over heated copper turnings.

4. Selection of the method to be used

Which of these three methods is to be employed will depend partly upon the probable frequency of use, the size of laboratory, the ability to take adequate safety precautions if hydrogen cylinders are to be used and the proximity to the laboratory of a supplier of hydrogen cylinders or nitrogen cylinders.

In addition there may be microbiological requirements to be met. At present comparatively little is known about the growth responses of most anaerobes to the different gaseous atmospheres which result from these three methods. Futter (1967) and Futter and Richardson (1970, 1971) found that the effect of different hydrogen/nitrogen mixtures on the recoverability of *Clostridium perfringens* spores depended upon the previous treatment of the spores. Recoverability of undamaged spores was independent of the hydrogen/nitrogen mixture used. Spores that had survived heat treatment showed poorest recovery (50–60 per cent) in a 1 : 1 hydrogen/nitrogen mixture, and good relative recovery in 100 per cent hydrogen or 100 per cent nitrogen. In the case of spores surviving either ethylene oxide treatment or γ-irradiation, progressively lower recoveries were obtained as the hydrogen content of the mixture was increased from 0 to 100 per cent. These results suggest that if *Clostridium perfringens* is being studied in heat-processed foods, method 1 or method 3 should be used in preference to GasPaks.

CULTIVATION IN A CARBON DIOXIDE-ENRICHED ATMOSPHERE

A few bacteria (e.g. some *Brucella*) will grow only in an atmosphere which contains a high concentration of carbon dioxide, and others, although capable of growing in ordinary air, can grow better in an atmosphere enriched with carbon dioxide. As already described on page 63 many anaerobes also prefer or require carbon dioxide in the oxygen-free atmosphere. It is relatively easy to provide a carbon dioxide-enriched atmosphere in the laboratory. Place the Petri-dish or test-tube cultures in a "half-size" biscuit tin (approximately $22\frac{1}{2} \times 20 \times 12\frac{1}{2}$ cm) together with a beaker containing about 15 ml of 2 N hydrochloric acid. Drop into the acid a marble chip of 1–1·2 g and put the tin lid in place. Seal the lid with Sellotape. The marble will react with the hydrochloric acid to give a concentration of about 5 per cent of carbon dioxide inside the tin.

Alternatively GasPak carbon dioxide generators are available (Baltimore Biological Laboratories).

BIOCHEMICAL TESTS FOR BACTERIA

The tests described in this section are a fairly limited selection of tests which are of general usefulness in the characterisation and identification of bacteria. Some organisms require the use of a medium specially adapted to particular nutritional or osmotic requirements. In other cases, a selective medium may be employed which not only allows the isolation of the organism under investigation but also incorporates one or more biochemical tests. Thus, where necessary, the appropriate special media are described in the various sections of Parts II and III of this manual.

It is important to check that each batch of a medium is satisfactory by using a strain of bacterium known to give a positive reaction. The eighth edition of *Bergey's Manual* (Buchanan *et al.*, 1974) lists the reactions of many type strains, and the sources of suitable cultures. In addition, where appropriate, incubate an uninoculated tube or plate with each test in order to detect false positive reactions due to impurities in, or a deterioration of, medium or reagents.

A. Reactions Involving Protein, Amino Acids and other Nitrogen Compounds, Including Tests for Proteolytic Activity

1. Hydrolysis of gelatin

(*a*) *Nutrient gelatin*

Medium. Nutrient broth with the addition of 10–15 per cent gelatin, fin a pH 7·2. Sterilise by autoclaving for 20 min at 115°C.

Stab inoculate a tube of nutrient gelatin and incubate at 20–25°C for up to 30 days. If growth is poor at 25°C, tubes may be incubated at the optimum temperature.

Recording result. Liquefaction of the test medium when an uninoculated tube has remained solid indicates that hydrolysis has occurred. Record the shape and extent of the liquefied portion of the gelatin. When tubes have been incubated above 25°C, the culture must be immersed in iced water for 5 min before being examined for hydrolysis.

(*b*) *Frazier's gelatin agar* (*modified*) (Smith, Gordon and Clark, 1952)

Medium. Nutrient agar + 0·4 per cent gelatin, final pH 7·2. Sterilise by autoclaving for 20 min at 115°C.

Inoculate a poured, dried plate of the medium by streaking *once* across the surface or placing a spot of inoculum and incubate at the optimum growth temperature for 2–14 days.

Test reagent. Mercuric chloride solution.

Recording result. Flood plates with 8–10 ml of test reagent. Unhydrolysed gelatin forms a white opaque precipitate with the reagent. Hydrolysed gelatin appears therefore as a clear zone. Record the width of the clear zone in mm from the edge of a colony to the limit of clearing.

(*c*) *Gelatin-charcoal discs* (Greene and Larks, 1955)

This method enables a result to be recorded usually after less than 24 h. The gelatin-charcoal discs used consist of formalin-denatured gelatin which contains charcoal. Although these discs can be prepared in the laboratory, they are also obtainable ready-made (Oxoid).

Medium. Peptone water dispensed in 1 ml amounts in 75 mm × 10 mm test-tubes. Sterilise by autoclaving for 15 min at 121°C. Immediately before use, add aseptically one gelatin-charcoal disc to each tube.

Pre-warm the test medium to the incubation temperature and then *heavily* inoculate from young nutrient agar slope cultures. Incubate at the optimum temperature in a water-bath. Examine at frequent intervals (e.g. hourly for 6 h and then less frequently after this period), extending the incubation time of negative reactors for up to 24 h.

Recording result. Hydrolysis of the gelatin causes particles of charcoal to become liberated and to settle to the bottom of the tube. On shaking, the charcoal particles are resuspended and can be readily seen.

It should be noted that formalin-denatured gelatin does not melt at the same low temperature as normal gelatin and consequently this test may be carried out at the optimum temperature of the organism being investigated or at any other incubation temperature required.

2. Hydrolysis of casein

Medium. Milk agar, which consists of agar with the addition of skim-milk to 10 per cent, sterilised by autoclaving for 20 min at 115°C. A more opaque milk agar medium may be made by mixing 10 ml of hot sterile 2·5 per cent water agar with 5 ml of hot sterile skim-milk (giving a 30 per cent milk agar) immediately before pouring the plates. This is best used in a double layer plate, with a thin layer of the 30 per cent milk agar overlaid onto 10 ml of water agar previously poured and set.

Inoculate a poured dried plate of the medium by streaking *once* across the surface and incubate at the optimum growth temperature for 2–14 days.

Test reagent. Mercuric chloride solution (as used for Frazier's gelatin medium), 1 per cent hydrochloric acid or 1 per cent tannic acid solution.

Recording result. Clear zones which are visible after incubation of the plates are presumptive evidence of casein hydrolysis. However, false positives may occur (Hastings, 1904) and to confirm that clearing is a result of casein hydrolysis, flood the plates with one of the test reagents which are protein precipitants. Record width of clear zone in mm. (See Smith *et al.*, 1952; Chalmers, 1962).

3. Hydrolysis of coagulated serum

Medium. Loeffler's serum, consisting of three parts serum + one part glucose nutrient broth. Since comparatively low temperatures are used in the preparation of this medium, the glucose nutrient broth should be sterilised by autoclaving before mixing with the serum (sterile sera are obtainable from, e.g. Wellcome Reagents, Oxoid Ltd.) in a sterile flask, and then the medium distributed aseptically into sterile small (5 ml) screw-capped bottles. The medium is then coagulated in the sloping position by heating slowly to 85°C (see also page 86). The medium may be sterilised by heating in the inspissator at 85°C for 20 min on each of three successive days.

Inoculate the serum slope by streaking across the surface in the usual way and incubate at the optimum growth temperature for 2–14 days.

Recording result. Visual examination will indicate whether hydrolysis (liquefaction) of the coagulated serum has occurred.

4. Production of indole from tryptophan

Medium. Peptone water (tryptone 1–2 per cent, sodium chloride 0·5 per cent, final pH 7·2) dispensed in 5 ml amounts in test-tubes and sterilised by autoclaving for 15 min at 121°C. Tryptone is the peptone to be recommended for this test, since it is rich in tryptophan, one of the amino acids destroyed by the usual methods of preparing peptones. Inoculate as for a broth culture, preferably from young agar slope cultures, and incubate at the optimum growth temperature for 2–7 days.

Test reagent. Several reagents and methods are available for the detection of indole. Kovacs's reagent is to be preferred, being rather more sensitive due to the higher solubility of the dye complex in the amyl alcohol layer.

Recording result. Add 0·5 ml Kovacs's indole reagent, shake tube gently and then allow to stand. A deep red colour develops in the presence of indole which separates out in the alcohol layer (see Report, 1958a).

5. Production of ammonia from peptone or arginine

(*a*) *Peptone water*

Inoculate a tube of peptone water (peptone 1 per cent, sodium chloride 0·5 per cent) and incubate with a sterile control tube at the optimum growth temperature for 2–7 days.

Test reagent. Nessler's reagent.

Recording result. Add a loopful of culture to a loopful of Nessler's reagent on a slide or glazed porcelain tile, or add 1 ml culture to 1 ml Nessler's reagent in a clean tube. The development of an orange to brown colour indicates the presence of ammonia. The sterile control tube should be tested at the same time and should turn pale yellow or show no colour reaction.

(*b*) *Arginine broth* (Abd-el-Malek and Gibson, 1948a)

This medium is used mainly for the differentiation of streptococci. Inoculate a tube of arginine broth and incubate with a sterile control tube at the optimum growth temperature for 2–7 days.

Test reagent and recording result. As for method (*a*).

(*c*) *Thornley's semi-solid arginine medium* (Thornley, 1960)

This medium contains phenol red as a pH indicator; it should be dispensed in 5 ml (bijou) screw-capped bottles to a depth of about 2 cm. It is used mainly for the differentiation of Gram-negative rods. Inoculate a bottle of medium and then seal the surface with sterile liquid paraffin or vaspar. Incubate with a sterile control at the optimum growth temperature for 2–7 days.

Recording result. Hydrolysis of arginine, with the formation of ammonia, results in alkalinity and is indicated by a change in colour of the medium to red from salmon pink. *Pseudomonas* and *Aeromonas* produce ammonia from arginine in the sealed medium, whereas members of Enterobacteriaceae do not. The former can grow *anaerobically*, by being able to use arginine for the generation of ATP without needing oxygen.

6. Production of hydrogen sulphide

This may result from the decomposition of organic sulphur compounds, e.g. cysteine and cystine, or from the reduction of inorganic sulphur compounds, e.g. sulphite.

(*a*) *Cystine or cysteine broth*

A basal medium of peptone water or nutrient broth is used, with the addition of 0·01 per cent cystine or cysteine. Indicator papers are used which consist of filter paper soaked in saturated lead acetate solution, dried, cut in strips and sterilised.

Inoculate, and then insert a strip of the indicator paper between the plug and the glass with the lower end above the medium. Incubate at the optimum growth temperature for 2–7 days, together with an uninoculated control.

Recording result. Production and liberation of hydrogen sulphide causes blackening of the lead acetate paper. If no blackening has occurred by the end of the incubation period, add 0·5 ml of 2 N hydrochloric acid and replace the plug and lead acetate paper immediately. If any sulphide has been produced but has remained in solution, the addition of the acid will cause the liberation of hydrogen sulphide (Skerman, 1959). Treat the uninoculated control tube similarly to check for possible false positives.

(*b*) *Ferrous chloride gelatin*

The medium is prepared by adding freshly prepared 10 per cent ferrous chloride solution to boiling nutrient gelatin to give a final concentration of ferrous chloride of 0·05 per cent, followed by dispensing into sterile narrow tubes, quick cooling and sealing with sterile air-tight (e.g. rubber) stoppers. Inoculate by stabbing. Incubate at 20–25°C for 7 days. Production of H_2S is shown by blackening of the medium.

N.B. This medium will also indicate liquefaction of gelatin. This is the recommended method for differentiation within the Enterobacteriaceae (see Report, 1958a).

(*c*) *Kliger's iron agar*

This is a complex medium containing 0·03 per cent ferric citrate and may be used for the differentiation of members of the Enterobacteriaceae. It is available in dehydrated form. After sterilisation, the medium is slanted with a deep butt (3 cm butt, 4 cm slant). Stab inoculate the butt of the medium, and streak the slant. Incubate at optimum temperature for up to 7 days. Production of H_2S is shown by blackening of the medium. In addition to the detection of hydrogen sulphide production, the medium also contains lactose and glucose for the differentiation of organisms on the basis of sugar fermentation.

7. Production of ammonia from urea

Christensen's urea agar (Christensen, 1946).

The basal medium is distributed in bottles or test-tubes, heat-sterilised and cooled to 50°C. Sufficient 20 per cent urea solution, previously sterilised by filtration, is then added to give a final concentration of 2 per cent. The medium is slanted, allowed to set, and is then ready for use.

Inoculate as for a slope culture. A control of basal medium containing no added urea should also be inoculated at the same time to check that ammonia is produced from urea and not from peptone. Incubate at the optimum growth temperature for 1–7 days.

Recording result. Urease production and subsequent hydrolysis of urea results in the production of ammonia, which increases the pH as shown by a change in colour of the medium from yellow to pink or red.

8. Reduction of nitrate

Nitrate peptone water is the medium employed in this test, consisting of peptone water with the addition of 0·02–0·2 per cent potassium nitrate (analytical reagent grade). The medium is distributed in tubes, each with an inverted Durham tube, and sterilised by autoclaving for 15 min at 121°C.

Inoculate as for broth culture and incubate together with a sterile control tube at the optimum growth temperature for 2–7 days.

Test reagents. Griess-Ilosvay's reagents (modified).

Recording result. Add 1 ml of each of the two reagents to the culture and to the control tube. The presence of nitrite is indicated by the development of a red colour within a few minutes. The control tube should show little or no coloration. Alternatively Merck nitrite test strips may be used (obtainable from Anderman & Co. Ltd.).

A negative result should be confirmed by the addition of a small quantity of zinc dust to the tube. This reduces to nitrite any nitrate still present. Thus the development of a red colour indicates that some nitrate remains. If the addition of zinc does not result in the development of colour, no nitrate remains, the nitrate having been reduced by the culture beyond the nitrite stage. The presence of gas in the Durham tube indicates the formation of gaseous nitrogen.

The details of nitrate reduction by bacteria are complex, and interested readers are referred to the review by Payne (1973).

9. Action on litmus milk

Inoculate as for a broth culture and incubate at the optimum growth temperature for up to 14 days.

Recording result. Examine tubes daily and record any changes in the medium. A number of different reactions and combinations of reactions may occur involving (1) lactose, (2) casein, (3) other milk constituents.

1.(a) Acid production shown by a change in the colour of the litmus to pink.
(b) If sufficient acid is produced the milk will clot. This is known as an acid clot.
(c) Reduction of the litmus and loss of colour may occur. This may precede or follow other changes.
(d) Gas may also be produced and show as gas bubbles in the medium, although normally this is only visible if clotting has occurred.

2.(*a*) Coagulation of the milk may occur as a result of proteolytic enzyme activity affecting the casein, the litmus colour remaining blue. This is known as a sweet clot.

(*b*) Hydrolysis of the casein as a result of proteolytic enzyme activity causes clearing and loss of opacity in the milk medium, usually referred to as peptonisation. Proteolysis may also result in an alkaline reaction due to ammonia production.

[3. Utilisation of citrate in the milk medium results in the production of an alkaline reaction – shown by the change to a deep blue colour in the litmus medium.]

B. Reactions Involving Carbohydrate and other Carbon Compounds

1. Hydrolysis of starch

Starch hydrolysis may be tested with solid or liquid media, although starch agar is perhaps more convenient. This medium consists of nutrient agar with the addition of 0·2–1 per cent soluble starch. The best results are obtained by preparing layer plates which are prepared by pouring 10 ml of nutrient agar into each plate, allowing it to set, and then overlaying this with 5 ml of starch agar.

Inoculate a poured dried plate of the medium by streaking *once* or spot-inoculating across the surface and incubate at the optimum growth temperature for 2–14 days.

Test Reagent. Gram's iodine solution as used for Gram's stain.

Recording result. Flood plates with 5–10 ml of iodine solution. Unhydrolysed starch forms a blue colour with the iodine. Areas of hydrolysis therefore appear as clear zones and are the result of β-amylase activity. Record the width of any clear zone in mm from the edge of the colony to the limit of clearing. Reddish-brown zones around the colony indicate partial hydrolysis of starch (to dextrins) which is the result of α-amylase activity.

2. Production of acids from sugars, glycosides and polyhydric alcohols

To peptone water or other basal medium add 0·5–1·0 per cent of substrate. An indicator is also incorporated in the medium to detect acid production. The indicator may be Andrade's indicator 1 per cent, phenol red 0·01 per cent, or bromcresol purple 0·0025 per cent. Durham tubes are included to detect gas production. Sterilise by steaming for 30 min on 3 successive days. Substrates which may be excessively decomposed by heat sterilisation should be sterilised as 10 per cent solutions by filtration and added aseptically to tubes of previously heat-sterilised basal medium to give the correct final concentration of substrate. (See pp 327–328.)

Incubate at the optimum growth temperature for up to seven days.

Recording result. Acid production is shown by a change in the colour of the indicator—Andrade's indicator to pink, phenol red to yellow, and bromcresol purple to yellow. Gas, if produced, accumulates in the Durham tube. On continued incubation some organisms can cause a pH reversion as a result of producing ammonia from the peptone. Total utilisation of glucose by bacteria which produce very little acid can be detected by the use of "Clinistix" (Miles Laboratories) (Park, 1967). The test strip changes colour if glucose is present.

3. Differentiation of oxidation and fermentation of carbohydrates

For this test Hugh and Leifson's medium (Hugh & Leifson, 1953) is employed. It is sometimes modified for particular groups of organisms – e.g. lactic acid bacteria (Whittenbury, 1963), and staphylococci and micrococci (see p. 343).

The basal medium, which contains a pH indicator, is dispensed in 5–10 ml amounts in 150 × 16 mm test-tubes and sterilised by autoclaving. The carbohydrate is prepared separately from the basal medium as a 10 per cent solution and sterilised by autoclaving or by filtration. The sterile carbohydrate solution is added aseptically to the sterile melted basal medium to give a final concentration of 1 per cent.

For each carbohydrate, stab inoculate two tubes of medium. Cover the surface of the medium in one tube with sterile liquid paraffin, vaspar or agar. Incubate at the optimum growth temperature for up to 14 days. One tube only can be used provided that it is examined daily.

Recording result. Acid production is shown by a change in the colour of the medium from blue to yellow (or from blue-green to red in the case of the double-indicator version). Fermentative organisms produce acid in both tubes. Oxidative organisms produce acid in the open tube only and usually, or at least initially, only at the surface of the open tube.

4. Production of carbon dioxide from glucose

The medium used is Gibson's semi-solid tomato-juice medium (Gibson and Abd-el-Malek, 1945). It consists of 4 parts skim-milk + 1 part nutrient agar with the addition of 0·25 per cent yeast extract, 5 per cent glucose and 10 per cent tomato juice, final pH 6·5. The medium is distributed in tubes to give a depth of 5–6 cm. Studies by Stamer, Albury and Pederson (1964) indicate that the requirement of lactic-acid bacteria for tomato juice may be met by manganese. This may be conveniently added as manganese sulphate (1–10 ml of 0·4 per cent $MnSO_4.4H_2O$ solution) to give a final concentration of 1–10 ppm Mn^{++} in place of tomato juice.

Dissolve the medium by heating at 100°C until molten, then cool to 45°C. Inoculate with *c.* 0·5 ml young broth culture, mix by rotation of the tube, then cool in tap water. When set, pour into the tube molten nutrient agar at *c.* 50°C to give a layer 2–3 cm deep above the surface of the medium. Incubate at the optimum temperature for up to 14 days.

Recording result. The semi-solid medium and agar seal trap any carbon dioxide gas produced in the medium. This is shown by disruption of the agar seal and by the presence of gas bubbles in the medium. It may be necessary to place the tube in hot water to release the gas.

5. Methyl red test

The medium used is glucose phosphate broth.

After inoculation, incubate at the optimum growth temperature for 2–7 days.

Test reagent. Methyl red solution (0·1 g methyl red in 300 ml of 95 per cent ethanol, made up to 500 ml with distilled water).

Recording result. Add about five drops of the indicator to 5 ml of culture. A red colour, denoting a pH of 4·5 or less, is described as positive. A yellow coloration is recorded as negative.

6. Voges-Proskauer test

This is a test for the production of acetylmethylcarbinol from glucose. To the inoculated medium after incubation, alkali is added, in the presence of which any acetylmethylcarbinol present becomes oxidised to diacetyl. The diacetyl will combine with arginine, creatine or creatinine, to give a rose coloration. The medium used is glocose phosphate broth, as in the methyl red test or other suitable medium, e.g. see p. 272.

After inoculation, incubate at the optimum growth temperature for 2–7 days.

Test reagents. (*a*) O'Meara's modification (O'Meara, 1931)—creatine and a 40 per cent solution of sodium hydroxide. (*b*) Barritt's modification (Barritt, 1936)—a 6 per cent ethanolic solution of α-naphthol and a 16 per cent solution of potassium hydroxide.

Recording result. (*a*) O'Meara's modification: add a knife-point of creatine to the culture, followed by 5 ml of 40 per cent sodium hydroxide. Shake the tube well. The development of a pink colour in the medium, usually within 30 min, indicates a positive reaction. (*b*) Barritt's modification: to 1 ml of culture in a test-tube, add 0·5 ml of 6 per cent α-naphthol solution and 0·5 ml of 16 per cent potassium hydroxide. Shake the tube. Development of a red coloration, usually within 5 min, constitutes a positive reaction. The Barritt procedure is rather more sensitive than the creatine method, being able to detect 1 ppm of acetylmethylcarbinol.

7. Utilisation of citrate as the sole source of carbon (see Report, 1958a)

(*a*) *Koser's citrate medium*

The liquid medium is dispensed in test-tubes previously cleaned scrupulously in an effective cleansing agent (see pages 82, 315), and sterilised by autoclaving.

Inoculate with a *wire needle* from a peptone water culture or from a saline suspension prepared from a young agar slant culture. A wire loop should *not* be used since this may provide an inoculum sufficiently heavy to make the medium turbid before incubation. In addition, a heavy inoculum will add appreciable quantities of organic compounds to the medium. Incubate at the optimum growth temperature for up to 7 days.

Recording result. Growth in the medium – involving utilisation of citrate as sole carbon source – is shown by turbidity in the medium. It is advisable to make several serial transfers into fresh sterile citrate medium to confirm that the result is a true positive, and to run controls on the same medium with the citrate omitted.

Several modifications of the original Koser's medium have been made including the addition of agar and the indicator bromthymol blue.

(*b*) *Simmon's citrate agar*

This medium contains bromthymol blue as a pH indicator, and agar. It is used as slopes, with a 1-inch butt.

The slope culture is inoculated by streaking over the surface with a loopful of peptone water culture or preferably with a wire needle of saline suspension as in (a) above. Incubate at the optimum temperature for up to 7 days.

Recording result. Utilisation of citrate and growth on the citrate agar results in an alkaline reaction, so that the bromthymol blue indicator in the medium changes from green to bright blue. When no growth occurs and citrate is not utilised, the colour of the medium remains unchanged.

8. Production of polysaccharide from sucrose (Evans *et al.*, 1956; Garvie, 1960)

The medium used, sucrose agar, consists of nutrient agar with the addition of 5–10 per cent sucrose and is sterilised by autoclaving.

Inoculate a poured dried plate by streaking to obtain separate colonies. Also set up a control on a medium which contains only 0·1 per cent sucrose. Incubate at 20–25°C or at the optimum growth temperature for 1–14 days.

Recording result. Synthesis of dextran or laevan from sucrose is indicated by the development of growth of a mucoid character.

9. Action on litmus milk

Inoculate as for a broth culture and incubate at the optimum growth temperature for up to 14 days.

Recording result. Examine tubes daily and record any changes in the medium. A number of different reactions and combinations of reactions may occur involving (1) lactose, (2) casein, (3) other milk constituents.

1.(*a*) Acid production shown by a change in the colour of the litmus to pink.

(*b*) If sufficient acid is produced the milk will clot. This is known as an acid clot.

(*c*) Reduction of the litmus and loss of colour may occur. This may precede or follow other changes.

(*d*) Gas may also be produced and show as gas bubbles in the medium, although normally this is only visible if clotting has occurred.

2.(*a*) Coagulation of the milk may occur as a result of proteolytic enzyme activity affecting the casein, the litmus colour remaining blue. This is known as a sweet clot.

(*b*) Hydrolysis of the casein as a result of proteolytic enzyme activity causes clearing and loss of opacity in the milk medium, usually referred to as peptonisation. Proteolysis may also result in an alkaline reaction due to ammonia production.

3. Utilisation of citrate in the milk medium results in the production of an alkaline reaction—shown by the change to a deep blue colour in the litmus medium.

C. Reactions Involving Fats and Related Substances, including Tests for Lipolytic Activity

1. Hydrolysis of tributyrin

The medium used is tributyrin agar which consists of yeast extract agar at pH 7·5, with the addition of tributyrin followed by emulsification, which is best carried out in an electrical mixer or blender. Sterilise by steaming for 30 min on 3 successive days. The medium may be used in the form of layer plates, 5 ml of molten tributyrin agar being overlaid on to previously poured and set plates of yeast extract agar.

Inoculate a poured dried plate of the medium by streaking *once* across the surface, Incubate at the optimum growth temperature for 2–14 days.

Recording result. Hydrolysis of tributyrin results in clearing of the medium and formation of a clear zone. Record the width of the zone in mm from the edge of the colony to the limit of clearing. This reaction is usually regarded as being specific for lipase but Sierra (1964) has reported a bacterial proteolytic enzyme which is capable of hydrolysing tributyrin, although not more complex fats.

2. Hydrolysis (lipolysis) of butter-fat, olive-oil and margarine

(*a*) *Butter-fat agar, olive-oil agar* (Berry, 1933)

The medium consists of yeast extract agar, pH 7·8, + butter fat or olive oil to 5 per cent. Emulsification of this medium can be achieved by shaking vigorously.

Inoculate a poured dried plate of the medium by streaking once across the surface and incubate at the optimum growth temperature for 2–14 days.

Test reagent. Saturated copper sulphate solution.

Recording result. Flood the plates with 8–10 ml of reagent and allow to stand for 10–15 min. Pour off the reagent and wash the plates gently in running water for one hour to remove excess copper sulphate. Where lipolysis has occurred, a bluish-green coloured zone appears, due to the formation of the insoluble copper salts of the fatty acids set free on lipolysis.

(*b*) *Victoria blue butter-fat agar* (Jones and Richards, 1952) *and Victoria blue margarine agar* (Paton and Gibson, 1953).

Use of margarine gives in a more stable medium than does butter. The media contain Victoria blue as an indicator of the presence of free fatty acids. Some strains of bacteria are inhibited by Victoria blue, so these media are not suitable for primary isolation of lipolytic organisms from mixed populations.

Recording result. Where lipolytic activity has occurred, the free fatty acids combine with the Victoria blue to form deep blue salts. Deep blue zones surrounding or beneath microbial growth are thus an indication of lipolytic activity. The background colour of the medium should be pinkish-mauve.

3. Hydrolysis of Tween compounds (Sierra, 1957)

In addition to the Tween, the medium must include a soluble calcium salt; released fatty acid is then detectable as the precipitated calcium salt.

Tween 20 (a lauric acid ester), Tween 40 (a palmitic acid ester), Tween 60 (a stearic acid ester), Tween 80 (an oleic acid ester), are amongst the Tweens which can be used, but Tween 80 is the one usually chosen. (Tweens are manufactured by Atlas Chemical Industries Inc., and are obtainable from Honeywill-Atlas Ltd.)

Inoculate a poured dried plate of the Tween agar medium by streaking *once* across the centre or by spot-inoculating and incubate at the optimum growth temperature for 1–7 days.

Recording result. Opaque zones surrounding microbial growth consist of calcium salts of the free fatty acids and are usually taken as being indicative of a positive lipolytic activity. It should be noted, however, that the Tween forms micelles which are pervaded by water. Current nomenclature defines a lipase as an enzyme which acts on water-insoluble esters or fats at the ester/water interface. Tween/water mixtures probably provide suitable conditions for the activation of both lipases and other esterases.

4. Hydrolysis of lecithin

(*a*) *Egg-yolk agar*

This consists of a nutrient agar with the addition of sodium chloride to 0·9 per cent and egg-yolk emulsion to 10 per cent.

Inoculate a poured dried plate of the medium by streaking *once* across the surface and incubate at the optimum growth temperature for 1–4 days.

Recording result. Lecithinase activity (i.e. hydrolysis of the lecithin of the egg-yolk medium) results in the formation of opaque zones around the region of microbial growth. Record the width of the opaque zone in mm.

(*b*) *Egg-yolk broth*

A nutrient broth with the addition of 0·9 per cent sodium chloride and 10 per cent (v/v) egg-yolk emulsion.

Inoculate and incubate at the optimum growth temperature for 1–4 days.

Recording result. Lecithinase activity results in opacity in the egg-yolk broth medium usually with a thick curd.

D. Miscellaneous Tests

1. Catalase test

Most organisms growing on aerobically incubated plates possess the enzyme catalase, although in differing amounts depending on species and strain (see, for example, Taylor and Achanzar, 1972). The lactic-acid bacteria (including *Streptococcus*, *Leuconostoc*, *Lactobacillus*) do not normally produce a detectable catalase, but Whittenbury (1964) reported a number of strains capable of giving a positive reaction in the catalase test (method (*a*) below) particularly when grown on media containing heated blood. On the other hand, obligate anaerobes (e.g. *Clostridium*) are usually catalase negative.

Test reagent. Hydrogen peroxide (10 vol. concentration). (This should be freshly prepared each day, and stored in the refrigerator between tests.)

Method. (*a*) Pour 1 ml of hydrogen peroxide over the surface of an agar culture. Alternatively, a loopful of growth may be emulsified with a loopful of hydrogen peroxide on a slide.

(*b*) Place 1 ml of hydrogen peroxide in a small clean test-tube and add 1 ml of culture withdrawn aseptically from a broth culture.

Recording result. Effervescence, caused by the liberation of free oxygen as gas bubbles, indicates the presence of catalase in the culture under test.

2. Oxidase test (Kovacs, 1956; Steel, 1961)

This test is particularly useful for differentiating pseudomonads from certain other Gram-negative rods.

Test reagent. 1 per cent aqueous solution of tetramethyl-*p*-phenylenediamine hydrochloride. This may be kept in a *dark* bottle in the refrigerator but auto-oxidation will cause the solution gradually to become deep purple, when it should be discarded. Auto-oxidation can be retarded by addition of ascorbic acid to 0·1 per cent (Steel, 1962).

Method. (*a*) Pour the reagent over the surface of the agar growth in a Petri-dish.

Recording result. Oxidase positive colonies develop a pink colour which becomes successively dark red, purple and black in 10–30 min.

Method. (*b*) Add a few drops of reagent to a piece of filter paper in a Petri-dish. With a platinum loop or glass rod (other materials may give fake positive results) smear some bacterial growth on to the impregnated filter-paper.

Recording result. A purple coloration is produced within 5–10 sec by oxidase positive cultures. A delayed positive is indicated by a purple coloration within 10–60 sec, any later reaction being recorded as negative.

3. Coagulase test

This test is used to differentiate pathogenic (*Staph. aureus*) from non-pathogenic staphylococci.

(*a*) *Slide method*

This test is carried out on 18–24 hour nutrient agar cultures.

Test reagent. Human or rabbit plasma. Dried rabbit plasma supplied by Wellcome Reagents Ltd. reconstituted and diluted 1 in 5 has been found satisfactory.

Method. Mark a slide into two sections with a grease pencil. Place a loopful of normal saline (0·85 per cent sodium chloride in aqueous solution) on each section and emulsify a small amount of an 18–24 hour agar culture in each drop until a homogeneous suspension is obtained. Add a drop of human or rabbit plasma to one of the suspensions and stir for 5 sec.

Recording result. A coagulase positive result is indicated by clumping which will not re-emulsify. The second suspension serves as a control.

(*b*) *Tube method*

In this modification, 18–24 hour nutrient broth cultures are used.

Test reagent. Human or rabbit plasma is used as above.

Method. Place 0·5 ml of diluted plasma into each of two small test-tubes. To one tube add 0·5 ml of an 18–24 hour broth culture. Incubate both tubes at 37°C and examine after 1 hour and at intervals up to 24 h.

Recording result. Clotting indicates that the strain under test is coagulase positive. Coagulation normally takes place within 1–4 h. The second tube serves as a control and should show no coagulation. The tube test is rather more reliable than the slide method.

4. Phosphatase test

It has been found that coagulase positive staphylococci produce the enzyme phosphatase (Barber and Kuper, 1951). Whereas the coagulase test must be applied to individual colonies, the phosphatase reaction provides a less time-consuming method of examining many colonies. The production of phosphatase can be detected by cultivation on a nutrient agar medium containing 0·01 per cent phenolphthalein phosphate. Since phosphatase positive organisms belonging to genera other than *Staphylococcus* could interfere with the test, polymyxin can be incorporated to make the medium more selective for the growth of staphylococci (Gilbert *et al.*, 1969). In this way the medium may be used to enumerate potentially pathogenic staphylococci in foodstuffs, etc., by using the Miles and Misra surface count technique or spread plates. (See also Report, 1972, for a comparison of the effectiveness of phenolphthalein phosphate polymyxin agar with Baird Parker's medium for isolating staphylococci from foods.)

Medium. Liquefy the nutrient agar, cool to 50°C and add aseptically sterile 1 per cent phenolphthalein phosphate solution to a final concentration of 0·01 per cent, mix, and pour plates.

Test reagent. Concentrated ammonia solution.

Method. Inoculate poured plates and incubate overnight at 37°C. Expose each plate to ammonia vapour by adding a few drops of ammonia to a filter paper inserted in the lid of the dish.

Recording result. Pink or red colonies indicate the presence of free phenolphthalein set free by phosphatase activity, and are therefore phosphatase positive.

5. Haemolysis of blood agar

Medium. Nutrient agar containing 0·85 per cent sodium chloride and 5 per cent (v/v) defibrinated or oxalated blood. Horse blood is suitable for streptococci, but for other organisms, e.g. staphylococci, the blood of other animals (e.g. sheep, rabbit, ox) may give better results. Liquefy the nutrient agar medium, cool to 50°C and add the sterile blood (0·5 ml to 10 ml agar) aseptically. This is mixed by rotation between the hands or inversion of the tube, and poured into a Petri-dish. Alternatively, 5 ml of blood agar may be poured on top of a thin layer (5–10 ml) of nutrient agar previously poured into the Petri-dish and allowed to solidify.

Method. In the case of staphylococci, streak a dried plate of the medium so as to produce separate colonies. If streptococci are being examined, either prepare pour plates or incubate streak plates anaerobically. Incubate at the optimum growth temperature for up to 2 days.

Recording results. Clear zones around the colonies indicate haemolytic activity. *β-Haemolysis* is the term given to this complete clearing of the blood agar when caused by streptococci. Zones of β-haemolysis possess sharply defined edges. *α-Haemolysis* is the term usually given to a greenish coloration produced around the colonies of some streptococci. These greenish zones have hazy outlines. It should be noted that the terms *α-haemolysin* and *β-haemolysin* have a significance which depends on whether they are used with reference to streptococci or staphylococci. In the case of staphylococci, α-haemolysin produces clear zones, β-haemolysin produces dark hazy zones. Nevertheless the haemolysis produced by staphylococcal α-haemolysin may sometimes be called β-haemolysis (see for example Breed *et al.*, 1957). Thus, to avoid confusion, it is recommended that the type of haemolysis is *described* rather than given a designation which may be misinterpreted. For a discussion of the characteristics of the haemolysins involved, see Wilson and Miles (1975) and Montie, Kadis, and Ajl (1970).

CLEANING OF GLASSWARE AND APPARATUS

A. Treatment of New Glassware

Borosilicate glass (e.g. Pyrex) or factory-washed soda-glass apparatus needs no special treatment before being used, other than normal washing up. New unwashed soda glass should be soaked in N hydrochloric acid overnight to partially neutralise the alkali contained in the glass.

B. Treatment of Used Glassware and other Apparatus

All glassware containing microbial cultures or otherwise contaminated by micro-organisms should be sterilised by autoclaving for 20 min at 121°C, which will also liquefy any solid media and allow easy removal.

The sterilised glassware should then be rinsed in tap-water followed by soaking in a *suitable* detergent solution and final cleaning with a brush. It is advisable to use a detergent specifically formulated for use in microbiological laboratories, since the detergent must (*a*) completely remove the most tenacious residues, e.g. proteins and fatty materials, (*b*) be capable of being removed easily and completely from the glass by rinsing, (*c*) neither cause deterioration of any of the materials from which the apparatus is constructed, nor be harmful to the skin. Glassware that has contained Vaseline, paraffin wax or liquid paraffin should be washed separately to avoid spreading greasy films.

Examples of such detergents are Pyroneg (Diversey (U.K.) Ltd.), and Decon 90 (Decon Laboratories Ltd.). If such a detergent is used it is possible to wash metal, rubber, and plastic components, e.g. screw-caps and tube closures, with the glassware. During washing, the rubber liners must be removed from metal screw-caps so that all surfaces receive adequate washing; for this reason autoclavable polypropylene screw-caps which do not possess separate liners (e.g. those made by Sterilin Ltd.) may be preferred.

After the detergent wash, rinse the apparatus five times in hot tap water and then three times in distilled water. Drain and dry in a heated drying cabinet.

Used pipettes

These should be discarded at the bench into jars containing a disinfectant solution. If the pipettes have been used for liquid cultures, or for pipetting water, ¼-strength Ringer's solution or other non-soiling liquids, a hypochlorite solution can be used in the jars. The available chlorine in the jar should at all times be sufficient to give a purple-black colour when a drop of the liquid is placed on a strip of starch-potassium iodide test paper (e.g. from Johnsons of Hendon). In the case of pipettes used for blood, serum, milk, and dilutions of many foods, hypochlorite is unsuitable. Hypochlorite must not be used if there is a danger of contact with formaldehyde (see page 308). It is then necessary to run tests on detergent-sanitisers which are readily available to determine: (*a*) their ability to remove residues, and (*b*) their bactericidal and bacteriostatic effects on the types of bacteria most often used or encountered. Suitability of preparations is best determined using the capacity test method (see page 93). We have found satisfactory 0·6 per cent Marinol D (Reddish Chemical Co.) in the used pipette jars, with the used pipettes then being transferred to 0·2 per cent Reddishquat (Reddish Chemical Co.) for soaking overnight before being placed in the pipette washer.

If pipettes are used for the transfer of dangerous pathogens it is best to use disposable pipettes placed into jars of Lysol or other suitable disinfectant, the pipettes being removed to autoclavable disposal bags and autoclaved before final disposal.

C. Disposable Apparatus

Plastic Petri-dishes, and pipettes and bottles which are disposable should be sterilised by autoclaving for 20 min at 121°C before discarding. They may be placed in autoclavable plastic bags (Sterilin Ltd.), the sealed bags with sterilised contents being more readily and hygienically discarded. Microscope slides and coverslips having been discarded into jars of disinfectant on the work bench can be disposed of without further sterilisation, but, if known pathogens are involved, the jars and contents should be autoclaved (in which case a disinfectant other than hypochlorite should be used).

STERILISATION

The sterilisation of culture media, containers and instruments is essential in bacteriological work for the isolation and maintenance of pure cultures. Bacteriological tests for sterility may be used to confirm the efficiency of the procedures used.

A. Sterilisation by Heat in the Absence of Moisture

Although heat is much more effective in the presence of moisture both dry heat methods and moist heat methods have their uses.

1. Red heat in the Bunsen flame

This is used for sterilising inoculating wires, loops and metal instruments which are not damaged by heat. If highly infective pathogens are being cultured, hooded burners or properly designed loop sterilisers are essential to contain spattered material.

2. Flaming after dipping in ethanol

This method is frequently used for scalpels, spatulae, etc., with the instruments not being heated to red heat. It does not necessarily achieve sterilisation.

3. Hot air oven

The hot air oven is heated by either electricity or gas and is thermostatically controlled. Sterilisation in the hot air oven is usual for dry glassware such as test-tubes, glass Petri-dishes, flasks and pipettes. Glass Petri-dishes and pipettes are packed most easily in copper, alloy or stainless steel containers made for the purpose. Glassware should be dry before placing in the oven. The hot air oven is also used for sterilising dry materials in sealed containers, e.g. chalk, and for mineral oils used in the preservation of stock bacterial cultures. Loading should take place when the oven is cold and spaces should be left between and around the items for circulation of air through the load. The holding time should be a minimum of 2 h beginning when the oven thermometer indicates 160°C. The oven should be allowed to cool before the door is opened as otherwise the glassware may crack.

The efficiency of sterilisation can be checked by using Browne's steriliser control tubes, Type III (green spot) (A. Browne Ltd.), which should be packed to simulate the most thermally protected material being treated.

B. Sterilisation by Heat in the Presence of Moisture

1. Boiling water bath

Boiling for 5–10 min is sufficient to kill non-sporing organisms, but some spores can survive. This method is useful where sterility is not essential or where better methods are not available, or are unsuitable. Distilled water should be used, particularly in districts where the water supply is hard, otherwise instruments etc. become covered with a film of calcium salts.

2. Koch's steam steriliser

Sterilisation in a Koch's steam steriliser ("steamer") in steam at atmospheric pressure and *c.* 100°C is used for media or constituents which are damaged by exposure to temperatures above 100°C, e.g. sugars, gelatin, and milk. The steamer may be used in two ways: (i) a single exposure at 100°C for 90 min; (ii) intermittent heating (Tyndallisation) by heating at 100°C for 30 min on each of 3 successive days interspersed with incubation under conditions in which the medium will be subsequently used. In (ii) the first exposure kills vegetative forms, and between the heat treatments the spores germinate which are thus killed in the subsequent heating.

3. Autoclave

Autoclaving is the most efficient method of sterilising culture media, and should be used for all media capable of withstanding the high temperature without decomposition. It is also used for glassware and instruments, and for sterilising cultures and contaminated material before washing. The actual temperature inside the autoclave depends on the steam pressure.

Gauge pressure (lb/in^2)	Gauge pressure (kN/m^2)	Temperature (°C)	Gauge pressure (lb/in^2)	Gauge pressure (kN/m^2)	Temperature (°C)
0	0	100	15	103·4	121
5	34·5	109	20	137·9	126
10	68·9	115			

Sterilisation in the autoclave is usually achieved by autoclaving at 121°C for 15–30 min in pure saturated steam at 15 lb/in^2 above atmospheric pressure (103·4 kilonewtons per m^2).

The total exposure time required at the desired temperature will depend upon a number of factors including:

(*a*) *the microbial load of the material being sterilised, and the nature of the contaminants.* Ingredients of culture media, and dehydrated media, usually will have been manufactured in a way which ensures very little contamination. But very heat resistant spores of thermophilic *Bacillus* spp., for example, *may* be encountered in dehydrated media or ingredients at levels so high that the media cannot be decontaminated without considerable thermal denaturation. This failure to obtain sterility would only be noticed when such media are used for thermophilic studies. Most manufacturers produce special media of guaranteed low spore counts for such purposes.

(*b*) *the size of the containers and the thickness of the wall.* About twice as long is required for heat to penetrate a 500 ml medical flat bottle full of medium, as for it to penetrate a 100 ml medical flat bottle full of the same medium. It requires a heating time about half as long again for the heat to penetrate 500 ml of water, diluent or nutrient broth contained in a thick-walled transfusion bottle as for the heat to penetrate the same amount of liquid in a medical flat bottle.

(*c*) *the nature of the contents.* The presence of agar increases the heat penetration time to nearly double that of water.

It is not possible to generalise on the process times necessary, partly because some heat penetration occurs during venting and whilst the steam pressure reaches the required level. The more efficient the autoclave in these respects, the more consideration which must be given to the factors just mentioned. Controlled experiments using spore suspensions, spore strips (obtainable from Oxoid Ltd. and other manufacturers of dehydrated media, etc.), thermocouples (which can be used in very few autoclaves), or ampoules of chemical indicators (e.g. Browne's tubes, manufactured by A. Browne Ltd.) may all help to establish the correct processing treatment.

It is important to ensure that all air is removed from the autoclave before sterilisation begins, since a mixture of air and steam results in a lower temperature for any given pressure. Also, the presence of air tends to prevent steam penetration and causes uneven heating in the different parts of the autoclave. If $\frac{1}{3}$ of the air remains in the autoclave as a result of inefficient venting, the temperature inside the autoclave will only reach 115°C at a pressure of 103 kN/m^2. In the absence of a thermometer in the steam drain, a check on the presence of air in the steam can be made by arranging for the vented steam to pass through a container of cold water. It should be noted that the design of some manually-operated autoclaves (especially those designed for connecting to a steam main) *may* be such that any air not vented by downward displacement at the beginning of the operation will remain in the inner chamber and *not* escape through the pressure valve. See also Rubbo and Gardner (1965), Report (1959).

4. The inspissator

This is used for the preparation of media such as Loeffler's serum medium. The medium is distributed in containers placed in a sloping position in special racks. The temperature is slowly raised to 85°C and maintained for two hours, causing the medium to be completely solidified.

When coagulation of the material is not required and lower temperatures (about 56°C) are used, it is necessary to repeat the process on several successive days.

C. Sterilisation by Filtration

This method is used for sterilising fluids and solutions which would be adversely affected by heat. Such components may then be added aseptically to other materials which have been heat sterilised. It is also used when a source of sterile air is required, e.g. for fermenters, although it may be difficult to preclude the entry of bacteriophage. There are various kinds of filter which may be used.

1. Earthenware candles

Earthenware candles such as the Berkefeld and the Chamberland filters were among the earliest filtration aids, but are now little used.

2. Asbestos filters

These are available in a number of graded porosities, some being used for clarification, some being used for sterilisation. Although these were until recently the most widely used type of filter for sterilisation of media and supplements of media, they are being rapidly superseded by membrane filters. Asbestos filters consist of a randomly arranged matrix of material, with a wide range of pore sizes. Even in the sterilisation grades of filter there will be many individual pores considerably larger than the micro-organisms to be stopped, and much of the retention of micro-organisms is caused by electrostatic attraction and adsorption. Thus we cannot rely absolutely on the asbestos filter always to achieve complete removal of all bacteria. The greater the difference in pressure on the two sides of the filter, the deeper will the bacteria penetrate the matrix. With a high enough pressure differential (perhaps even as low as 140 kN/m^2) "breakthrough" of the organisms can occur when the forces of attraction between micro-organisms and filter are overcome. With filtration under negative pressure (using a vacuum pump) loss of sterilising efficiency from this cause will not occur since the maximum pressure differential theoretically attainable is one atmosphere, and in practice much lower pressure differentials are used. However, if positive pressure filtration apparatus is employed there will be a temptation to use as high a pressure differential as possible in order to shorten the filtration time.

Further disadvantages of asbestos filters are that their adsorbent nature can cause the depletion of certain constituents from media or solutions, and that they can be a source of unwanted and unknown contaminating substances—

these aspects are particularly important when chemically defined media are being prepared. They may also present a health hazard.

3. Membrane filters (see also pages 32–33)

These are usually made of highly porous cellulose acetate, and are available in a wide range of different porosities. In contrast to an asbestos filter, a membrane filter retains particles above a certain size *on its surface* because the size of the largest pore is smaller than the smallest of the retained particles. Particle retention is effected by pore size and not by electrostatic attraction or adsorption. Thus the pore size rating given by the manufacturer is an absolute one and although the pores will vary somewhat in their size, the manufacturer can quote a range of pore size outside which no pores will occur. For example the grade GS Millipore filter has a quoted pore size range of 0·22 μm ± 0·02 μm. This means that the employment of high pressure differentials in positive pressure filtration to achieve short filtration times does not affect the sterilising efficiency of the filter.

For the bacteriological sterilisation of media etc., the final sterilising filter in any procedure will be Millipore Grade GS (pre size 0·22 μm ± 0·02 μm) or the equivalent grade from another manufacturer, although grades with much smaller pore sizes are available from most manufacturers of membrane filters (e.g. Millipore Grade VS has a pore size of 0·025 μm ± 0·003 μm). Note that the 0·45 μm filter (e.g. Millipore Grade HA) used for counting and isolation work in the microbiological analysis of samples is not suitable for the provision of a filtrate of guaranteed bacteriological sterility.

Filtration can be achieved by the application of negative pressure (i.e. suction) or positive pressure. In the microbiological analysis of samples negative pressure is nearly always used. When media, sugar solutions, sera, etc., are to be sterilised by filtration, it is better to employ positive pressure for two reasons. Firstly the use of negative pressure requires either the receiving vessel to possess a side arm to which the suction is applied (e.g. a filter flask) or use of a bypass filter attachment (e.g. as made by A. Gallenkamp & Co. Ltd.). In the former case the sterile liquid will need to be transferred aseptically to the final sterile container. A bypass filter attachment will allow filtration directly into the final sterile container, but there must still be adequate protection against blowback of non-sterile air, water or oil from the vacuum line or leakage of non-sterile air into the container through improperly sealed joints—and this last is quite difficult to ensure. The second drawback associated with the use of negative pressure is that the absolute (and unattainable) maximum pressure differential is one atmosphere. With a positive pressure filter unit rapid filtration can be obtained by the use of much higher pressure differentials. (This advantage of positive pressure filtration does not obtain with the normal ranges of *asbestos* filters since these should normally be used at pressure differentials of only about 140 kN/m^2 or less.)

There is available a wide range of membrane filtration apparatus for sterilising filtration by positive pressure. In most laboratories engaged in bacteriological and mycological analytical work sterilising filtration will be used for carbohydrate solutions, alcoholic solutions, yeast extract solutions, sera, etc., which are all mostly required in fairly small quantities, say up to 100 ml batch size. One of the most useful pieces of equipment for the small microbiological laboratory is the Swinnex filter-holder (Millipore Filter Corporation) for use with 13 mm, 25 mm, or 47 mm membrane filters; these holders attach to any hypodermic syringe employing the usual Luer type of connection. Such filters are perfectly capable of sterilising 1, 10, or 100 ml amounts of expensive media supplements or of rarely used media with practically no wastage. The Swinnex holder is also extremely useful for in-line sterilisation of air or media being supplied, for example, to fermenters.

Carbohydrate (not polysaccharide) solutions, alcohols, salts solutions, vitamin solutions and water can be fairly readily filtered through the sterilising grade of filter with or without (but preferably with) a disposable fibreglass prefilter pad preceding it. Media incorporating proteins, and some other liquids can be filtered more readily if two membrane filters are used after the fibreglass prefilter—the final sterilising grade of filter being preceded by a coarser grade (e.g. Millipore DA, pore size $0 \cdot 65\ \mu m \pm 0 \cdot 03\ \mu m$). Liquids such as serum and plasma can be filter-sterilised by employing a number of filters with progressively smaller pore sizes; for example, a fibreglass pre-filter followed by a filter of pore size $1 \cdot 2\ \mu m$, a filter of pore size $0 \cdot 65\ \mu m$, a filter of pore size $0 \cdot 45\ \mu m$, and finally the sterilising grade of filter with a pore size of $0 \cdot 22\ \mu m$. Because the membrane filters are so thin, a stack of filters can quite easily be placed in a single filter holder; the use of Terylene net separators between each pair of filters is recommended to improve flow rates. The stack of filters with separators is sterilised *in situ* in the filter holder by autoclaving.

D. Chemical Disinfectants

Chemical disinfectants are used mainly for disinfecting the skin, floors, buildings, apparatus, and for articles which cannot be effectively heated without damage. In the laboratory, pipettes and slide preparations containing living cells should be discarded into jars containing suitable disinfectants (see page 82). Any cultures spilled in the laboratory should be covered with absorbent cotton-wool soaked in disinfectant before removal. An iodophor such as 0·4 per cent Wescodyne (Mirfield Agricultural Chemicals Ltd.) is suitable for these purposes.

For certain purposes, volatile antiseptics (e.g. chloroform, toluene) may be used to prevent growth in nutrient solutions, the antiseptic later being allowed to evaporate by heating.

E. Preparation of Clean Glassware and Materials for Sterilisation Prior to Use

Before sterilisation, test-tubes, flasks and bottles should be stoppered with non-absorbent cotton-wool so as to form a firmly fitting plug. Alternatively, loosely fitted metal caps may be used. Pipettes should be plugged with cotton-wool and the ends singed.

Pipettes and Petri-dishes to be sterilised in the hot-air oven are placed in metal containers or may be wrapped in paper individually or in small lots of up to 5 or 6.

Culture media and aqueous solutions to be sterilised in the autoclave or steamer should be covered with grease-proof paper to protect the cotton-wool plugs from excessive wetting by steam. Any screw-caps should be slightly loose to prevent breakage during heating. After heating and removal from the autoclave, the caps may be firmly tightened.

F. To Test the Sterility of Laboratory Equipment

In the event of difficulties arising in the laboratory through contamination from some unknown source it will be necessary to test that laboratory equipment is effectively being sterilised.

It is essential that very careful aseptic procedures are followed in tests for sterility since, obviously, any extraneous contamination can lead to false positive results.

1. *Media.* After sterilisation, media should be tested for sterility by incubating at 30°C or 37°C (or for work with thermophiles, at 55°C) for 48 h, followed by storage at room temperature for 48 h prior to use. Contaminated or non-sterile media will show growth provided, of course, that conditions are suitable for the growth of any contaminants present; absence of growth implies that the medium is sterile.

2. *Pipettes.* Rinse two pipettes in a tube of nutrient broth by sucking the broth up and down ten times. Discard the pipettes, and incubate the broths at 30°C for 48 h. In the absence of growth, the pipettes are presumed sterile.

3. *Petri-dishes.* Pour in 10 ml of molten sterile nutrient agar medium at 45°C. Incubate at 30°C for 48 h. In the absence of growth, the Petri-dishes are presumed sterile.

4. *Rinses.* Rinses are usually quarter-strength Ringer's solution and are used, for example, in rinsing laboratory and food processing equipment.

Inoculate 2 ml from each rinse into a Petri-dish. Add 10 ml of molten nutrient agar at 45°C, mix and allow to set. Incubate at 30°C for 48 h. In the absence of growth the rinses are presumed sterile.

5. *Sample bottles, test-tubes and other containers.* Pour in 10 or 25 ml rinse (according to size of container) and shake ten times. Inoculate 2 ml of rinse into a Petri-dish. Add 10 ml molten nutrient agar at 45°C, mix and allow to set. Incubate at 30°C for 48 h. In the absence of growth, the sample bottles or containers are presumed sterile. Alternatively nutrient agar can be added directly to the sample bottles, etc., and allowed to set after swirling round the inside surfaces. A third method is to add nutrient broth directly to the container, and to rinse the internal surfaces. The broth may be incubated *in situ.*

N.B. The procedures described in paragraphs 2–5 obviously are only capable of detecting contaminants which are able to grow in nutrient agar or nutrient broth and to produce detectable growth under the given incubation conditions of temperature, gaseous atmosphere and time.

LABORATORY EVALUATION OF DISINFECTANTS

A. The Rideal-Walker Test

In this test, the activity of the test disinfectant is compared with that of phenol as a standard; the result obtained is known as a phenol coefficient. The Rideal-Walker test, however, is only valid in comparing disinfectants similar in chemical composition to phenol, and is therefore less useful in assessing the efficiency of disinfectants such as hypochlorites and quaternary ammonium compounds. For the procedure refer to Cruickshank (1965). Sykes (1962) has discussed the rationale of disinfectant evaluation, and the status and applicability of the various testing methods.

B. The Suspension Test

(International Dairy Federation, 1962b)

In principle, this method (also known as a survivor-curve method) consists of adding a known number of micro-organisms in suspension to a solution of disinfectant at the required concentration and then determining the number of survivors after given time intervals. An advantage of the suspension test is that concentrations of disinfectant and exposure times can be chosen which simulate those used in practical conditions. The suspension test has therefore proved extremely useful in studying the activity of disinfectants used in the food and dairy industries. The effect of additional organic matter on disinfectant activity can be studied by adding organic matter, e.g. milk, to the disinfectant at the time of adding the cell suspension.

1. *Preparation of cell suspension*

A suitable test organism may be chosen as appropriate, e.g. *Escherichia coli*, *Streptococcus lactis*, and incubated for 24 h at the optimum growth temperature in a suitable medium before harvesting. Cell suspensions are prepared from solid media by washing off the growth with sterile diluent (quarter-strength Ringer's solution). Cells can be separated from liquid media by centrifuging, discarding the supernatant liquid and resuspending the sedimented cells in sterile diluent. The resulting cell suspension is shaken to disintegrate clumps and can then be standardised to the desired strength by diluting with sterile diluent. A concentration of approximately 10^{10} cells per ml is recommended, and can most conveniently be found for routine purposes by first determining the relation between the numbers of micro-organisms and the optical density of the suspension. In subsequent work, it is then sufficient merely to adjust the suspension to the required opacity, e.g. with Brown's opacity tubes (see page 44).

2. *Test procedure*

(*a*) Distribute the disinfectant solution at the required concentration in 99 ml quantities in 250 ml conical flasks. Concentrations tested should include those likely to be encountered under practical conditions.

(*b*) Add 1 ml of cell suspension to the disinfectant solution, taking care that the tip of the pipette is held just above the surface of the disinfectant while delivering the suspension. Note the precise time of adding the suspension and mix well by rotating the flask.

(*c*) After exposure periods of 30 sec, 2 min, 5 min, and 10 min, remove 1 ml from the flask and transfer to tubes containing 9 ml of a sterile solution of inactivator. Mix well by rotation.

The inactivator must be appropriate to the disinfectant under test. Sodium thiosulphate (0·5 per cent) is used for tests with hypochlorite and iodophors. For quaternary ammonium compounds the following inactivators may be satisfactory but suitability should be checked for the particular disinfectant (see page 95) (1) a mixture of 2 per cent egg lecithin in a 3 per cent aqueous solution of Cirrasol ALN-WF; or (2) 2 per cent Tween 80; or (3) 10 per cent serum. (Cirrasol ALN-WF is obtainable from Honeywill-Atlas Ltd.; Tween 80 is made by Atlas Chemical Industries Inc., and obtainable from Honeywill-Atlas Ltd.) Organic mercurial compounds can be inactivated by the addition of 0·25 per cent sodium thioglycollate solutions (Sykes, 1965).

(*d*) Determine the numbers of survivors for each exposure period by taking out 1 ml and 0·1 ml from each tube of inactivator solution onto a suitable agar medium and incubating at the optimum growth temperature of the test organism.

C. The Capacity Test

This provides information on the capacity of a use-dilution of a disinfectant to be soiled with micro-organisms and organic material without losing disinfectant activity (International Dairy Federation, 1962a; Kelsey, Beeby and Whitehouse, 1965; Kelsey and Sykes, 1969; Bergen and Lystad, 1971).

This test procedure is appropriate for example for testing disinfectants or detergent-disinfectant mixtures to be used in used-pipette containers in microbiological laboratories, and determining the required concentrations and required intervals for renewal of the disinfectant in such containers. The capacity test is more suitable for this purpose than the type of use-dilution test described by AOAC (Association of Official Analytical Chemists, 1975) which determines the ability of a disinfectant to render sterile artificially contaminated surfaces (e.g. stainless steel cylinders on which *Salmonella* broth culture has been dried). The AOAC use-dilution method does not distinguish between disinfection and detergency—organisms removed from the test surfaces may still remain viable.

Outline test procedure

The details of the experiment should be determined by the conditions under which the disinfectant is used.

The test organisms are grown in nutrient broth or other suitable medium at optimum temperature for 24 h. (Alternatively suspensions washed from agar slopes may be used.) Suitable test organisms include:

Pseudomonas fluorescens
Escherichia coli
Klebsiella aerogenes
Proteus sp.
Staphylococcus epidermidis

Clumps of organisms are broken by shaking with sterile glass beads for 1 min. Prepare an appropriate dilution of the culture or suspension (about 10^8 organisms per ml) using an inorganic diluent. Add 1 ml of sterile nutrient broth or 1 ml of sterile (UHT) milk to 10 ml of the dilution to be used, to provide a standard organic content. Determine the viable count of this dilution. The test temperature employed is commonly 22°C, but other temperatures can be used. Six ml of the required concentration (e.g. the recommended use-dilution) of disinfectant are placed in a sterile jar. At 10 min intervals 1 ml of the bacterial suspension is added, and the mixture mixed by swirling (avoid foam or bubble formation). Eight min after each addition, a sample is withdrawn for determining the presence or absence of viable organisms.

Five single drops (0·02 ml) are placed on the surface of a poured, dried nutrient agar plate; and one drop placed in each of two nutrient broths containing an appropriate inactivator. This procedure is continued for 1 h, which allows six additions of organisms with the concentration of disinfectant being cut to approximately half (actually 0·49).

The end point of the test with respect to the activity of the disinfectant is the highest number of additions that gives fewer than 5 colonies from 5 drops or fewer than 2 positive broths. The concentration of disinfectant, number of bacteria added, and concentration of organic matter at this point can be calculated. In practice a disinfectant use-dilution can be regarded as satisfactory if 3 or more increments can be added before a positive culture is obtained.

Preparation of disinfectant dilutions

It is suggested that 3 dilutions be examined: the manufacturer's recommended concentration, one half the recommended concentration, and $1\frac{1}{2}$ times the recommended concentration.

Standard hard water

Kelsey and Sykes (1969) recommend using a standard hard water to prepare the dilutions of the disinfectants.

> 17·5 ml of 10 per cent (w/v) solution of $CaCl_2.6H_2O$ and 5 ml of 10 per cent (w/v) solution of $MgSO_4.7H_2O$ are added to 3·3 litres of distilled water. Sterilise by autoclaving.

Inactivators

These should be incorporated in nutrient broth at the concentrations given for the suspension test (see page 93).

To test the suitability of an inactivator

Prepare (*a*) inactivator + disinfectant in test ratios and concentrations
(*b*) inactivator + water
(*c*) water alone

Inoculate 10 ml of each with 1 ml of suspension containing 10^3 to 5×10^3 organisms per ml. Immediately plate 1 ml of each with nutrient agar. Repeat after 30 min and 60 min contact. The inactivator is suitable if the counts obtained in the three systems are not significantly different.

THE EFFECT OF HEAT ON MICRO-ORGANISMS: THE DETERMINATION OF DECIMAL REDUCTION TIMES (D-VALUES) AND *z*-VALUES

Heating a bacterial culture for increasing periods of time results in a progressive reduction in the viable population. The more organisms present initially, the longer the time required at a given temperature to kill the whole population.

Plotting the logarithms of the numbers of survivors against time, gives a curve which tends to a straight line. The reciprocal of the slope of this line is the D-value, which is defined as the time taken at a given temperature (T) to effect a reduction of 90 per cent in the microbial population. The higher the temperature, the smaller will be the D-value. If the logarithms of the D-values obtained at various temperatures are plotted against temperature, and the best straight line drawn through the points, the reciprocal of the slope of this line is the *z*-value. The *z*-value is defined as the number of degrees by which the temperature has to be raised or lowered to bring about 90 per cent reduction or 10-fold increase in the D-value.

Theoretically (if these straight line relationships were to be maintained over the whole lethal temperature range) the effect on a given bacterium of a typical heat treatment, including heating, holding and cooling times, can be determined provided that the temperature: time curve of the heat treatment, the *z*-value, and at least one D-value for the organism be known. A discussion of these mathematical calculations and of the limitations involved are outside the scope of this book but readers are referred to Stumbo (1973) for details.

Two examples of the experimental procedures which may be used are described here. The choice of media, suspending liquids, temperatures and times of heat treatment, and incubation temperatures and times depend upon the organism and the heat treatment system under investigation, and may require a simple trial experiment to determine the details of the experimental procedure.

Method 1. Using test-tubes to determine D-values and *z*-value of an asporogenous culture

1. Dilute a 24-h broth culture, e.g. of *Escherichia*, to give a total count of *c*. 3×10^8 per ml. (This can be determined by microscopic count, or by nephelometry or opacity tubes if these have been calibrated using microscopic counts.)

2. Put 10 ml of this dilution into 90 ml of sterile diluent (e.g. 0·1 per cent peptone water) which contains a number of glass beads. Shake well to mix

and to break up clumps. Distribute in 5 ml amounts into five sterile 150 × 16 mm test-tubes (preferably with polypropylene closures).

3. Put the tubes prepared in (2) into a water-bath at 58°C with a control test-tube containing a thermometer and 5 ml of diluent. When the temperature of the contents of the control tube reaches 1°C below the temperature of the water-bath, remove one of the tubes and simultaneously start timing from this moment (t_0).

4. Rapidly cool the removed tube in cold water. Prepare six decimal dilutions and perform a plate count using all of these dilutions.

5. Withdraw another tube and repeat step (4) after each of the following times: 2½, 5, 7½ and 10 min.

6. Incubate the plates and determine the number of survivors after the various heating periods. Draw a graph of $\log_{10}$ (number of survivors) : time. From this determine the $D_{58°C}$ (in minutes) as the reciprocal of the slope of the best straight line.

7. Repeat steps (1) to (6), using a water bath set at each of the following temperatures and removing tubes after heating for each of the times shown:

56°C for 0, 10, 20, 30, 40, and 50 min
54°C for 0, 10, 20, 30, 40, and 50 min
52°C for 0, 20, 35, 50, 65, and 80 min
51°C for 0, 20, 40, 60, 80, and 100 min
50°C for 0, 20, 40, 60, 80, and 100 min.

8. Plot $\log_{10}D_T$: T, to estimate the *z*-value.

Method 2. Using sealed tubes to determine the D-values and *z*-value of spores

Method 1 represents the simplest method for determining D-values and *z*-values. It has many sources of inaccuracy, amongst which are the long heat penetration times (for both heating and cooling), the temperature gradients which exist within the tubes, the likely presence of bacteria (*a*) on aerosol droplets in the contained air and (*b*) on the test-tube walls above the bulk liquid level (these bacteria can become reintroduced into the bulk liquid from time to time). These inaccuracies become more significant the shorter the heating times studied. The following method is described for use with spores of *Bacillus stearothermophilus*.

1. Cut glass tubing of 4 mm external diameter, 2 mm internal diameter, into 100 mm lengths. Using a suitable gas flame seal one end of each tube to give a hemispherical end. Plug with cotton wool and sterilise by autoclaving.

2. Prepare a spored culture of *Bacillus stearothermophilus* by culturing on fortified nutrient agar (Gould, 1971) in large sterile medical flat bottles, incubated at 55°C for 6 days. The incubation temperature and period to

provide a high ratio of spores to vegetative cells should be determined by experiment for the strain under investigation (the incubation conditions described were found to be satisfactory for *B. stearothermophilus* NCIB 8923 (ATCC 7954).

3. Rinse the spore crop from the slopes by the addition of a small amount of sterile phosphate buffer at pH 7·0, with a few glass beads to allow this to be effected using gentle agitation.

4. Add papain to the spore suspension to a concentration of 1 mg per ml, and incubate at 37°C for 48 h, to destroy vegetative cells.

5. Centrifuge the spore suspension, pour off the supernatant liquid and resuspend the pellet in phosphate buffer. Repeat this centrifuging and washing process 6 times. (The use of sterile disposable centrifuge tubes with screw caps is recommended.)

6. Finally resuspend the spores in 0·85 per cent sodium chloride to a concentration of *c.* 10^8 spores per ml (using a counting chamber to determine the count microscopically) and store in a refrigerator.

7.(*a*) Use 1 ml of this working spore suspension to prepare a dilution series and carry out a plate count to determine the viable spore population. Incubate the plates at 55°C with daily counting of colonies, avoiding contamination, until the count does not increase. This indicates the viable spore population *relative to the medium used.* (*N.B.* Plates incubated at 55°C or above should be placed in sealed bags or other containers to minimize dehydration of the medium.)

(*b*) Prepare a similar plate count after the suspended spores have been exposed to 80°C for 10 min. This brief heat treatment may be found to activate the spores and by thus encouraging germination, to result in a higher count than that obtained on the unheated suspension.

8. This and following steps are carried out concurrently with step (7); that is, all the plate counts produced by steps (7) to (13) are incubated together.

Introduce the well-mixed spore suspension into eight of the tubes prepared in step (1), filling each tube to approximately $\frac{1}{4}$ of its length. The amount of spore suspension introduced into each tube can be predetermined by using a sterile microlitre syringe, or by weight if uncalibrated Pasteur pipettes are used.

Rapidly seal the open end of each tube by rotation in a Bunsen flame, taking care as far as possible to protect the spore suspension from heating.

9. Use one of the tubes to ascertain the count at t_0. The surface of the tube should be swabbed with 70 per cent ethanol and a sterile glass cutter used with aseptic precautions to weaken the glass at several points. Aseptically place the tube in a sterile Universal container with 10 ml of sterile phosphate buffer. Break the tube using a sterile glass rod. Mix the contents well and prepare a dilution series and a plate count in the usual way.

10. Immerse seven of the tubes in an oil bath set at 110°C. Remove one tube after each of the following heating periods: 1, 3, 5, 10, 15, 25, and 30 min.

11. Each tube should be cooled rapidly in a beaker of cold water, and a plate count made as described in step (9).

12. Incubate the plates at 55°C and determine the number of survivors per ml of suspension after the various heating periods. Draw a graph of $\log_{10}$ (number of survivors) : time, and ascertain the $D_{110°C}$ (see above).

13. Repeat the experiment using an oil bath set at each of the following temperatures and determine the number of survivors per ml of suspension after heating for each of the following periods:

115°C for 1, 3, 5, 10, 15, 20, and 25 min
118°C for 1, 3, 5, 10, 15, 20, and 25 min
121°C for 1, 3, 5, 7, 10, 13, and 16 min.

These times and temperatures have been found satisfactory in experiments involving *B. stearothermophilus* NCIB 8923 (ATCC 7954) (T. Player and N. Marsden, unpublished work, 1973).

14. Plot $\log_{10}D_T$; T to estimate the *z*-value.

Use of the D-values and z-value in process calculations

The D-values and *z*-value for an organism may be used to determine the destructive effect of a heat treatment on that organism. For any heating and cooling curve there may be calculated a heating time at a given single temperature (the F-value) (assuming instantaneous heating to and cooling from that temperature) which has an equivalent destructive effect on a population of the organisms (see Stumbo, 1973).

It should be emphasised that the reliability of such calculations depends on the extent to which the *z*-value *is* constant with temperature, i.e. whether there is a straight line relationship of the form

$$\log D = a - \frac{T}{z}.$$

If, however, the thermal destruction of organisms is described by an Arrhenius relationship (log D is proportional to the reciprocal of the temperature) then *z* will *vary* with temperature. In such a case, the *z*-value determined experimentally will provide an estimate of the effect of temperature on the D-value, *only over the range of temperatures used in the experiment*. The D-values and *z*-value should not be used therefore for process calculations relating to temperatures very far outside the experimentally used range – for example D-values and *z*-value determined by heat resistance experiments performed at 80°, 90°, 100° and 110°C should not be used in process calculations relating to UHT processes at temperatures of around 130–140°C (see Gillespie, 1951; Cowell, 1968).

SEROLOGICAL METHODS

A. Introduction

The *in vitro* reactions between antigens and antibodies are of great value in many instances, the following being particularly useful to the food microbiologist.

1. The classification and identification of micro-organisms

In this case the bacterium is unknown, the antibodies known. This type of study is known as *antigenic analysis* and is used most often in systematic bacteriology. For example:

(*a*) In the investigation of an outbreak of food poisoning, *Salmonella* may have been isolated from a food which is under suspicion. The strain of *Salmonella* isolated from the food can be tested against a series of antisera containing antibodies against antigens of known species of *Salmonella*. Cultures of *Salmonella* isolated from faeces of patients can be similarly tested, and the species of *Salmonella* in the food and faeces identified. Agglutination reactions are used for this purpose.

(*b*) Precipitin reactions are used in the identification and classification of *Streptococcus* spp.

2. The detection and identification of antibodies in blood, milk and other body fluids

In this application, the antibodies are unknown and known bacterial cultures are used. This type of serological investigation is used most frequently in clinical and veterinary bacteriology. For example:

(*a*) In the diagnosis of brucellosis in cattle a culture of *Brucella abortus* is used to provide known antigen and tested against the milk from the animal. If antibodies are present in the milk there will be a (positive) reaction.

(*b*) The Widal agglutination test for the diagnosis of typhoid fever. A culture of *Salmonella typhi* is used to provide the known antigens and tested against the patient's serum. If the serum contains antibodies against this particular pathogen, there will be a positive reaction. This indicates either that the patient may have (or have had) typhoid fever, or that he has been vaccinated against the disease.

B. The Agglutination Reaction

This is used very largely for the Gram-negative intestinal bacteria, which include the organisms causing typhoid fever, food poisoning and dysentery.

Agglutination results from mixing together cellular antigen and homologous antiserum. As a result of the antigen-antibody reaction, the bacterial cells clump together and form flocculent masses or dense granules. The reaction between "H" (i.e. flagellar) antigen and its homologous antiserum results in flocculent clumping, whereas the reaction involving the "O" (i.e. somatic) antigen results in a more dense and granular clumping.

Antisera commercially available are of two types: polyvalent antisera which react with organisms of a particular genus or with groups of serotypes, and which are suitable for preliminary screening; and specific antisera which allow an identification of a particular serotype. Bottles of antisera should be stored in the dark at 4–7°C, and under these conditions they may remain usable for many months longer than the indicated shelf life, although the titre (or concentration) of effective antibody should be checked periodically.

1. The slide agglutination test

Procedure

1. Using a clean, grease-free slide, mark it into three with a wax pencil labelling the three sections "H", "O" and "C" (control).
2. Place one or two loopfuls of physiological saline in section "C".
3. Place a drop of a formaldehyde-treated suspension of bacteria (see below) in the section marked "H", and a drop of a heat-treated bacterial suspension (see below) in the section marked "O".
4. Mix one loopful of undiluted polyvalent "H" antiserum with the bacterial suspension labelled "H".
5. Mix one loopful of undiluted polyvalent "O" antiserum with the bacterial suspension labelled "O".
6. Rock the slide gently backwards and forwards and observe for three minutes over a dark background. If agglutination occurs there will be a clumping of the bacteria, usually within 30 sec. There should be *no* change in the control suspension.

N.B. It is important to flame the wire loop *and* allow it to cool between each transfer.

Slide agglutination tests may be performed with antigen preparations obtained as described below, but the suspensions should be made more turbid than when they are to be used in tube tests.

Bradstreet *et al.* (1961) recommend that a presumptive *Salmonella* should be tested by slide agglutination using a polyvalent *Salmonella* "O" antiserum. If negative, repeat the test using polyvalent "H" phase 1 and 2 antisera. If both are negative, repeat using *S. typhi* "Vi" antiserum. For further identification of agglutinating cultures, "Rapid Salmonella Diagnostic Sera" and, if necessary, single factor sera can be used (see McCoy and Spain, 1969). For a further discussion of single factor reactions, absorption, etc. see Oakley (1971). Harvey and Price (1961, 1974) have described a simple method of inducing H-antigen phase change in *Salmonella* cultures, when this is necessary for identification of particular serotypes.

2. Tube agglutination test

This enables an estimation of the titre (or concentration) of *antibody* to be made. This test method can be used to check on the efficacy of antisera used in slide tests for identification purposes. In this method, drops of bacterial suspension, antiserum and saline are dispensed in agglutination tubes. A Pasteur pipette with a rubber teat is used for adding the drops, the accuracy of the method depending on the drops being of constant volume. This is achieved best by the use of graduated micropipettes but can also be achieved by using the same ordinary Pasteur pipette for all reagents. The pipette must be held vertically and the drops expelled slowly (about one drop per sec).

Tube number:	(1)	(2)	(3)	(4)	(5)
		Number of drops added:			
Saline	0	5	8	9	10
Antiserum (1:10 dilution in saline)	10	5	2	1	0
Bacterial suspension	15	15	15	15	15
Final dilution of antiserum	1:25	1:50	1:125	1:250	Control

Procedure

1. Label the tubes.
2. Using a Pasteur pipette and rubber teat, add the appropriate number of drops of saline to the tubes.
3. With the same pipette, add the appropriate number of drops of a 1 in 10 dilution of antiserum to the tubes.
4. Rinse the pipette in saline at least three times, discarding the rinsings.
5. With the rinsed pipette, add 15 drops of bacterial suspension (prepared as described below) to each tube.
6. Discard the pipette into a jar of a suitable disinfectant solution.
7. Tap the tubes gently to remove air bubbles and incubate in a water bath, at 37 or 48°C (see note below) for 1–18 h. The contents will become mixed due to the action of convection currents, which are maximised if the level of the water in the water bath is maintained a little below the level of the liquid in the agglutination tubes. "H" agglutination will occur usually within

1 h; "O" agglutination often requires overnight incubation. If incubation is followed by storage in the refrigerator for 12–18 h the reactions may become more marked.

8. When a tube agglutination is read, each tube (starting with the control) is tapped gently with the finger until the cells are resuspended. The control tube should show uniform turbidity in contrast to the clumping seen in a positive reaction. The *titre* of the antiserum is the highest dilution at which clumping can be detected.

N.B. Occasionally antigen preparations have a tendency to clump spontaneously, hence the incubation of an antigen control. Standard antisera are obtainable from Wellcome Reagents Ltd. Incubation at 48–50°C is suitable for rabbit antibodies. Reactions involving human antibodies should be incubated at 37°C.

Preparation of antigens for agglutination tests

1. The organism is subcultured onto a freshly made nutrient agar slope to examine reactions involving "O" antigens, and into nutrient broth for "H" antigens. Incubate for 4–18 h at 37°C.

2.(*a*) For a preparation to contain "H" antigens, a 5 ml broth culture should be treated by the addition of 2–3 drops of commercial formalin and diluted to an opacity equivalent to a concentration of *c.* 5×10^8 organisms per ml.

(*b*) For a preparation containing "O" antigens wash the growth from the slope with *c.* 3 ml saline. The bacterial suspension should be heated in a boiling water bath for 10 min to destroy the less heat-stable "H" and "Vi" antigens. Dilute the suspension to an opacity equivalent to a concentration of 5×10^8 organisms per ml.

N.B. Disinfectants based on hypochlorite or other chlorine-containing compounds should NOT be used for discard jars, as this introduces a hazard of toxic fumes resulting from the action of formaldehyde.

3. The milk ring test for *Brucella abortus*

The milk ring test ("MRT" or "ART") is an agglutination reaction used to diagnose brucellosis in dairy herds, *and* to detect from bulk milk supplies herds in which brucellosis is present.

It is essentially a test which demonstrates the presence of agglutinins (antibodies) in the milk, by the addition of stained bacteria to the milk sample followed by incubation for 30–60 min. The stained *Brucella* are clumped by any agglutinins present and, in cow's milk or sheep's milk, the stained agglutinin-antigen complex rises with the fat globules. This causes the cream layer at the top to become deeply coloured. (In the case of goat's milk, the agglutinated stained bacteria go to the bottom of the test-tube, owing to the different creaming properties of the milk.)

The milk ring test is extremely sensitive and is very suitable for testing cans or churns of milk because the milk of one infected cow can be detected when mixed with that of many uninfected cows.

Procedure

1. Add one drop of stained *Brucella abortus* antigen (Wellcome Reagents Ltd.) to a 25 mm high column of a well-mixed milk sample in a small, narrow 75 × 10 mm test-tube. Mix thoroughly by shaking but *avoid frothing*.
2. Incubate the mixture at 37°C for 30–60 min and then examine.

Result

If the cream layer is deeply coloured and the milk beneath the cream layer is white or nearly so, the test is regarded as positive, indicating the presence of agglutinins in the milk. If the cream layer is white and the milk beneath deeply coloured, or the cream layer is the same colour as the milk layer, the test is recorded as negative.

C. The Precipitin Test

Whereas, in the agglutination test, the antibody causes clumping of an already particulate antigen material (e.g. bacterial cells), in the precipitin test a *clear* solution of a protein or polysaccharide antigen extracted from bacterial cells is used. The clear antigen solution when mixed with the appropriate clear antiserum will produce a mixture which first turns cloudy and then precipitates.

Preparation of the antigen solution for Lancefield's method of streptococcal grouping

1. Grow the strain of *Streptococcus* under investigation overnight in 50 ml of yeast glucose lemco broth.
2. Centrifuge the broth and discard the supernatant liquid.
3. Add 2 ml of 0·05 N hydrochloric acid containing 0·85 per cent w/v sodium chloride to the sediment of bacteria and transfer the suspension to a small test-tube, using a pipette.
4. Place the tube in boiling water and leave for 10 min.
5. Cool the tube quickly under running water.
6. Add one drop of 0·04 per cent phenol red solution. The liquid should turn the phenol red to yellow.
7. Slowly, *drop by drop*, add 0·3 N sodium hydroxide until the fluid is alkaline (red in colour). Then add, *drop by drop*, 0·1 N hydrochloric acid until the liquid is neutral (the colour should be salmon-pink).

8. Centrifuge and remove the *supernatant liquid* by means of a Pasteur pipette to another test-tube, and discard the sediment. The liquid constitutes the antigen solution, which is ready for use. If it is desired to keep the antigen extract, 0·5 per cent phenol should be added, and the extract kept refrigerated.

The precipitin ring test

This is the most common method of performing a precipitin test and is the test usually employed in streptococcal grouping. It can be made roughly quantitative by using a series of tenfold dilutions of antigen ranging from 1 : 10 to 1 : 10^7.

Procedure

1. 0·1 ml of antiserum is placed in a 6 mm × 50 mm tube.
2. 0·1 ml of prepared antigen or of a serial dilution of antigen solution is layered over the antiserum. This must be done slowly and with care so that a sharp interface forms between the two solutions. A Pasteur pipette with a finely drawn end is used and the tip is placed against the inside of the tube just above the surface of the antiserum so that the antigen solution runs on to the antiserum.
3. Set up, in the same way, a saline-antiserum control and an antigen-normal serum control.

Formation of a white precipitate at the interface of the reagents within 30 min indicates a positive reaction.

If economy of reagents is necessary, capillary tubes may be employed instead of 6 mm × 50 mm tubes. The capillary tubes can be held vertical in a block of "Plasticine".

When a precipitin test is set up using serial dilutions of antigens, one tube often shows a precipitate before the others. The ratio of dilution of antigen to dilution of antiserum in this tube is called the *optimal ratio*.

N.B. Standard antisera are obtainable from Wellcome Reagents Ltd.

MOULDS AND YEASTS

This section describes very simple basic techniques which can be used in preliminary studies of the microfungi encountered by the food microbiologist. More detailed procedures such as physiological tests for yeasts, and slide culture techniques will be found in Parts II and III.

A. General Conditions for the Growth of Moulds and Yeasts

Mycological media differ from bacteriological media in a number of ways because of the differing requirements for growth. Most fungi have an optimum pH much lower than that of most bacteria. Also, fungi are more capable of growing on media of inorganic salts with the addition of carbohydrate as an energy source, although some fungi possess a growth requirement for B-group vitimins or other growth factors which may need to be fulfilled by adding 0·1 per cent yeast extract. Almost all moulds, particularly the saprophytes, are obligate aerobes.

B. Media for the Growth of Moulds and Yeasts

There are many media used for the culture of moulds and yeasts, the following being among the more useful.

Malt extract agar for the isolation, counting and cultivation of moulds and yeasts. The pH may be at 5·4 or 3·5 depending on the purpose of the medium.

Czapek-Dox agar and *Potato dextrose agar* are general purpose media for the cultivation of both moulds and yeasts.

Davis's yeast salt agar (Davis, 1958) for counting moulds and yeasts. *Buffered yeast extract agar* is a simpler version of this medium which is available in dehydrated form (Oxoid). The pH of these media can be lowered to pH 3·5 to inhibit bacterial growth.

Orange serum agar contains clarified orange juice, yeast extract and glucose, and is a medium suitable for culture and counting of microfungi and aciduric bacteria.

Osmophilic agar (Scarr, 1959; Beech and Davenport, 1969) is a wort agar with a high concentration of sucrose and glucose, for the growth of osmophilic and osmotolerant organisms.

Rose bengal agar for the isolation of moulds from samples containing large numbers of bacteria. In addition to rose bengal, chlortetracycline (Jarvis, 1973) or streptomycin can be added to make the medium even more inhibitory against bacteria.

All but the last two media are readily available in dehydrated form.

As an alternative to acidification, media can be rendered selective against bacteria by the addition, immediately before pouring the plates, of sterile penicillin and streptomycin solutions to give final concentrations of 20 units and 40 units per ml respectively (Buckley, Campbell and Thompson, 1969). For a comprehensive survey of media used to culture yeasts and moulds see Booth (1971).

C. Examination of Moulds

1. Record the colonial characteristics and examine the colonies under the ×10 (low-power) objective of the microscope.

2. Prepare slides of the mould growth for microscopic examination in the following way. Pick off a portion of the growth with a needle and tease it out in a drop of lactophenol-picric acid or lactophenol-cotton blue placed on a microscope slide. Cover with a clean cover slip, taking care to exclude air bubbles.

Examine the prepared slide under the microscope, first using the low-power objective, and then using the ×40 (high-power dry) objective for a closer examination of a selected field.

Moulds are microscopically examined in *wet* preparations as described above. Water or aqueous solution of stains are not usually employed for mounting as many moulds are inadequately wetted by water and become enclosed in air bubbles. Consequently a mountant such as lactophenol is employed, usually containing a stain such as picric acid or cotton blue. Lactophenol also has the advantage that wet preparations do not spoil by rapid evaporation as is the case with water-mounted preparations, although a gradual deterioration will occur from other causes.

3. Moulds can also be grown and examined by the use of slide cultures (see pages 279, 287).

D. Examination of Yeasts

1. Colonial characteristics should be noted.

2. Prepare wet mounts by suspending a portion of culture in a drop of water. Add a small drop of Gram's iodine and cover with a coverslip. Observe with the ×40 (high-power dry) objective. Loeffler's methylene blue can be used instead of iodine.

3. Prepare a heat-fixed smear in the usual way and stain by Gram's method.

E. The Identification of Moulds and Yeasts

Moulds are identified on the basis of morphological and cultural characteristics including:

1. The colonial characteristics—size, surface, appearance, texture and colour of the colony.

2. The vegetative mycelium—presence or absence of cross-walls, and diameter of hyphae.

3. The asexual and sexual reproductive structures—e.g. sporangia, conidial heads, zygospores, arthrospores.

Yeasts may reproduce by budding, binary fission, ascospores, or less commonly by other methods. Identification is more difficult than is identification of multicellular moulds partly because yeasts rarely produce ascospores on ordinary media. Sporogenous yeasts can be induced to form ascospores by subculturing twice on a nutrient agar containing 5 per cent glucose and 0·5 per cent tartaric acid, followed by subculture on to an agar medium containing 0·04 per cent glucose and 0·14 per cent anhydrous sodium acetate only. Other media are described in Part III.

THE ISOLATION OF BACTERIOPHAGES

Bacteriophages (i.e. bacterial viruses) may be found in many natural environments, but bacteriophages capable of infecting a particular species of bacterium would most likely be found in an environment capable of supporting the growth of the host bacterium. For example, coliphage (a phage active against *Escherichia coli*) could be sought in faeces, and therefore also in sewage or in polluted water, whereas the most likely source of a phage against a soil bacterium would be the soil. Nevertheless, sewage is a useful source of phage against a wide range of bacteria, particularly if an enrichment stage is used. The number of phage particles present in the environment would, in any event, be quite small, so the first step in the isolation of a bacteriophage should be enrichment. Bacteriophages will multiply only in growing bacteria, so culture techniques must aim at providing an active culture of the host bacterium.

The following procedure is given merely as an example of a typical simple method of isolation. For further details of isolation, concentration and purification of phages see Billing (1969).

Isolation of Bacteriophages Active Against *Escherichia coli*

Preliminary enrichment

To a sample of 5 ml of crude sewage or polluted water, add 1 ml of chloroform. Mix and centrifuge. Transfer 1 ml of the supernatant liquid to 5 ml of a broth culture of *E. coli* in the exponential phase (this can be prepared by adding 1 ml of an overnight culture to 4 ml of broth and incubating for 2 h). Incubate at 37°C for 24 h.

Isolation of phages

After incubation, remove the bacteria by centrifugation. Transfer the supernatant liquid to another centrifuge tube, add 1 ml of chloroform and mix well. Centrifuge again, and aseptically transfer 1 ml of the supernatant liquid to 9 ml of quarter-strength Ringer's solution. Prepare a decimal dilution series.

The dilutions can now be examined for the presence of phages by using a Miles and Misra surface drop technique (see page 31) to inoculate "lawn" cultures of *E. coli*. A "lawn" culture is prepared by spreading 0·1 ml of a 24-h culture of *E. coli* over the surface of a nutrient agar plate. Incubate the plates at 37°C overnight. The phages will be able to develop in these organisms as they grow.

After incubation the presence of phage is shown by a clear area or by several small clear areas (i.e. plaques) where the supernatant liquid has been placed on the bacterial inoculum, the bacteria in these areas having been lysed by the phage. Occasionally, single colonies of phage-resistant mutants of the bacteria may develop inside this clear area. The use of decimal dilutions allows plaques to be obtained which have developed from a single phage particle.

Preparation of phage culture

In order to isolate a pure phage culture from the plaques so obtained, material from an isolated plaque can be transferred with a sterile needle to an actively growing broth culture of *E. coli*, and the culture incubated until clearing occurs. It is advisable at the same time to incubate a broth culture of *E. coli* not infected with phage as a control since sometimes complete clearing will not occur due to the presence and subsequent development of phage-resistant mutants. In such cases, noticeably less turbidity in the case of the inoculated culture can be taken as indicative of the presence of phage. Purity can be achieved by a repetition of plating to obtain well-isolated plaques and liquid culturing. The pure phage stock so obtained should be freed of bacteria and cell debris by centrifugation followed by filtration through a bacteriological filter (preferably a membrane filter).

Storing phage cultures

Most phage suspensions remain active in broth or serum. Some types of phage are inactivated at gas-liquid interfaces during shaking of suspensions; such phages can usually be protected from inactivation by using suspending liquids which contain a small amount of protein (by addition of gelatin or use of serum).

Further information on the concentration, purification and storage of phage cultures is given by Adams (1959) and Billing (1969).

THE MICROBIOLOGICAL ASSAY OF GROWTH FACTORS

A number of vitimins can be satisfactorily assayed using chemical procedures. Amongst the vitamins which are still frequently assayed microbiologically are biotin, pantothenic acid, para-aminobenzoic acid, and nicotinic acid and nicotinamide. The following typical procedure is that used for the assay of nicotinic acid and its analogues. It is based on the assessment of the growth response of a strain of *Lactobacillus plantarum* when various amounts of sample extract are added to a chemically-defined medium deficient in the vitamin. This growth response is compared with the growth obtained in media to which known amounts of vitamin have been added.

Procedures have been described for the microbiological assay of other vitamins by Barton-Wright (1963) and the Difco Manual (1953 and Supplement); for the assay of available amino acids in foods using *Streptococcus zymogenes* by Ford (1962, 1964); and for the assay of available amino acids and assessment of the nutritional value of protein using *Tetrahymena* by Stott and Smith (1966), Stott, Smith and Rosen (1963), and Shorrock and Ford (1973).

Procedure for microbiological assay of nicotinic acid (niacin) and nicotinamide

(*a*) *Test culture*

Use a culture of *Lactobacillus plantarum* NCIB 6376 (equivalent to ATCC 8014). The stock culture should be carried as a stab-inoculated tube of Micro-Assay Culture Agar (Difco), which is incubated at 30°C for 24–48 h, and then stored refrigerated. Subcultures should be made monthly.

To prepare an inoculum, transfer with a straight wire from the stock culture to a tube of Micro Inoculum Broth (Difco). Incubate at 30°C for 24 h. Dilute to 1 in 100 with sterile quarter-strength Ringer's solution. One drop of this dilute suspension is used to inoculate each of the assay tubes.

(*b*) *Preparation of food sample extracts for assay* (Barton-Wright, 1963)

Grind the dry materials and suspend 5·0 g in 50 ml N hydrochloric acid. In the case of niacin-rich materials, e.g. dried yeast or wheat germ, use 1·0 g samples. (High fat materials should first be defatted by a preliminary extraction for 16–18 h with analytical-grade light petroleum (boiling point 40–60°C).) Autoclave for 20 min at 121°C.

Cool, add 2·0 ml of 2·5 M sodium acetate solution, and then adjust pH to 4·5 with N sodium hydroxide solution (using bromocresol green as an external indicator). Filter, and then adjust to pH 6·8 with the sodium hydroxide solution. This procedure ensures that no acid-precipitable material remains in the extract to cause turbidity as a result of acid production by the lactobacilli during incubation of the assay tubes.

Make up the solution of the extract to a volume which gives an estimated probable 0·04–0·05 μg of niacin per ml.

(*c*) *Preparation of basal medium*

Use the dehydrated medium supplied by Difco Ltd. (Bacto-Niacin Assay Medium 0322) or by Oxoid Ltd. (Niacin Assay Medium CM 211). Add 7·5 g to 100 ml of distilled water, and heat to boiling for 2–3 min. A slight precipitate will be seen – this should be kept evenly distributed while the medium is being dispensed in 5 ml amounts into chemically clean 150 × 16 mm test-tubes (which must be optically standardised if the assay is to be assessed nephelometrically) (see page 44).

(*d*) *Preparation of standard response curve for niacin*

A standard curve must be established for every assay. The suggested range of concentrations of niacin for establishing the standard curve are:

μg per assay tube (10 ml): 0, 0·025, 0·05, 0·10, 0·15 and 0·20.

This range is obtained using a standard working solution of niacin containing 0·05 μg/ml (this being obtained by dissolving 0·05 g of niacin in 1 litre of distilled water, 1 ml of this stock solution being then made up to 1 litre of distilled water to give the standard working solution).

Twelve tubes each containing 5 ml of basal medium have added to them (each concentration being set up in duplicate):

(1) 5·0 ml of distilled water
(2) 0·5 ml of standard niacin solution + 4·5 ml of distilled water
(3) 1·0 ml of standard solution + 4·0 ml of distilled water
(4) 2·0 ml of standard solution + 3·0 ml of distilled water
(5) 3·0 ml of standard solution + 2·0 ml of distilled water
(6) 4·0 ml of standard solution + 1·0 ml of distilled water

(*e*) *Preparation of assay test range*

The assay of test samples should be set up at three or more concentrations with duplicate tubes at each concentration. The number of concentrations of extract depends on whether an approximate niacin content can be assumed.

All tubes, both standard range and test range are closed with suitable closures: cotton wool is best avoided as it is almost impossible to prevent fibres from entering the medium, and these will affect turbidity measurements. If nephelometry is being used as a method of assessment and it is intended to take the readings with the closures *in situ*, e.g. if incubation is to be continued, the test tube closure may affect the turbidity reading by reflecting a (variable)

amount of light back down the tube. In this case check the effect of the test tube closures which it is intended to use (see page 319–320).
Sterilise by autoclaving at 115°C for 10 min.

(*f*) *Incubation and reading results*

After inoculating each tube with 1 drop of the dilute suspension prepared in (*a*), incubate at 35–37°C for 72 h. After 24 h incubation take a series of readings using an EEL nephelometer (see page 45), or other suitable instrument. After 72 h incubation take a second set of turbidity measurements, and finally determine the growth response acidimetrically by titration with 0·1 N sodium hydroxide solution to pH 6·8 using bromothymol blue as an internal indicator.

(*g*) *Computation of results* (see Barton-Wright, 1963)

Graphical plots of the response data should provide curves with reasonably linear portions. If this be the case, the nicotinic acid content of the sample can be determined by the ratio of the slopes of the two lines provided by the test series and by the standard response series. (The slope provides the increase in response per unit of standard solution or test solution.) Ideally the two lines should intersect on the vertical axis.

If the plots do not provide straight lines, plot the logarithms of both doses and responses. The resulting lines should be approximately parallel. Measure the horizontal distance from the test line to the standard line—this represents the logarithm of the number of ml of standard solution equivalent to 1 ml of test extract (Barton-Wright, 1963), so the vitamin content of the sample can be calculated.

GROWTH RATE DETERMINATIONS ON PURE CULTURES

It may occasionally be useful in food spoilage investigations to obtain more precise information about the rate of growth of particular organisms than can be achieved by routine keeping quality tests or incubation of whole food samples. The following procedure which is similar to that described by Barnes and Impey (1968) can be modified to simulate particular food environments.

Procedure

Whenever possible use recent isolates from the foodstuff concerned. The inoculum may be either a culture in a liquid medium when this medium is also the test medium, or it may be a washed cell suspension. Dispense and sterilise 100 ml amounts of culture medium in conical flasks and equilibrate at the desired incubation temperature prior to inoculation. Add sufficient test culture to give a concentration of cells in the medium of *c.* 10^4 per ml, mix the flask by rotation, and withdraw a 1 ml sample for the preparation of decimal dilutions and determination of the viable count at the start of the experimental period (t_0).

Further 1 ml quantities should be withdrawn and viable counts determined at frequent intervals, so that the number of results obtained prior to the start of the stationary phase is adequate for the construction of meaningful growth curves and the calculation of mean generation time. In any particular system, preliminary outline experiments are likely to be required in order to establish the most satisfactory detailed procedure. The results obtained may be presented graphically by plotting $\log_{10}$ counts as ordinate against time (minutes) as abscissa.

The mean generation time may be calculated using the formula

$$G = \frac{t \log 2}{\log b - \log a}$$

where G is the mean generation time, a and b are the numbers of bacteria at two points in the logarithmic growth phase separated by the time interval t.

This technique can be particularly helpful when, for example, comparing the effects of temperature or the presence of food preservatives on the growth of the component organisms in a microflora. In the case of some foodstuffs (e.g. milk, sugar syrups) the sterilised foods can be used as the growth medium. Other foodstuffs (e.g. cooked meats) will require to be prepared as sterile homogenates, although in the case of raw foods filter-sterilised aqueous extracts may be suitable. However, it will be appreciated that such processing may affect significantly the growth characteristics of the organism under investigation, compared with the growth obtained in the original food environment.

It is also possible to study microbial interactions by the determination of counts and generation times when the isolates are grown in the test system as mixtures of the strains under investigation.

PART II

TECHNIQUES IN APPLIED MICROBIOLOGY

INTRODUCTION

There are a number of aspects to the role of the food microbiologist in the food industry: these include quality control on incoming raw materials, quality control of production, hygiene training for production staff, the development of suitable Codes of Practice for hygienic food production, assessment of detergent-disinfectants and the establishment of efficient cleaning regimes, the examination of samples of finished products, and the investigation of customer complaints of a microbiological nature. Often, undue emphasis is placed on the microbiological examination of many and frequent samples of the finished product only, with over-simple tests being performed which provide very little real information about the microbiological status of the food. ICMSF (1974) drew attention to the unsatisfactory nature of casual sampling. They cautioned against deriving an unjustified sense of security from the interpretation of results obtained in unsatisfactory sampling plans. They recommended that sampling procedures should be statistically based: improved sampling plans of the type suggested by them are able to provide a basis for statistically valid conclusions about the microbiological quality of the foods tested. Readers involved in the design of sampling plans and in the choice of microbiological standards are strongly recommended to consult ICMSF (1974).

It may be possible for a manufacturer or caterer to require that the suppliers of their "raw" materials meet certain specifications for the supplies to be accepted. For example, in the production of many low-acid canned foods, the heat treatment given needs to be far in excess of any legally required "botulinum cook", in order to achieve an acceptable low level of spoilage in the finished product. This is because the spore-forming spoilage bacteria will be present in the foodstuff before heat treatment in much greater numbers than will *Clostridium botulinum* spores, and very many of such spores will have much higher D-values than will the *Cl. botulinum* spores. If the raw materials (e.g. rice, spices) which are used in the manufacture of the canned foods and which are known to contribute the majority of the bacterial spores are required to meet a specification relating to a low spore count, a less severe heat treatment will be needed to obtain an acceptably low level of spoilage, with a consequent improvement in organoleptic quality.

Quality control on production is likely to operate at three levels: hygiene control on the processing (and this includes cleaning regimes for equipment, etc.); detection of possible hazards from pathogenic organisms in the product; and the assessment of the potential shelf-life (or storage-life) of the product. The application of a repressive system of quality control based only on the sampling of finished products and rejection of batches which fail the standard is unlikely to succeed in its aim since microbiological testing is destructive, and in consequence only a relatively small sample of the entire batch can be taken. The passing of the standard by the samples tested can provide no absolute assurance that the rest of the batch would also pass if similarly tested (see Ingram and Kitchell, 1970; ICMSF, 1974). An example of a repressive system which can provide a good measure of assurance (except in the case of deliberate sabotage!) is the in-line detection of foods which contain adventitious metal particles, which by means of electronic metal detectors can be applied non-destructively to every item in a production run for a very wide range of food products. There is no way in which a food company can protect itself against a statutory zero-tolerance on pathogens by microbiological testing of samples of the finished product. The only possibility of protection would come from extremely rigorous on-line control of the hygienic precautions taken (see also ICMSF, 1974). For this to be successful the co-operation of the food handlers and production staff would be necessary.

Thus a more positive approach to quality control is one based on the provision of training courses in hygiene for production staff, and the application of agreed Codes of Practice which, although they may be based on officially published Codes of Practice, will be more helpful and easier to understand if they are written for the specific production situation. In this case results obtained on samples of the finished products are used as an indicator of the success of the hygiene codes. In factories with fairly stable operating conditions in which a processing line produces a few lines only, it may be possible to use control chart procedures to provide early warning of emerging problems. Interested readers are referred to the useful discussion by Steiner (1967) of control chart procedures in quality control. An interesting account of the application over two decades of a quality control programme based on Codes of Practice, the training and involvement of food handlers and production staff, and the use of non-legal specifications is provided in the papers by Goldenberg (1964), Stephens (1970), Goldenberg and Elliott (1973), and Goldenberg and Edmonds (1973).

Detailed microbiological examinations of samples taken during and after the production and processing of food products *can* provide useful information, allowing an assessment of the probable shelf life, a check of the process in order to correct or to anticipate any deterioration in the production methods, and some indication of potential public health hazard. However, the choice of the methods of examination of a particular food product requires, amongst other things, an intimate knowledge of the preparation, storage and distribution of the raw constituents. The ICMSF (1974) suggested that in deciding whether to test a food sample for a given pathogen, the known food-borne disease record of that food should be considered. However, the lack of a record of implication in food poisoning may be merely due to the non-recognition of the existence of the hazard. Space only allows two examples to be given here to indicate the possibilities for misjudgment.

Until the late 1950s many cakes and confectionery products incorporated in their external decoration raw shredded desiccated coconut. Such products frequently were decorated after cooking so that the coconut received no heat treatment. Outbreaks of salmonellosis occurred, in which uncooked desiccated coconut was the common epidemiological factor. It was discovered that the desiccated coconut was peculiarly liable to contamination by *Salmonella* because of the method of preparation, and the drying process involved (Wilson and MacKenzie, 1955). Until the outbreaks of salmonellosis occurred in which the coconut was implicated, few people would have considered routine examination for *Salmonella* to be a test which had any significance in the microbiological analysis of desiccated coconut. Yet this *Salmonella* hazard may have existed for some time and may have resulted in a number of outbreaks of food poisoning. Without knowledge of the hazard the questions asked of patients concerning their food consumption histories could easily have been insufficient in these hypothetical earlier outbreaks to detect the fact that desiccated coconut was the common factor.

In the second example, until the laboratory investigations which resulted from the Aberdeen typhoid fever outbreak in 1964 (Report, 1964; Howie, 1968) it had been considered that unspoiled cans of heat-processed meats such as corned beef could present no significant hazard as a vehicle for heat-sensitive pathogens such as *Salmonella*. On investigation it was found that when gas-producing coliform organisms were introduced into a can of corned beef together with *S. typhi*, the latter could easily outgrow the coliforms and prevent their producing gas. It was demonstrated *in vitro* that the anaerobic growth of *S. typhi* was enhanced by the presence of nitrate in concentrations in the medium equivalent to those found in corned beef (Meers and Goode, 1965). The Aberdeen typhoid outbreak in 1964 caused the Committee of Enquiry (Report, 1964) to review an outbreak of typhoid fever which occurred in Oswestry in 1948. When the Oswestry outbreak had been first investigated it had been decided that circumstantial evidence pointed to milk as the vehicle but later investigations had shown this to be unlikely. The re-examination of the results of the epidemiological investigations (Report, 1964) strongly suggested the possibility of corned beef being the vehicle. The *S. typhi* strain involved in the Oswestry outbreak was then re-typed and found to be phage-type 34—the same type as that responsible for the Aberdeen outbreak.

Thus, in placing reliance on known records of involvement when trying to decide which tests should be performed on a food, it should be recognised that these records are in themselves imperfect indicators of hazards. ICMSF (1974) point out that "most food control efforts should be directed to the areas of greatest risk". Nevertheless, whilst *priority* in testing should be given to known hazards, when time, expense and other considerations allow, it may be salutary to use tests given a lower priority.

The two examples just given concern contaminated materials which themselves cause the outbreaks of disease. However, the fact that a contaminated raw ingredient is to be incorporated in a product receiving a heat treatment sufficient to destroy the pathogens is not necessarily a sufficient safeguard, since in most factories, restaurants and shops (and indeed homes) cross contamination can occur all too easily. For example, outbreaks of paratyphoid fever in which cakes filled or decorated with imitation cream were implicated, were found by Newell (1955) to be due to contaminated frozen egg. The frozen egg was being incorporated into the cake or pastry. While the baking may be expected to have destroyed the salmonellae in the cake mixture, the imitation cream, which did not contain frozen egg as an ingredient, was liable to contamination with *Salmonella* because in each of the bakeries involved the same mixing machine was used for mixing both cake and filling. In some cases the bowls and machines were not even rinsed between mixing the cake mix and the imitation cream.

The investigation of complex food products and their constituents for the presence of potential spoilage or food poisoning organisms also requires care. For example, certain spices may give very high viable counts. Black pepper normally contains a large number of aerobic spore-bearers. The introduction of pepper into a food product towards the end of its preparation (e.g. in *Pommes de terre duchesse*) may therefore introduce significant numbers of *Bacillus*, which may, under certain circumstances, be able to multiply sufficiently to cause spoilage or *B. cereus* food poisoning. In the preparation of the product already mentioned—namely *Pommes de terre duchesse*—cooked sieved potato is mixed with butter, egg, pepper, and nutmeg and piped into attractive shapes which are then flashed under a grill, both to brown the outside and to reheat to serving temperature. If this were being prepared in a large restaurant, the situation might arise where large amounts of the product are made at one time and kept until required. Immediately before serving, it may or may not be warmed under a grill or in an oven. Such a situation could result in numbers of *Bacillus cereus* sufficient to cause food poisoning. *Bacillus cereus* food poisoning seems to be very common in Hungary, and Ormay and Novotny (1969) have suggested that this may be due to the use of large amounts of spices customary in the production of traditional Hungarian meat and vegetable dishes. The outbreaks of *B. cereus* food poisoning caused by fried rice dishes (Vernon and Tillett, 1972) provide another example of the potential impact of modifications to kitchen practice caused by problems resulting from scaling up production in catering establishments (see Gilbert, Stringer and Peace, 1974).

Thus the more complex a food product, the more care is required in the correct selection of samples to be tested, in the correct selection of tests and counts to be made, and in the interpretation of the results obtained.

METHODS OF SAMPLING AND INVESTIGATION

It is necessary to take large and representative samples of foods. In the case of a packaged food product, the package should be opened only in the laboratory and sampling must be performed aseptically. All apparatus used for sampling should have been previously wrapped and sterilised. Batches of foods which are in packets, cans, bags or other containers should be examined by individually testing a number of units which have been selected at random. ICMSF (1974) discuss in detail the criteria for determining the stringency of sampling which may be required for any given combination of food, type of processing, method of storage and type of micro-organisms under consideration.

A. Liquid Samples

It is usually relatively easy to obtain representative samples of liquids. Frequently the liquid (e.g. milk, ice-cream mix, sugar syrups) will be held in a vat and will be subject to continuous or periodic mixing; otherwise the liquid should be mixed thoroughly up and down (e.g. using a sterile ladle) before the sample is taken. A large sample (e.g. 100–500 ml) should be withdrawn into a sterile container for transport to the laboratory. In the laboratory the liquid should be mixed thoroughly once again before pipetting the amounts required for investigation (see page 25).

B. Solid Samples and Sampling of Surfaces

Sampling of solids may be performed using sterile scalpels, spoons or cork-borers depending on the nature of the material to be sampled. A particulate food such as flour or dried milk is capable of sufficient mixing to enable a single, fairly small sample, e.g. 100 g to be taken. If, however, the product is in bulk larger samples from more than one location should be taken. The large samples should then be treated separately, each being mixed thoroughly in the laboratory before smaller samples are removed for testing.

Meat, fish and similar foods should be examined by taking deep samples as well as surface samples. Deep samples should be taken with care to minimise contamination from superficial levels. Some foodstuffs such as fresh and cooked meats can be sampled using sterile carving knife, scalpel and forceps. In the case of frozen foods a cork borer or even an electric drill fitted with a bore-extracting bit can be used to obtain deep samples without the need for thawing.

Surface Samples

Surface slices may be removed, or alternatively surfaces may be examined by transferring the micro-organisms from sample to microbiological medium with the aid of a supposedly inert carrier that neither causes death nor allows the multiplication of the micro-organisms. Such carriers include rinses, swabs and adhesive tape. Since the continued long-term viability of the micro-organisms on the carrier without death or multiplication is difficult if not impossible to achieve, the micro-organisms must be inoculated into suitable media at the earliest opportunity. The longer the delay before inoculation the less reliable the quantitative assessments.

A further method is to transfer the micro-organisms directly from the sample surface to the surface of the medium by impressing the one upon the other. The two principal impression techniques which achieve this are the agar sausage (Ten Cate, 1965) and the impression plate.

Impression techniques, and adhesive tape transfer do not allow dilution series to be prepared, and therefore colony counting is possible only when the microbial load is small. A great advantage of impression techniques, adhesive tape, contact slides and swabs is that they allow a non-destructive examination of the food samples.

The seven methods described below are not quantitatively equivalent and sometimes may not even rank samples in the same order according to their apparent bacterial load.

(1) Surface slices Remove very thin slices of the superficial layers of the food using sterile scalpels and forceps. In the case of table poultry, for example, the skin provides suitable material for sampling. Homogenise the slices in a suitable diluent to obtain an initial 1 : 10 dilution.

(2) Rinses and washes Rinse or wash the food (1 part by weight) in sterile diluent (10 parts by weight) and then consider the washings to be the initial 10^{-1} dilution. This procedure is applicable to foods such as sausages, dried fruits, vegetables and salad vegetables. It should be borne in mind that frequently the micro-organisms on the surface of the food will not be detached merely by agitation in the diluent so it is advisable also to obtain samples from which dilutions can be prepared by comminution. The extent of agitation and consequent removal of micro-organisms should be standardised. In reporting the results it must be recorded that the count represents bacteria on the surface only.

(3) Swabs When quantitative results are required, the area to be examined should be defined by the use of a previously sterilised template.

(*a*) *Cotton wool swabs* are prepared from non-absorbent cotton wool wound to a length of 4 cm and a thickness of 1–1·5 cm on wooden sticks or stiff stainless steel wire. They should be placed in alloy tubes which are then plugged and sterilised.

Method of use: Moisten the swab with sterile quarter-strength Ringer's solution, and rub firmly over the surface being examined, using parallel strokes with slow rotation of the swab. Swab the surface a second time, using parallel strokes at right angles to the first set. Care must be taken that the whole of the predetermined area is swabbed. Replace the swab in the tube. To prepare counts, add 10 ml of quarter-strength Ringer's solution. Agitate the swab up and down in the tube 10 times to assist the rinsing of the bacteria from the surface of the swab. Prepare plate counts from 1 ml amounts of the swab washings and from dilutions prepared as required. Counts should be recorded as the number per square centimetre of surface swabbed.

(*b*) *Alginate swabs* are an alternative type of swab which can be used (Higgins 1950), and are prepared from calcium alginate wool ("Calgitex" wool, obtainable from Roussel Laboratories Ltd.; swabs obtainable from Medical Wire and Equipment Co. (Bath) Ltd.). It is recommended that the amount of alginate wool used for each swab should not exceed 50 mg. The alginate wool is wound on a 1·5 mm diameter wooden stick to give a swab 1–1·5 cm long by 7 mm in diameter, moistened *very slightly* in quarter-strength Ringer's solution, placed in a 10 cm by 1 cm tube and sterilised by autoclaving at 121°C for 15 min. It is convenient to close the tube with a cork instead of cotton wool, and to mount the wooden stick in the cork. The cork may then be used as a handle.

Method of use: swab a predetermined area of the surface to be examined by rubbing firmly over the surface in parallel strokes, with slow rotation of the swab. Since alginate wool is very smooth, it is necessary for the rotation of the swab to be in the direction which will prevent the swab from unwinding. Then swab the same surface a second time, using parallel strokes at right angles to the first set. After swabbing the surface to be examined, and when it is required to carry out the laboratory examination, break off the swab aseptically into a screw-capped bottle containing 10 ml of a sterile 1 per cent solution of Calgon (sodium hexametaphosphate) in quarter-strength Ringer's solution. Replace the cap and shake the bottle vigorously. This causes the alginate wool to disperse and dissolve, giving a suspension of all the bacteria present on the swab. Prepare plate counts from 1 ml and 0·1 ml amounts (or further dilutions as required) using appropriate media. BBL-trypticase soy agar (Baltimore Biological Laboratories) in place of nutrient agar has been recommended by Post and Krishnamurty (1964), since it appears to nullify to some extent the slightly bacteriostatic nature of the Calgon (Post, Krishnamurty and Flanagan, 1963). BBL-trypticase soy agar may be replaced by Oxoid Tryptone Agar, and plate count agar should be an acceptable alternative. In order to reduce the final concentration of Calgon to a less inhibitory level, 15–20 ml of medium should be added to each plate. Counts should be recorded as the number per square centimetre of surface swabbed.

(4) Adhesive tape This method of surface sampling involves the use of a suitable self-adhesive tape (e.g. "Sellotape", "Scotch Tape") or of self-adhesive labels (e.g. Ryman). Self-adhesive labels have the advantage that the sampling details can be written on the back of the label and that they are already attached to a non-adhesive mount; self-adhesive tape needs to be transferred to a suitable sterile mount. Thomas (1961) described the use of self-adhesive tape and labels for investigations on the microbial flora of the human skin, but food surfaces and equipment surfaces also can be examined in this way. Tapes and labels are frequently self-sterilising for a short time after manufacture, by the action of volatile solvents, etc. Trials should be run on any tape which it is intended to use in this way, in order to obtain assurance on (*a*) the sterility of the tape, and (*b*) the absence of residual bactericidal activity.

The adhesive strip or label should be turned back on itself at one end for about 1 cm to form a tab for holding it. The strip is removed from its mount, pressed against the surface to be examined, pulled off immediately and replaced on its mount. In the laboratory, the strip is removed from the mount, pressed against the surface of an appropriate culture medium, and then removed and discarded.

(5) Agar sausages The agar sausage consists of sterile agar medium solidified inside a sterile cylindrical plastic casing (Ten Cate, 1965). To use, the end of the agar and casing is cut off aseptically and the exposed sterile agar surface pressed against the sample surface. A slice of agar is then removed using a sterile scalpel, placed aseptically into a Petri dish, impressed side uppermost, and incubated.

(6) Impression plates Impression plates are plates of the form shown below; they are available in a sterile disposable plastic form from a number of manufacturers (including Sterilin Ltd.)

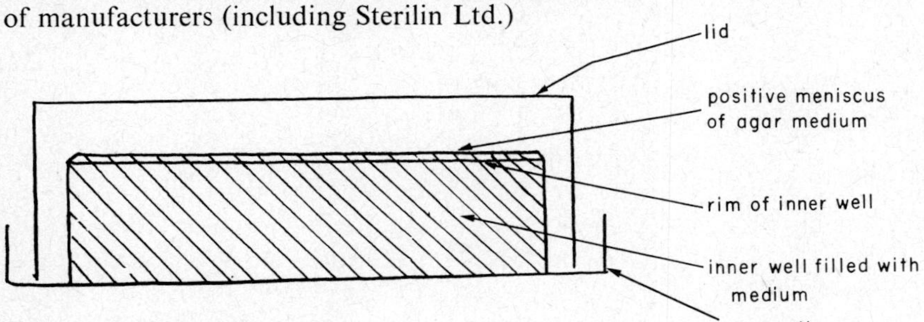

Cross section of impression plate

The centre well is filled with sufficient of the required agar medium to produce a convex meniscus. When set, the agar surface can be pressed against the surface to be sampled.

(7) **Contact slides** (Thomas, 1966) Press a sterile glass slide against the food sample to be examined. Transport the slide back to the laboratory. It may then be examined microscopically after fixing and staining (e.g. by Gram's method). Alternatively, the slides may be impressed on to poured plates of media, to achieve some transfer of the organisms to the agar surface. After removal of the slides (using sterile forceps) the plates may then be incubated. Although this method does not allow quantitative estimates to be made, it is useful for rapidly determining the main types which comprise the dominant microflora, especially on such foods as raw meat, poultry, and soft cheeses.

C. Sampling for Anaerobic Bacteria

If an examination for anaerobic bacteria is to be undertaken it is important that food samples likely to contain little free oxygen, e.g. deep tissues of meat, should not be exposed to normal atmospheric concentrations of oxygen as would occur if small samples were taken. When small samples are unavoidable, and also when swabs are used, a suitable transport medium (e.g. Stuart's transport medium) which is capable of maintaining reduced conditions must be employed. Thus, if an alginate wool swab is used it should not be reinserted into its original tube but placed in a bottle of Stuart's transport medium. The swabs may also be moistened with reinforced clostridial medium before use.

D. Attributes Sampling Plans

1. Sampling to detect and count low concentrations of organisms

When it is desired to count concentrations of viable organisms in foodstuffs which are below the limit of sensitivity of colony count techniques, it is possible to do so using a multiple tube technique, employing successive dilutions as described on page 34. If necessary large amounts (e.g. 100 g, 10 g and 1 g) of food can be used as the inocula into suitable large containers holding appropriate amounts of medium.

An alternative approach—known as an attributes sampling scheme—is to take a number of samples of the same size, and to determine whether any contain the organisms in question. If the organisms are not detected it is then possible to calculate the maximum proportion of units of that size which will, with a given probability, contain at least one of the organisms in question.

If the foodstuff be distributed already in discrete units (e.g. as is the case with canned foods) it is preferable to adopt this unit as the basis for each sample. Thus such attributes sampling schemes are readily applied to problems such as the determination of the proportion of cans of food, or packets of UHT milk, which contain spoilage organisms. It is not so obvious that attributes sampling schemes can be applied to foodstuffs which are not distributed in a large number of discrete units (e.g. a vat of ice-cream mix), and also to foods in which the unit size is inconveniently large (e.g. a large can of frozen egg albumen).

In such cases, it is assumed that the distribution of the organisms is random throughout the batch, and that the batch of food is made up of units of a given size (e.g. 10 g or 25 g). A number, n, of such units is examined for the presence of the organisms being sought. Then the proportion, in the entire batch, of positive units (i.e. units containing at least one organism) which may be detected with a given probability by at least 1 positive unit being found amongst the n units examined is given by

$$d = 100(1 - \sqrt[n]{1 - p}) \quad *$$

For example, suppose that we take 6 units each of 25 g and examine them for the presence of bacteria. Then the percentage of positive units in the whole batch which will be detected with a 95 per cent probability, by at least 1 of the 6 units being found positive is given by

$$d = 100(1 - \sqrt[6]{1 - 0{\cdot}95})$$

Hence
$$\frac{d}{100} = 1 - \sqrt[6]{0{\cdot}05}$$

$$1 - \frac{d}{100} = \sqrt[6]{0{\cdot}05}$$

$$\log\left(1 - \frac{d}{100}\right) = \tfrac{1}{6}(\log 0{\cdot}05) = \tfrac{1}{6}(\bar{2}{\cdot}6990) = \tfrac{1}{6}(-1{\cdot}3010)$$

$$1 - \frac{d}{100} = \text{antilog}\,(-0{\cdot}2168) = \text{antilog}\,(\bar{1}{\cdot}7832) = 0{\cdot}6070$$

$$\frac{d}{100} = 1 - 0{\cdot}6070 = 0{\cdot}393$$

Therefore $d = 39{\cdot}3$ per cent.

That is, if all the units are negative then, with $p = 0{\cdot}95$, the number of

* This is derived from the formula $p = 1 - \left(1 - \frac{d}{100}\right)^n$ which shows the probability of at least 1 defective unit being in the sample of n units taken, when d is the percentage defective in the entire batch. This relationship is true provided that n is a small fraction (less than $\frac{1}{4}$) of the entire batch.

micro-organisms in the batch is such as to give less than 39·3 per cent of 25 g units positive (in other words there are fewer than 40 bacteria per 2·5 kg in the long run).

If we wish to find the number of units which need to be sampled in order to detect a given proportion of defective units, we can do so from

$$n = \frac{\log(1 - p)}{\log\left(1 - \frac{d}{100}\right)}$$

For example, suppose that we wish to apply a standard for a food such that, with a 95 per cent probability, the foodstuff contains less than 1 *Salmonella* per 500 g. How many 25 g units must be sampled, and 1 found to be positive, in order to detect a level of contamination of 1 organism per 500 g? There are 20 units of 25 g in 500 g. Thus the standard we wish to apply is equivalent to there being less than 5 per cent of defective 25 g units (i.e. less than 1 unit in any 20 units being positive in the batch as a whole), with a 95 per cent probability. The number of units, n, which needs to be taken is given by

$$n = \frac{\log(1 - 0{\cdot}95)}{\log\left(1 - \frac{5}{100}\right)} = \frac{\log 0{\cdot}05}{\log 0{\cdot}95} = \frac{\bar{2}{\cdot}6990}{\bar{1}{\cdot}9777} = \frac{-1{\cdot}3010}{-0{\cdot}0223} = 58{\cdot}3$$

That is, we need to sample 59 units, each of 25 g, all of which must be negative, in order to ensure that in the long run the food contains (with a 95 per cent probability) less than 1 *Salmonella* per 500 g.

2. Use of attributes sampling plans in conjunction with colony or MPN counts

In the above section we have considered the use of attributes sampling plans for detecting small numbers of organisms. The ICMSF (1974) recommended the application of 3-class attributes plans to microbial counts in general, and used such plans as the basis for their recommended microbiological standards for a wide range of foodstuffs. In a 3-class plan, instead of the yes-no decision of the 2-class plan discussed above, a further class is distinguished, the three classes being: acceptable, marginally acceptable, and wholly unacceptable. Whereas any count above the limit for unacceptability will cause a rejection of the batch, a certain proportion of counts in the "marginally acceptable" range is allowable. Thus the use of a 3-class attributes plan in conjunction with counts such as general viable counts, coliform counts, etc., will require the testing of multiple samples taken at random. One of the stated advantages of 3-class plans is that the probability of acceptance or rejection of a batch is less affected by the nature of the distribution of the organisms within the batch than is the case with 2-class plans. For further details see ICMSF (1974).

E. Choice of Samples on a Non-random Basis

Usually attempts are made to take samples randomly. The samples to be taken from a production line or from a store may for example be chosen using a table of random numbers to identify the time of sampling or the location of samples. However, a conveyor line for a cooked food may show a gradual build up in microbial numbers over the period of a run. On the other hand, a pipeline system for food distribution, if inadequately cleansed and if at least a rinse is not performed immediately before starting a production run, may cause higher counts to be obtained at the beginning of the run. If the food is then distributed through the pipeline at a temperature within microbial growth range, a build up of micro-organisms may later occur towards the end of the run.

In the case of a batch of food in store, taking samples on a random basis may be justified if the storage conditions are the same for the entire batch. Often, however, temperature and other gradients will occur across the stack of food. Obviously in such cases most information would be obtained by the choice of samples from specific locations in the stack. The microbiological effects of such gradients can then be determined, particularly if there is opportunity for recording the environmental parameters involved (e.g. by use of a multichannel recorder and a number of thermocouples to detect temperature gradients in batches of food kept in cold stores).

F. Transport and Storage of Samples

Whenever possible, the original state of the sample should be maintained until the laboratory tests are carried out. For example, frozen foods should be kept frozen by using solid carbon dioxide (with precautions to protect the frozen food samples against exposure to the gaseous CO_2) or "cold packs", in conjunction with insulated containers for transport to the laboratory, followed by storage in a deep freezer until tested.

Perishable but unfrozen food samples should not be frozen before testing (unless the testing *must* be long delayed) but, if storage for a short period is unavoidable, the samples should be chilled and kept refrigerated at about 4°C until tested. It should be remembered, however, that refrigeration of samples for periods of three days or more will result in the multiplication of any psychrotrophic micro-organisms present in the food, and may cause the death of some mesophiles or thermophiles.

The history of sampling, transport and storage should be recorded for all samples.

PREPARATION OF DILUTIONS

Throughout the methods for food sampling it is suggested that initial 1 : 10 dilutions of food samples be prepared from 10 g amounts of the food suspended or homogenised in 90 ml amounts of diluent, since frequently this is the largest sample that can conveniently be handled in many laboratories. Nevertheless, it is recommended that, if possible, 1 : 10 dilutions are prepared by suspending or homogenising 50 g in 450 ml of diluent, as this gives much more reliable and representative results, whilst remaining a quantity reasonably easy to handle.

A. Choice of Diluent

1. General purpose diluents

For many purposes sterile 0·1 per cent peptone water at pH 6·8 – 7·0 (Straka and Stokes, 1957), phosphate buffer or quarter-strength Ringer's solution are satisfactory diluents. 0·1 per cent peptone water appears to have a greater protective effect than phosphate buffer or quarter-strength Ringer's solution, but in any case the inoculation of media should be carried out within 15–30 min of the preparation of the dilutions.

The possible modification of the nature of the diluent by the sample at the lowest dilutions should not be overlooked. In particular, if the food sample contains a high proportion of undissolved water-soluble material, what will be the effect of its solution in the diluent? Will the pH or available water (a_w) be greatly affected? If in doubt the pH of the first dilution should be checked, and some indication of a possible change in a_w may be obtained by a simple determination of the soluble solids. It may be necessary to add phosphate buffer to the diluent to counter the pH drift. For example, Dixon and Wilson (1960) succeeded in isolating *Salmonella* from superphosphate-containing horticultural fertilisers by suspending the samples in 0·5 M phosphate buffer at pH 7·0–7·2. Such fertilisers had been thought previously to be free from *Salmonella*.

If dilutions of a highly soluble dry specimen (e.g. milk powder, baby food preparations) are being prepared, so that a low a_w occurs at the lowest dilution, the preferred diluent may even be distilled water (Silverstolpe, Plazikowski, Kjellander and Vahlne, 1961): the most suitable diluent should be chosen by testing a range of diluents and the one which gives the highest recovery rates used subsequently.

2. Diluents for anaerobes

When examining a sample for anaerobic bacteria either qualitatively or quantitatively, reinforced clostridial medium or a similar formulation must be used as the diluent in order to maintain reduced conditions. Dispersion of the sample in the diluent should be achieved by a method which provides least opportunity for introducing oxygen into the mixture. For example, the Colworth Stomacher (see below) is better than a top-driven homogeniser or a bottom-driven blender unless the sample container is flushed with oxygen-free nitrogen before blending.

3. Diluents for osmophiles and halophiles

A suitable diluent for osmophilic counts is sterile 20 per cent sucrose solution. For investigations of halophilic organisms (e.g. in samples of curing brines), sterile 15 per cent sodium chloride can be used as the diluent.

B. Liquid Samples

Dilutions of liquids can be prepared in a manner similar to that described in Part I, page 25. Pipette aseptically 10 ml of the thoroughly mixed sample into a sterile glass bottle with a ground glass stopper and add 90 ml of diluent to give a 1 : 10 dilution v/v. Alternatively, weigh, with aseptic precautions, 10 g of the thoroughly mixed sample into the bottle and add 90 ml of diluent to give a 1 : 10 dilution w/v, which, for all practical purposes, is equivalent to a 1 : 10 w/w. Prepare further decimal dilutions as necessary in the usual way. Inoculation of media should be carried out within 15–30 min of the preparation of the dilutions.

C. Fine Particulate Solid Samples

The preparation of the initial dilution of fine particulate solid samples such as flour and milk powder is accomplished easily. Weigh aseptically 10 g of sample into a sterile glass bottle which is marked at 100 ml capacity and fitted with a ground glass stopper. Add sterile diluent to the 100 ml mark to give a 1 : 10 dilution w/v. Shake the suspension 25 times with an excursion of 30 cm. Prepare further dilutions as necessary in the usual way. Care should be taken with highly soluble samples that the counts obtained by distribution of measured volumes can be related accurately to the original sample (expressed as the count per gram). In order to achieve this it is necessary to determine the volume of the first dilution. The possibility of alteration of pH and a_w must also be considered as already mentioned above. Inoculation of media should be carried out within 30 min of the preparation of the dilutions.

D. Other Solid Samples

1. Comminution

Tests of foods which may contain micro-organisms below the surface should be carried out by blending or comminuting at least 10 g of the food in the appropriate amount of sterile diluent using suitable apparatus.

(*a*) *Using a bottom-driven food blender*

One of the best types readily available in the U.K. is the "Ato-Mix" (made by M.S.E. Ltd.). This uses an autoclavable stainless steel container with blending blades attached to a leak-proof drive through the base of the container; the lid of the container is also leak-proof.

The container and its lid can be autoclaved at temperatures up to but not exceeding 121°C. For autoclaving, the lid should be loosely seated (i.e. with the knob rotated anticlockwise so that the white plastic adjuster is not expanding the sealing gasket) with a piece of paper separating the lid from the container. After autoclaving, and when cool, the paper can be removed carefully, the lid placed in position and sealed by rotating the knob clockwise to keep the interior sterile. (If autoclaved with the lid in direct contact with the container, the gasket will stick to the container and will be damaged.)

Between similar samples the interior of the container and inner face of the lid should be washed well with hot water, and swabbed with 70 per cent ethanol. After excess ethanol has been drained from the container, the remaining ethanol may be flamed quickly from the interior; the lid must not be flamed. Before the next sample is placed in the container, the container should be cooled to about 15°C by placing in a refrigerator.

The sample and sterile diluent (10 g in 90 ml in the 200 ml container, or 50 g in 450 ml in the 1 litre container) are blended for 30 sec at low speed and 2 min at high speed followed by a final 15 sec at low speed.

One disadvantage of bottom-driven blenders is the temperature rise which occurs and which can be substantial with some types of sample. Unfortunately the use of refrigerated diluent when the sample is not at refrigerator temperature may cause a drop in count as the result of cold-shock (Meynell, 1958; Gorrill and McNeil, 1960; Strange and Dark, 1962). Most workers have been in agreement that it is usually Gram-negative bacteria which show cold-shock, that exponential phase cells are most susceptible (stationary phase cells being most resistant) and that the effect is more marked in quarter-strength Ringer's solution than in sucrose solutions or complex diluents. However, there has been at least one report (MacKelvie, Gronlund and Campbell, 1968) of a *Pseudomonas* which became more susceptible to cold shock as the cells aged.

(*b*) *Using a top-driven homogeniser*

The "Ultra-Turrax" (made by Janke and Kunkel KG) is an example of a laboratory-scale top-driven homogeniser suitable for processing samples for microbiological examination. The homogeniser shaft rotates at high speed and creates a considerable vortex. If the sample and diluent are placed in too small a container most of the contents will be ejected on to the bench!

The method of clearing and sterilising the homogeniser shaft between samples is as follows: remove any gross debris; operate the mixing shaft (20 sec) in a beaker of warm water with a little low-foam detergent; wipe with clean tissue; operate (20 sec) in a beaker of rinsing water; operate (20 sec) in a beaker of Wescodyne iodophor solution; operate (20 sec) in a beaker of sterile sodium thiosulphate solution; finally operate (20 sec) in two rinses of sterile distilled water. If in doubt about the efficacy of the cleaning regime, 1 ml amounts of the last rinse can be plated out.

The motor should not be operated except when under load, and the running time for homogenisation should be less than three minutes.

The principal disadvantage of the smaller Ultra-Turrax is that it will only blend efficiently food samples which already consist of fairly small fragments (approximately 5 mm in diameter).

(*c*) *Using the Colworth Stomacher*

The Colworth Stomacher was developed at the Unilever Research Laboratories (Sharpe and Jackson, 1972) and is available from A. J. Seward.

The sample plus diluent are placed inside a sterile, disposable, thin and flexible polythene bag. The bag is placed into the blender chamber with a few cm of the bag projecting above the top of the door of the chamber, closing of the door firmly seals the bag while blending occurs. Starting the Stomacher causes two large flat stainless steel paddles alternately to compress the bag plus its contents against the flat inner surface of the door. Operation for 30 sec is sufficient for dispersing most samples, although samples containing high concentrations of fat should be processed for 90 sec.

Amongst the advantages of this apparatus are that the samples do not come into contact with the blender but are contained in cheap disposable bags; practically no temperature rise occurs; good dispersal is obtained even with deep-frozen samples; in the case of many foods very low dilutions can be prepared if low counts are suspected. It is also quiet compared with other forms of homogeniser. The apparatus is highly recommended. Tuttlebee (1975) in a comparative study of the use of Stomacher and Atomix concluded that in all respects the Stomacher was to be preferred.

2. The examination of surfaces

The methods of examining surfaces of food samples have already been described (pages 124–127).

GENERAL VIABLE COUNTS

General viable counts are determined usually by colony counting methods (see page 25ff) although the multiple tube technique may be used if low concentrations of bacteria are expected. The choice of medium and incubation conditions is difficult when general viable counts are attempted on the mixed microflora usually found in foods. Frequently viable counts are required of populations for which there is little knowledge of the types of organisms present, and in these circumstances, because of the variety of nutritional and physical requirements represented, it is impossible to obtain counts which truly indicate the number of viable organisms present.

A sample may contain some organisms which are obligate aerobes, some which are obligate anaerobes and some which are facultative with respect to oxygen; some of the organisms may be capable of good growth at 37°C but not at 20°C, some may grow at 20°C but not at 37°C and others may grow at both temperatures. Even if four sets of plates were to be incubated—one set aerobically at 20°C, another aerobically at 37°C, a third anaerobically at 20°C, and the fourth anaerobically at 37°C—it would not be possible to determine the *total* viable population of these organisms, because an unknown proportion of the population would grow and be counted on more than one set of plates.

Similar problems occur when attempting to choose media. Colony counts and multiple tube counts assess viability as the proportion of the population capable of continued multiplication in the incubation conditions provided. In other words, the development of a colony (on or in an agar medium) or of turbidity indicates clonal viability in the incubation environment and *not* clonal viability in the environmental conditions found either in the normal post-processing state of the food or in the consumer. This distinction is especially important when stressed or damaged organisms are encountered in, for example, processed or preserved foods. Thus if a food sample be examined using both total (microscopic) count and viable count, a high total count in conjunction with a low viable count does not necessarily indicate that the majority of the organisms observed microscopically are dead, but perhaps merely that they are unable to multiply in the particular incubation environment.

The food microbiologist has to choose one or more sets of incubating conditions on empirical grounds, and as a general rule those incubation conditions are chosen which have been found to give the highest counts in the largest number of samples. In most test situations more than one incubation temperature is likely to be employed:

0–10°C	for counts of psychrotrophs and psychrophiles
20–32°C	for counts of saprophytic mesophiles
35–37 (−45)°C	for counts of parasites and commensals of homothermic animals
55–63°C	for counts of thermophiles.

Micro-organisms capable of growth at low temperatures may be either psychrophilic micro-organisms with a low optimum temperature for growth, or mesophilic psychrotrophs which, according to the definition of Eddy (1960a), are organisms capable of growth at refrigeration temperatures of +5°C or lower but with an optimum growth temperature typical of mesophiles. Psychrotrophic and psychrophilic counts are best determined using surface count techniques in order that the bacteria are not exposed to molten agar. True psychrophiles, which are probably rare in foods except those of marine origin, may be killed by exposure to ambient temperatures, and if their presence is suspected, samples, diluents and plates must be kept at temperatures below 10°C (Farrell and Rose, 1967).

The more complex the medium, the greater the range of types *likely* to be capable of growth (but see page 225). Thus, glucose tryptone yeast agar (plate count agar) allows the growth of more types than does nutrient agar. One of the authors obtained a viable count of less than 10,000 per gram of a frozen pre-cooked meat dish when nutrient agar was used, whereas the same sample gave a count of several millions per gram when plate count agar was used as the plating medium. The discrepancy occurred because the predominant component of the microflora of the food was a streptococcus, which requires glucose for growth. Supplementation by the addition of serum, egg yolk, or milk may provide even greater counts with certain types of sample, but such supplementation is not used routinely because of the extra cost and the increased complexity of medium preparation (serum and egg yolk needing to be added immediately prior to pouring the plates).

It is not uncommon for a significant component of the microflora of a food or other habitat to be unable to grow unless media are supplemented with an extract of the food or other material comprising the habitat. For example, Varnam and Grainger (1973) reported that certain Gram-negative, oxidase-positive, polarly flagellate bacteria (*Vibrio*-like organisms) common in bacon curing brines "required" a filter-sterilised pork extract in the growth medium for good growth to occur. This extract however, appeared to be acting not by providing an essential nutrient, but by removing peroxide from the medium, since it was subsequently found (Varnam and Grainger, 1975) that substitution of the pork extract by manganese dioxide (Meynell and Meynell, 1969) would also allow growth of these organisms.

Many marine bacteria are unable to grow unless at least part of the distilled water in the medium is replaced by seawater.

When attempting to study the constituent microflora of a habitat, the effects of similar supplementations should always be tested. In attempting to assess the efficacy of the isolation medium in recovering the constituent microflora, the use of direct microscopic examination of the samples and dilutions can be useful. If a particular morphological type of organism is seen microscopically but is not recovered on isolation media, there is a strong possibility that the isolation environment is proving unsuitable for the growth of that microbial type. However, care must be taken in interpreting and comparing microscopic and cultural examinations since some organisms (e.g. coryneform bacteria) may exhibit different morphological appearances in the sample and on the medium.

Bacterial counts in the presence of moulds and yeasts may be determined by using media incorporating the antifungal antibiotic cycloheximide (Actidione, obtainable from Upjohn Ltd.) to 10 p.p.m. (see page 219).

Yeast and mould counts are usually obtained by employing media such as orange serum agar, Davis's yeast salt agar, or malt extract agar, in which the acidity has been adjusted to around pH 3·5 by use of sterile lactic acid or citric acid immediately prior to pouring the plates. Alternatively, penicillin and streptomycin may be incorporated in the media to final concentrations of 20 units and 40 units per ml respectively (Buckley *et al.*, 1969). It must be realised however that whereas counts of yeasts present similar problems to bacterial counts (yeasts being unicellular organisms), mould counts may be difficult or impossible to interpret. Since most moulds will grow in a mycelial form in foods, the number of colonies obtained on a plate will depend on the degree of homogenisation and the extent of the consequent fragmentation of the hyphae. If the mould growth has resulted in formation of conidia or other spores, vast numbers of these may be present, each potentially capable of forming a colony. Thus the number of colonies obtained in a plate count does not reflect the amount of mould growth in the foodstuff.

DETECTION AND ENUMERATION OF INDICATOR BACTERIA

The term "indicator organisms" can be applied to any taxonomic, physiological or ecological group of organisms whose presence or absence provides indirect evidence concerning a particular feature in the past history of the sample. It is often associated with organisms of intestinal origin but other groups may act as indicators for other situations. For example, the presence of members of the set "all Gram-negative bacteria" in heat-treated foodstuffs is indicative of inadequate heat-treatment (relative to the initial numbers of these organisms) or of contamination subsequent to heating. Coliform counts, since the coliforms represent only a sub-set of "all Gram-negative bacteria", provide a much less sensitive indicator of problems associated with heat treatment, but are still frequently used in the examination of heat-treated foodstuffs.

Although most counts of indicator bacteria are based on the use of selective media, it is possible that higher recovery rates may be obtained with a non-selective resuscitation stage preceding the use of the selective medium (see page 148).

A. Coliform Organisms and *Escherichia coli*

In water analysis the presence of *Escherichia coli* indicates faecal pollution of the water, there being a positive correlation between the concentration of the organisms and the amount and/or recency of the pollution (see page 226). Thus the presence of *E. coli* implies that the water may contain pathogens of intestinal origin. *Escherichia coli* may be used as an indicator in food microbiology. In addition certain serotypes of *E. coli* are enteropathogenic. They are a significant cause of infantile gastroenteritis (Taylor, 1961; 1966), but may also cause outbreaks of food poisoning amongst the adult population (Marier *et al.*, 1973; Report, 1974).

Raw food or foods containing uncooked ingredients will frequently contain coliform organisms including *E. coli*, but multiplication of the organisms may have occurred in the food, so that there may be no correlation between numbers and the level of initial contamination. However, the presence of *E. coli* and other enteric commensals in foodstuffs does indicate the possible presence of enteric pathogens (quite apart from the possibility of these commensals themselves causing food poisoning). Therefore, care must be taken in the interpretation of counts.

For example, consider a piece of freshly prepared sterile food which is then handled by a worker with faecally contaminated hands. Immediate examination of a sample of the food will detect *E. coli* in numbers proportionate to the amount of faecal contamination of that food, with the corresponding implication of the possibility of other enteric pathogens. If, however, sampling and examination does not take place immediately after the contamination then, depending on the nature of the food, and the conditions of storage one of the following courses of events may ensue:

(i) The *E. coli* may die off so that the occurrence of the initial faecal contamination is undetected. In this case we may be incorrect in assuming that other enteric pathogens will have died at the same rate (see page 121).

(ii) The numbers of *E. coli* remain approximately the same. Again, this cannot be taken to indicate that other enteric organisms will react in the same way to the environments in which they find themselves.

(iii) *Escherichia coli* grows and increases in number. In this case we must assume that the conditions which were favourable for the growth and multiplication of *E. coli* probably also favoured the growth and multiplication of other bacteria of enteric origin.

Thus although large numbers of *E. coli* do not indicate recent or heavy faecal pollution, they must be taken to indicate a possible hazard from enteric pathogens.

Nevertheless the ability of *E. coli* to survive and/or multiply cannot be used as a reliable indicator of the ability of pathogens such as *Salmonella* to survive and/or multiply (Anderson and Hobbs, 1973; Akman and Park, 1974). In consequence, it has been proposed by Mossel and his colleagues that a better measure of the hygienic quality of foods would be a "total Enterobacteriaceae count" (Mossel, 1957; Mossel, Mengerink and Scholts, 1962). Such counts are easily made by adding glucose to the selective coliform counting medium. If violet red bile agar is used then glucose should be added to the recipe to a final concentration of 1 per cent. An excellent review of the rationale of the use of indicator organisms in general and of the "total Enterobacteriaceae count" in particular has been made by Mossel (1967).

Throughout this book the following definitions relating to bacteria in this group have been assumed (note that each group constitutes a subset of the previously defined group):

Total Enterobacteriaceae. Bacteria which, in the presence of bile salts, will grow and produce acid from glucose (as determined by use of violet red bile glucose agar).

Coli-aerogenes bacteria. Bacteria which, in the presence of bile salts or other equivalent selective agents, can grow and produce acid *and* gas from lactose when incubated at 30°C.

Coliform bacteria. Bacteria which, in the presence of bile salts or other equivalent selective agents, can grow and produce acid *and* gas from lactose when incubated at 35 or 37°C.

Faecal coliform bacteria. Bacteria which, in the presence of bile salts or other equivalent selective agents, can grow and produce acid *and* gas from lactose when incubated at 44–45·5°C. Note that the incubation temperature is critical, and a water bath should always be used for this test. The choice of temperature within the indicated range depends on the choice of medium, if equivalence of counts is used as the criterion, but it does not necessarily follow that when the same counts are obtained by two different procedures the same bacterial types are being detected.

Escherichia coli. Bacteria which, in addition to showing the above characteristics, are also methyl red +, Voges-Proskauer −, and cannot utilise citrate as a sole carbon source. Indole positive strains are termed *E. coli* type 1, and are presumed to have the intestine as their primary natural habitat.

The presence of *any* of the above groups in heat-treated foods is indicative of one or more of the following:

(i) the initial concentration of the bacteria was so high that the heat treatment was inadequate to reduce the concentration to an indetectably low level;
(ii) post-heating conditions allowed multiplication of survivors until their numbers became detectable;
(iii) contamination occurred subsequent to the heat treatment.

Rarely will survival be due to the presence of relatively heat-resistant strains, and unless isolates are tested for their heat resistance by D-value determinations unusual heat resistance should not be hypothesised in preference to one of the three explanations given above.

The counting procedures described below can be modified by use of glucose or choice of incubation temperature as indicated to count any of the categories defined above. In general, the more inclusive categories should be used in examining heat-treated or shelf-stable foods. The less inclusive ones are used in the examination of raw meats and similar foods in which faecal contamination is likely to have resulted in the presence of enteric commensals or when it is appropriate to use the counts to assess levels of innate contamination, rather than their being used to assess efficacy of processing and preservation procedures.

1. Coliform counts by the multiple tube technique

The presence of coliform organisms in media described below is detected by the production of acid and gas from lactose. Since many foods contain significant amounts of other carbohydrates (e.g. sucrose, glucose, fructose) it is especially important that *presumptive* positive results are confirmed by subculture as described when high concentrations of the food sample have been incorporated into the medium.

(a) Lauryl sulphate tryptose broth at 35°C

Pipette aseptically 1 ml of each of the prepared dilutions of the food into each of three or five tubes of lauryl sulphate tryptose broth. If very low concentrations of organisms are expected, 10 ml or even 100 ml amounts of the lowest dilution may be added to equal volumes of double-strength medium, three or five bottles being prepared at each dilution (see pages 34–36).

Incubate at 35°C and examine after 24 and 48 h for growth accompanied by the production of gas. (The production of acid may be detected by the addition of a pH indicator to the tubes after incubation.) A tube showing the production of acid together with sufficient gas to fill the concave of the Durham tube (i.e. the meniscus should be at a point where the sides of the Durham tube are parallel) is recorded as being *presumptively positive*. A presumptive positive result is also recorded if the Durham tube contains less than the stated amount of gas, but effervescence occurs when the side of the test-tube is tapped. In routine testing gas production is usually regarded as indicating a presumptive positive result without the need for testing for acid production.

Presumptive positive tubes should be confirmed by subculturing 2 or 3 loopsful from each tube into a tube of brilliant green lactose bile broth. These are incubated at 35°C for 48 h, gas production (judged on the criteria described above) being taken as indicating a *confirmed positive* result. By comparing the combination of positive and negative tubes obtained at each stage with probability tables (see Appendix 2, page 383), the most probable number of "presumptive coliforms" and of "confirmed coliforms" can be determined.

(b) MacConkey's broth

An alternative procedure involves the use of MacConkey's broth with incubation for 48 h at 30°C (for "coli-aerogenes" counts) or at 37°C (for coliform counts). Presumptive positive results (acid and gas produced) should be confirmed as above by subculture into brilliant green lactose bile broth or by streaking across MacConkey agar plates. Counts are determined by reference to probability tables (see p. 383).

(c) Brilliant green lactose bile broth

A third multiple tube technique which may be used involves the incubation at 35°C of inoculated tubes of brilliant green lactose bile broth to establish presumptive positive results (by gas production). Confirmation of presumptive positive results is by streaking a loopful from each such tube across a suitable medium (e.g. MacConkey's agar, Endo agar).

2. Faecal coliform counts by multiple tube technique

An extension of the coliform count made at the time of subculturing for confirmation of positives can provide the additional information needed for counting those faecal coliform organisms which are able to produce acid *and* gas from lactose at 44–45°C (and which, for most purposes, can be regarded as equivalent to *Escherichia coli*). The test for this characteristic is known as the modified Eijkman test.

Presumptive positive tubes from the foregoing tests are subcultured into either

(i) *E.C. medium*, incubated at 44·5 $\pm$ 0·2°C (American Public Health Association, 1971) or at 45·5 $\pm$ 0·2°C (Thatcher and Clark, 1968) for 24 h.

or (ii) *Brilliant green lactose bile broth*, incubated at 44·0 $\pm$ 0·2°C for 24 h.

It is advisable to prewarm the tubes to the incubation temperature before inoculation.

(The use of MacConkey's broth at 44–45°C for this test has been largely superseded by the media listed above.)

Escherichia coli type 1 is indole positive at 44–45°C and therefore the Eijkman test can be supplemented by inoculating a peptone water and incubating at the elevated temperature for 24 h and then testing for indole production (see page 68). The confirmation of *E. coli* by IMViC tests is time-consuming: since cultures for these tests are incubated at 35–37°C, it is first necessary to obtain pure cultures by streaking the appropriate broth cultures on to a solid medium (preferably one that is differential but non-selective).

A more rapid count of *E. coli* may be attempted by inoculating the original samples and dilutions into tubes of media which are incubated for 4 h at 30°C, and then transferred to a water bath at 44·0 $\pm$ 0·2°C for 18–20 h.

3. Colony counts of coliform bacteria and *E. coli*

The usual medium employed is violet red bile agar. Prepare duplicate sets of plates using 1 ml amounts of the chosen range of dilutions of the food. Add to each plate 15 ml of violet red bile agar, melted and cooled to 45°C, mix well and allow to set. Finally overlay with another 5 ml of violet red bile agar. After allowing to solidify, invert the plates and incubate at 35°C for 24 h for coliform counts. After incubation, count the number of dark red colonies. Normally typical coliform colonies will be 0·5 mm or more in diameter and show evidence of the precipitation of bile salts in the medium immediately surrounding the colonies, but of course overcrowding of the colonies will cause a reduction in size. In the case of foods containing carbohydrates other than lactose, media to which low dilutions of the food have been added may give anomalous results as already mentioned. With plates of 10^{-2} or lower dilutions it is therefore necessary to confirm presumptive coliform colonies either as coliforms by picking them off and inoculating lactose broth or MacConkey's broth and incubating at 35°C or as *E. coli* by the Eijkman test (see above).

B. Faecal Streptococci

The faecal streptococci include the species *Streptococcus faecalis* (and the subspecies *Strep. faecalis* subsp. *liquefaciens*, and *Strep. faecalis* subsp. *zymogenes*), *Strep. faecium*, *Strep. bovis*, and *Strep. equinus*. All belong to the Lancefield group D (see page 271). These organisms are useful indicators of the possible presence of enteric pathogens since, compared with *E. coli*, they are more resistant to freezing, low pH, and moderate heat treatment. Thus faecal streptococci may frequently be detected in frozen foods, fruit juices, or foods which have received a cursory heat treatment, even when *E. coli* has been killed by the inimical conditions. Faecal streptococci may help in assessing the standard of hygiene in factories (Slanetz *et al.*, 1963) and it is possible that they may sometimes be responsible for food poisoning (Moore, 1955).

Three procedures for isolation and enumeration are listed here: that using Packer's crystal violet azide blood agar has been recommended by the IAMS Committee (Thatcher and Clark, 1968) because the medium is amongst the least selective against some of the species in this group. The presence of azide helps to inhibit staphylococci, as a result of its inhibitory action against the cytochromes. As with all selective techniques, a compromise must be reached between selectivity and sensitivity. In order to have the maximum sensitivity against the widest range of strains there may be a reduction in the selectivity against organisms not belonging to the faecal streptococci; time-consuming confirmatory tests have great importance in such a case—in just those circumstances when many colonies develop which need to be screened, yet which prove not to be of the types which it is desired to enumerate. *Streptococcus bovis* and *Str. equinus* in particular are sufficiently different from the other Group D streptococci that a single selective medium suitable for both these physiological groups of species would be very difficult to devise without losing sensitivity against non-streptococci.

1. Glucose azide broth

This procedure employs the multiple tube technique, so that small numbers of faecal streptococci can be detected.

Pipette aseptically 1 ml of each of the prepared dilutions of the sample into each of three or five tubes of glucose azide broth. If very low concentrations of organisms are expected, 10 ml or even 100 ml amounts of the lowest dilution may be added to equal volumes of double-strength medium, three or five bottles being prepared at each dilution. Incubate the inoculated glucose azide broths at 35°C for 72 h and examine for the production of acid. Record those tubes which are positive (i.e. in which acid is produced) and subculture a loopful from each positive tube into a fresh single strength glucose azide broth (5 ml per tube) and incubate at 45°C in a water bath for 48 h, examining the tubes after 18 and 48 h. The production of acid at 45°C within 18 h indicates faecal streptococci. If acid is produced after 18 h but within 48 h, it is presumptive evidence of faecal streptococci, and can be confirmed rapidly by microscopic examination for the presence of short-chained streptococci. The most probable number of faecal streptococci can be determined using probability tables (see Appendix 2).

2. Packer's crystal violet azide blood agar

This medium is used with a pour plate technique. Prepare duplicate sets of plates using 1 ml amounts of the chosen range of dilutions of the sample. Add to each plate 15 ml of Packer's crystal violet azide blood agar, melted and cooled to 45°C, mix well and allow to set. Incubate at 35°C for 3 days. After incubation count all small violet-coloured colonies as presumptive faecal streptococci.

Representative colonies of presumptive faecal streptococci should be subcultured by streaking across poured plates of Barnes' thallium acetate tetrazolium glucose agar (Barnes, 1956). Incubate at 35°C for 24 h. Typical colony forms are: *Strep. faecalis* – colonies with a red centre and with or without a white periphery; other Group D streptococci – white or pink colonies. Morphology and Gram-staining reaction can be determined at this stage to provide reasonable confirmation.

3. Maltose azide media

(*a*) *Pour plate count*

Pipette 1 ml amounts of the chosen range of dilutions into a set of Petri-dishes. Add to each plate 15 ml of molten maltose azide tetrazolium agar at 45°C, mix thoroughly with the inoculum and allow to set. Incubate at 35°C for 48 h. After incubation, count the number of colonies which are dark red or have a red or pink central area.

(*b*) *Multiple tube technique*

If very small numbers of faecal streptococci are expected to be present in a food sample, use a multiple tube count with maltose azide broth as the selective medium. Pipette aseptically 1 ml of each of the prepared dilutions into each of five tubes of maltose azide broth. Incubate at 35°C for 48 h. After incubation examine the tubes for acid production, which is indicative of faecal streptococci. Calculate the most probable number of streptococci using probability tables (see Appendix 2).

THE DETECTION AND ENUMERATION OF PATHOGENIC AND TOXIGENIC ORGANISMS

A. Introduction

In addition to the micro-organisms which may be present in food and cause a "food poisoning", a number of diseases other than so-called food poisoning may be spread via food (see Kaplan, Abdussalem and Bijlenga, 1962; Riemann, 1969). The possibility of foods playing an important part in the spread of some virus diseases is being given an increasing amount of attention.

Among virus diseases which may be food-borne are infectious hepatitis and poliomyelitis. There is considerable controversy over the importance of foods as vehicles for the transmission of virus diseases. One reason for conflicting and inadequate information is the complex and expensive procedure necessary for the detection of viruses in foods. In our opinion the current evidence is such as to indicate that providing sufficient care is taken over bacteriological examinations and the interpretation of results, and appropriate action taken, there is little justification for the examination for viruses being attempted in a quality control laboratory. For further information consult Hermann and Cliver (1968), Sullivan and Read (1968), and Shuval and Katzenelson (1972).

"Food poisoning" can be caused either by toxins produced by bacteria which may be non-viable when the food is consumed, or by the multiplication of bacteria in the intestine after consumption. In the former case, when the organisms cause an intoxication, a normal condition for toxin production is the multiplication of the organism in the food resulting in a large total count (not necessarily a large viable count) at the time of testing, and this implies that the food is, or has been, in a state such as to allow the growth of bacteria. In the latter case, where the organisms cause an infection, they need only be present in relatively small numbers in the food but they must be viable. Food which does not support the growth and multiplication of the organisms concerned may nonetheless constitute a health hazard if it is not of a nature which causes the death of the organisms since the food may be used in a situation which allows transfer of the organisms to another food that *will* support growth. An example of such a hazard was provided by the outbreaks of salmonellosis caused by contaminated desiccated coconut (Wilson and MacKenzie, 1955).

As already mentioned there is evidence that large numbers of coliforms (Taylor, 1961, 1966; Marier *et al.*, 1973; Report, 1974) may cause gastro-enteritis, and also streptococci (Moore, 1955) may cause food poisoning.

Most procedures for the detection of viable pathogenic or toxigenic bacteria involve the use of highly selective media. The selective and differential media so far discussed (e.g. pages 54–58, and 139–146) are normally employed in circumstances in which the organisms sought are expected to constitute a relatively large proportion of that section of the microflora which shows physiological and/or ecological similarities. If this assumption be untrue then an unacceptably high level of false positive (or false negative) results may be obtained: for example, anaerobic spore-bearing organisms may cause misleading results in coliform tests performed on samples of chlorinated water (Report, 1969).

In the case of pathogenic and toxigenic organisms it is necessary to use selective and differential media and methods which can detect the organisms when they are in the presence of perhaps very large numbers of physiologically and/or ecologically similar organisms. Most selective pressures are only *relatively* selective and usually selectivity is improved only at the expense of sensitivity. This problem is exacerbated in food microbiology by the likelihood of the organisms having suffered some metabolic injury from a food-processing procedure so that the organisms are inhibited by the very media which are expected to select for them. Yet because *Salmonella* in a frozen or dried food cannot recover and grow in selenite broth, for example, it does not follow that the organisms could not recover in a non-selective environment (for example, trifle) (Corry, Kitchell and Roberts, 1969). Thus selective procedures are relatively complicated and lengthy—liquid enrichment usually precedes the use of solid selective media, and non-selective resuscitation may precede the liquid enrichment stage.

The composition of the non-selective resuscitation medium will depend on the type of damage most likely to have been suffered (Tomlins and Ordal, 1976). If organisms in lag or late stationary phase are frozen or dried, then the metabolic injury is likely to result in a temporary nutritional requirement for substances to be supplied exogenously which normally can be synthesised endogenously, so that higher recoveries are likely in rich media (see, for example, Straka and Stokes, 1959; Fisher, 1963). On the other hand if logarithmic phase cells are heated the damage suffered is likely to include nucleic acid damage; such organisms may recover in a minimal medium but not in a rich medium (Gomez, Sinskey, Davies and Labuza, 1973; Wilson and Davies, 1976). It is advisable therefore to use *both* rich medium resuscitation *and* minimal medium resuscitation until it can be ascertained which gives the best recovery rates from a particular food-processing-storage system.

Confirmatory and identification tests which follow the use of selective media are usually specially designed and offer confirmation only in the context of the prior selection of presumptive positive results. It is essential to ensure that the organisms submitted to confirmatory tests are obtained as pure cultures. For example, consider the case of a mixed microflora containing some coliform organisms and some salmonellae which is cultured on to desoxycholate citrate agar, a medium selective for *Salmonella* which differentiates on the basis of lactose fermentation. Coliform organisms grow only reluctantly on this medium and if in association with a developing colony of *Salmonella* they may not show their lactose-fermenting ability. The presumptive *Salmonella* colony transferred directly to confirmatory physiological test media is very likely to show positive results characteristic of both the *Salmonella* and the coliform organism. The purification of the *Salmonella* culture on a non-selective but differential medium would have demonstrated the presence of the coliform organisms.

Therefore the stages in isolation and identification are likely to be:

1. Pretreatment of the sample if necessary (see page 132)
2. Resuscitation in a "non-selective" non-differential medium
3. Enrichment in a liquid selective medium
4. Detection on solid selective and/or differential media
5. Purification of cultures on a differential, non-selective medium
6. Confirmatory physiological tests
7. Serological tests and possibly phage typing.

Such a procedure may occupy two weeks, so that the incentive to introduce rapid alternative procedures will be readily appreciated. One such method involves the use of fluorescent antibody techniques to detect the organisms by direct microscopic examination of the liquid enrichment cultures obtained at Stage 3. Such methods are beginning to be applied, especially in the case of *Salmonella* detection procedures (see page 155).

Unfortunately standardisation of isolation procedures irrespective of the foodstuffs to be examined may be impracticable in the recovery and enrichment steps: there is increasing evidence that the effectiveness of a given enrichment technique (in terms of selectivity and sensitivity) may depend as much on the unintentional modification of the medium caused by the addition of (usually) appreciable amounts of food as it does upon the recipe used to prepare the medium. Obviously different foods will have different effects. A possible approach to the standardisation of detection methods which has much to commend it is to specify for each pathogen the resuscitation stage and the first liquid enrichment stage for the particular type of food being examined, with later stages being common to the examination of all categories of sample. In the present state of our knowledge specifying the most effective first steps is possible only with a very few foods. Consequently generalised procedures must be recommended with the important proviso that such procedures should not be applied uncritically but should be regarded as open to modification to suit particular circumstances.

B. Quantification of Selective Isolation Techniques

If the procedure does not involve a liquid enrichment technique, counts may be performed on the solid selective media by the usual methods outlined in pages 31–32 (e.g. Miles and Misra surface drop count). Procedures which incorporate liquid resuscitation or liquid enrichment stages can be made quantitative by a modification of the multiple tube count technique, using the MPN tables in Appendix 2. This necessitates the use of replicate food samples at three consecutive "dilution levels" (e.g. 100 g, 10 g and 1 g amounts of food) during the *first* culture stage.

An alternative method of rendering the isolation procedure quantitative is to follow the suggestion made by the Committee on Salmonella (1969) and use an attribute sampling scheme (see page 128 ff.). The Committee on Salmonella (1969) proposed the use of a unit size of 25 g, and also suggested the following simple approximation to the formula given on page 130:

$$n = (-2{\cdot}3/d) \log (1 - p)$$

where n is the number of units that, if examined and all found to be negative, will imply with confidence p that the population of positive units in the whole lot averages less than d.

For a 95 per cent confidence in the result:

$$n \approx 3/d$$

It has been suggested (ICMSF, 1974) that comparable reliability can be obtained by compositing the samples taken. The units, n, are chosen randomly from the batch, but in the laboratory they may be composited before examination. For example, for $n = 30$, and a unit size of 25 grams, the units may be combined so that three 250-gram amounts are examined; although the same total weight is examined, savings are made in time and media for isolation and identification.

C. *Salmonella* and *Shigella*

Shigella dysentery may be spread by food as a result of contamination of the food by food handlers, etc. Unfortunately the minimum infective dose is very low, and completely satisfactory enrichment methods have yet to be described; in consequence the real significance of foods as vehicles of infection is unknown. However, it would seem that of the liquid media devised for enrichment of enterobacteria, Hajna's GN broth (Hajna, 1955) is the best for this purpose (Taylor and Harris, 1965; Taylor and Schelhart, 1967). Isolation of shigellae on solid media is probably best accomplished using Taylor's xylose lysine desoxycholate (XLD) agar (Taylor, 1965). The following procedure for the isolation of *Salmonella* therefore includes these two media in the hope that the possible presence of shigellae in food samples will not be overlooked.

1. Non-selective resuscitation

(*a*) In those cases in which any salmonellae present may have suffered metabolic damage when in lag or stationary phase (e.g. dried foods, frozen foods) add aseptically 25 g (or 10 g) of the food product to 250 ml (or 100 ml) of lactose resuscitation broth (North, 1961a) or to 250 ml (or 100 ml) of nutrient broth. Thoroughly mix, and incubate for 24 h at 37°C.

(*b*) If a dried food has a very high soluble solids content (e.g. dried milk, dried baby food), also prepare a non-selective resuscitation stage by adding 25 g (or 10 g) of the food to sterile distilled water, making up the volume to 250 ml (or 100 ml). Incubate for 24 h at 37°C.

N.B. If the pH of the suspension is outside the range pH 6·8–7·0, adjust to pH 6·8–7·0 with sterile N sodium hydroxide solution (or sterile N hydrochloric acid).

(*c*) If it be suspected that any salmonellae present may have suffered damage (e.g. from heating or freezing) whilst in a growth phase add 25 g (or 10 g) of the food product to 250 ml (or 100 ml) of minimal nutrients recovery medium.

2. Selective enrichment

(*a*) Add 10 ml of each pre-enrichment resuscitation culture to 100 ml of selenite broth; similarly add 10 ml to 100 ml of tetrathionate broth and 10 ml to 100 ml of Hajna's GN broth.

(*b*) In the case of those foods not requiring a non-selective resuscitation stage add 25 g (or 10 g) of the food to 250 ml (or 100 ml) of selenite broth; similarly add 25 g (or 10 g) of the food to 250 ml (or 100 ml) of tetrathionate broth, and 25 g (or 10 g) of the food to 250 ml (or 100 ml) of Hajna's GN broth.

Incubate the enrichment broths for 72 h at 37°C preparing streak plates on solid selective media after 18–24 h and again after 72 h.

Harvey and Price (1968, 1974) considered that incubation of selenite broth at 43°C following non-selective resuscitation at 37°C provided a better chance of isolating the salmonellae, but if 43°C is chosen as the incubation temperature for selenite broth, the medium must be sterilised by filtration as heat-sterilised selenite broth shows toxicity to salmonellae at 43°C. In addition it is recommended that an incubation temperature of 43°C is not used for enrichment media other than selenite broth, and that it is not used unless a non-selective resuscitation stage is used first.

3. Plating on solid selective media

Streak loopfuls from each enrichment broth on to plates of Wilson and Blair's bismuth sulphite agar, Taylor's XLD agar, and brilliant green agar. Incubate the plates for 24 and 48 h at 37°C and examine for the presence of typical colonies. Typical *Salmonella* colonies on Wilson and Blair's bismuth sulphite agar appear jet black with or without a metallic sheen, or have jet black centres with colourless translucent peripheries, or are grey to greyish-brown. The medium surrounding the colonies usually turns to brown to black.

Colonies of *Salmonella* species (including *S. arizonae*) on Taylor's XLD agar typically are red with black centres; whereas colonies of *Shigella* typically are uniformly red. (*N.B.* Some strains of *Pseudomonas* and *Proteus* also produce uniformly red colonies on this medium.) Typical *Salmonella* colonies on brilliant green agar are pink or red (occasionally colourless) surrounded by a zone of bright red medium. Brilliant green agar has a lower selectivity than the other media named. Because of this, *Salmonella* colonies on brilliant green agar may be rendered atypical by the presence of lactose-fermenting organisms.

4. Subculture of presumptive salmonellae and shigellae

Pick off all suspect colonies and streak on a non-selective lactose agar (neutral red chalk lactose agar may be used) to confirm the purity of the cultures. (MacConkey's agar can be used as an alternative differential medium, although its selective properties are not required and are a disadvantage at this stage.) As already explained (page 49) this step is most important to prevent physiological tests being performed on mixed cultures.

It is worth remarking that outbreaks of gastro-enteritis caused by lactose-fermenting variants of *Salmonella* have been reported (Poelma, 1968) and that *S. arizonae* and some strains of *Shigella* will slowly attack lactose (see pages 250–251) so that the purification stage should not be employed for *rejecting* cultures, especially if these have been derived from Wilson and Blair's bismuth sulphite agar.

Harvey and Price (1967) found that salmonellae could be recovered more frequently if an additional stage was introduced, which they termed a selective motility technique. The method (Harvey and Price, 1967, 1974) consists of removing the total growth from a selective agar medium with a short swab. The swab is placed in a short length of glass tubing placed inside a screw-capped bottle containing a semi-solid nutrient agar, the level of which is sufficient to immerse the swab. The swab is prevented from touching the base of the bottle by a short piece of glass rod placed within the glass tube. The whole is incubated at 37°C for 24 h. The growth from the surface

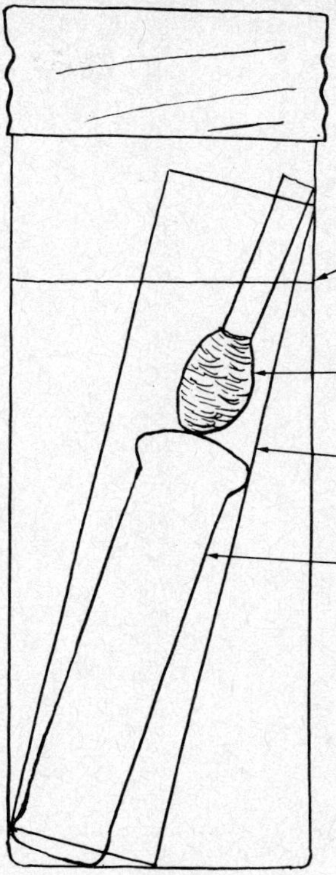

Secondary enrichment by selective motility technique
(Harvey and Price, 1967)

of the agar *outside* the inner tube is subcultured on to brilliant green MacConkey's agar (see p. 348) which is then incubated at 37°C for 24 h. Presumptive *Salmonella* and *Shigella* are picked off, purified, and subjected to physiological and serological tests.

5. Physiological test media for identification of salmonellae and shigellae

Inoculate each culture for identification into the two tubes of Kohn's media and on to a nutrient agar slope moistened with one drop of nutrient broth. Inoculate Kohn's medium No. 1 by streaking the slope and stabbing the butt, and Kohn's medium No. 2 by stabbing to about half the depth of the medium. Into the mouth of the tube containing Kohn's medium No. 2 insert a strip of lead acetate paper and a strip of indole test paper so that they do not touch the surface of the medium and do not touch each other. The paper strips are held in place by the plug. Incubate at 37°C for 18–24 h. The reactions on Kohn's media of *Salmonella*, *Shigella* and *Proteus* are as follows:

	Medium No. 1			Medium No. 2			
	Fermentation of glucose	Fermentation of mannitol	Urease production	Fermentation of sucrose and/or salicin	Motility	Production of H_2S[1]	Production of indole[2]
Salmonella typhi	A	A	—	—	+	+	—
Other *Salmonella* spp. and *S. arizonae*	AG	A	—	V(—)[3]	+	V	—
Shigella sonnei	A	A	—	—	—	—	—
Sh. flexneri, *Sh. boydii*	A	A	—	—	—	—	V
Sh. dysenteriae	A	—	—	—	—	—	V
Proteus	(—)	(—)	+	AG or —	+	V	V

AG: acid and gas. A: acid only. +: positive. V: variable reaction. (—): urease activity masks this reaction. —: negative or no reaction.

1 Detected with lead acetate paper inserted before incubation.
2 Detected with indole test paper inserted before incubation.
3 Sucrose-fermenting salmonellae have been described (Poelma, 1968).

Alternatively the culture can be picked off into a proprietary rapid multimedia testing system. One system which is extremely rapid to use is the "Enterotube" (Roche Products Ltd.)—using only a single inoculation this tests for 11 characteristics in an 8-compartmented tube, namely: fermentation of glucose, lactose, and dulcitol, and gas production from glucose; decarboxylation of lysine and ornithine; production of hydrogen sulphide; production of indole; deamination of phenylalanine; hydrolysis of urea; and utilisation of citrate.

Roche Products provide a manual ("ENCISE-System") for identifying the Enterobacteriaceae by means of the "Enterotube" system, which is based on the work of Edwards and Ewing (1972), but the basic probable combinations of reactions for *Salmonella* and *Shigella* are as follows:

Test	*Salmonella*	*Shigella*
Glucose fermentation	+	+
Glucose, gas from	+/—	+/—
Lysine decarboxylase	+/—	—
Ornithine decarboxylase	+/—	+/—
Hydrogen sulphide	+/—	—
Indole	—	+/—
Lactose fermentation	—	—
Dulcitol fermentation	+/—	+/—
Phenylalanine deaminase	—	—
Urea hydrolysis	—	—
Citrate utilisation	+/—	—

Moussa (1975) in an evaluation of the Enterotube and two other systems, found that it gave an overall accuracy of 99·2 per cent in 1540 tests, compared with accuracies of 92·2 and 80·5 per cent for the API and PathoTec systems respectively.

6. Serological differentiation of *Salmonella*

When the identity of a presumptive *Salmonella* culture has been confirmed by biochemical reactions, the nutrient agar slope culture should be used for a slide agglutination test (see page 101).

7. Rapid detection of *Salmonella* using fluorescent antibody techniques

The time taken to confirm the absence of *Salmonella* from a food product can be reduced by using fluorescent antibody techniques for the direct microscopic screening of the selective enrichment broths. This obviates the necessity for solid selective media, and biochemical identification of pure cultures being set up on *all* samples. However, there is a possibility of organisms other than *Salmonella* giving a positive reaction so that immuno-fluorescent-positive enrichment broths should be streaked on solid media, and the test procedure performed on some isolates.

In order to use this technique the laboratory must be equipped with a UV-fluorescence microscope (such as that manufactured by Wild Heerbrugg). Reagents for the test are available in a useful kit form as the C.S.I. "Fluoro-Kit" (Clinical Sciences Inc.) (see also Insalata, Dunlop and Mahnke, 1973). The reagents for immunofluorescence techniques are at present expensive and the technique is likely to be considered only when large numbers of samples are being processed. Further details of fluorescent antibody techniques are given by Georgala and Boothroyd, 1964, 1968; Georgala, Boothroyd and Hayes, 1965.

D. *Clostridium perfringens* (*Cl. welchii*)

Food poisoning caused by *Clostridium perfringens* appears to be an infection rather than an intoxication, if by intoxication we mean symptoms caused by the ingestion of pre-formed toxin. There is some evidence that the symptoms are produced as the result of formation in the gut of toxin during sporulation of a large number of ingested vegetative cells (Duncan, 1973). Large numbers of viable organisms must be present in the food to cause symptoms. Food poisoning caused by *Cl. perfringens* therefore normally requires conditions favourable for the multiplication of the bacteria in the food. Consequently large numbers of *Cl. perfringens* may be expected in a food that has caused food poisoning, whereas in routine microbiological quality control the emphasis is on detecting the presence of *Cl. perfringens* in small numbers if, due to subsequent storage and treatment of the food, bacterial multiplication is possible.

1. Colony count method

Prepare serial dilutions of the food to be examined, using reinforced clostridial medium as the diluent. Carry out plate counts on the dilutions using sulphite polymyxin sulphadiazine agar (Angelotti *et al.*, 1962) and the Miles and Misra surface count technique (see page 31). Incubate the plates anaerobically (see pages 59–64) at 37°C for 24 h. Colonies which are coloured black (due to the reduction of sulphite causing a precipitation of iron sulphide) are presumptive sulphite-reducing clostridia. Pick off such colonies and stab into tubes of nitrate peptone water +0·3 per cent agar and also streak on plates of Willis and Hobbs's lactose egg-yolk milk agar (Willis, 1962), having previously spread half of each plate with a few drops of *Cl. perfringens* antitoxin and having allowed it to dry. Streak the organism under examination over the whole plate.

Incubate the tubes of nitrate agar anaerobically at 37°C for 24 h, and incubate the plates of Willis and Hobbs's medium anaerobically for 24 h at 37°C. After incubation, examine the nitrate agar for the type of growth (indicating motility or lack of it) and test for nitrate reduction (see page 71). Examine a Gram-stained smear prepared from the growth on Willis and Hobbs's medium. Black colonies on sulphite polymyxin sulphadiazine agar which give cultures of Gram-positive non-motile rods, capable of reducing nitrate, are *Cl. perfringens*. This is confirmed by the reaction on Willis and Hobbs's medium: *Cl. perfringens* gives cloudy zones around the colonies on the section of the plate without antitoxin, but no zones on the section with antitoxin, and also ferments lactose (see also Willis, 1965).

2. Multiple tube count method

If very small numbers of clostridia are expected to be present in a food sample, a multiple tube count may be carried out using differential clostridial medium (DRCM) as the selective medium. Pipette aseptically 1 ml of each of the chosen dilutions (prepared using reinforced clostridial medium as the diluent) into each of five (or three) bottles of DRCM.

Incubate at 30°C for 7 days. Tubes showing blackening should be confirmed for *Cl. perfringens* as described under method 1. Calculate the most probable number of *Cl. perfringens* using probability tables (see Appendix 2).

E. *Clostridium botulinum*

Whereas food poisoning caused by *Cl. perfringens* is an infection, botulism, which is caused by *Cl. botulinum*, is an intoxication. The bacteria multiply in the food before it is consumed, producing an extremely potent exotoxin. Most outbreaks of botulism have been caused by smoked, cured, pickled or canned food, which in Europe is usually meat or prepared meat products (e.g. sausage, pâté, etc.) which may or may not be canned, and in America is usually canned fruit or vegetable. Recently, certain vacuum-packed foods (including smoked fish) have also been implicated in outbreaks of botulism. In the investigations of these last mentioned outbreaks it was found that *Cl. botulinum* Type E was capable of significant toxin production in the packaged food when stored for as little as 5 days at 10°C.

The spores of *Cl. botulinum* are relatively heat-resistant but are more sensitive to heat at low pH. Since canned foods present suitably anaerobic conditions for the growth of *Cl. botulinum* if the pH is above 4·5, heat processing is designed to destroy the *Cl. botulinum* spores. In Acts of Parliament and in Statutory Regulations, canned food is defined as "food in a hermetically sealed container which has been sufficiently heat processed to destroy any Clostridium Botulinum in that food or container or which has a pH of less than 4·5".

The enumeration of *Cl. botulinum* in foods is difficult and dangerous, and identification usually involves at some stage confirmation of toxin production by the bacteria. Alternatively, and more commonly, an examination of the suspected food for the toxin may be made. This involves immunodiffusion tests or the injection of mice, and will usually be carried out by the Public Health authority or a similar body in the event of an outbreak of botulism. Details of the techniques which may be used have been described by IAMS Committee (Thatcher and Clark, 1968), but work must be undertaken only by adequately trained staff (in whom high antitoxin titres are maintained) in specially equipped laboratories.

F. *Staphylococcus aureus*

Staphylococcal food poisoning is an intoxication which depends on the ability of the food concerned to support the growth of the staphylococci which produce the toxin. The staphylococcal toxin can withstand heating at 100°C for 30 min and therefore the absence of viable organisms in the food is not proof of safety if there is a high total (microscopic) count of cocci. *Staph. aureus* may grow and multiply in pickled and cured foods, although there is evidence that toxin production is inhibited by salt concentrations which allow the growth of the organisms. However, this appears to depend on the composition of the food in which the organisms are growing, and there have been reports of toxin production in habitats containing salt concentrations typical of many cured foods.

Procedure

(1) Prepare serial dilutions of the food and transfer 0·1 ml amounts to duplicate plates of poured, well-dried Baird-Parker's medium. Spread the inoculum evenly over the surface with a sterile, bent glass rod. Incubate for 24 h at 37°C. If no colonies have developed, reincubate for a further 24 h. *Staph. aureus* typically forms colonies which are 1·0–1·5 mm in diameter, black, shiny, convex, with a narrow white entire margin and surrounded by clear zones extending 2–5 mm into the opaque medium. Pick off a number, or all, of the colonies which are typical of *Staph. aureus* and test for coagulase production (see page 79). The absence of coagulase should not automatically exclude a diagnosis of staphylococcal food poisoning since there have been reports of enterotoxin production by coagulase-negative strains (e.g. Breckinridge and Bergdoll, 1971).

(2) If low concentrations of *Staph. aureus* are suspected, a liquid enrichment stage can be included, adding 10 g amounts of food to 100 ml of salt meat broth. Incubate at 37°C for 24 h and then streak loopsful on to Baird-Parker's medium. The enrichment technique can be rendered quantitative as described on page 150. Note, however, that a salt enrichment broth may be insufficiently selective when applied to high-salt food samples. (e.g. cured or pickled foods), and also that in refrigerated foods *Staph. aureus* may have become salt-sensitive as a result of the refrigeration. For the enrichment of *Staph. aureus* from high salt foods Giolitti and Cantoni's medium (Giolitti and Cantoni, 1966) may prove more effective.

G. *Bacillus cereus*

Bacillus cereus food poisoning (see Hauge, 1955; Davies and Wilkinson 1973; and also page 122) is caused by consuming foods containing very large numbers of bacteria (usually of the order of 10^7 per gram) (Goepfert *et al.*, 1973). Therefore when examining suspected foodstuffs during epidemiological investigations of outbreaks of food poisoning, a selective medium is not necessary. A count of aerobic phospholipase producers can be obtained by the surface inoculation of eggy-yolk agar (page 78), followed by microscopic confirmation of Gram-positive rods and biochemical tests for *B. cereus* (see below).

Very many foods can be expected to contain small numbers of *B. cereus*, because it is such a common environmental contaminant. In consequence selective enrichment techniques will not usually provide information of much value. At intermediate population concentrations, in which a foodstuff may represent a potential hazard because some proliferation of *B. cereus* has occurred, it may be useful to employ a selective and differential medium such as mannitol egg-yolk phenol red polymyxin (MEPP) agar (Mossel, Koopman and Jongerius, 1967) with direct plate counts.

Procedure

Prepare serial dilutions of the food (up to 10^{-4}) and transfer 0·1 ml amounts to duplicate plates of poured, well-dried MEPP agar. Spread the inoculum evenly over the surface of each plate with a sterile, bent glass rod. Incubate for 48 h at 30°C. Presumptive *B. cereus* colonies are rough and dry with a violet-red background and a halo of dense white precipitate. These reactions result from absence of attack on mannitol, and presence of lecithinase activity against the egg yolk.

To confirm colonies as *B. cereus* purify by streaking on egg-yolk agar plates, and prepare pure cultures on nutrient agar slopes. Examine microscopically for the presence of Gram-positive rods which will usually be capable of sporulation, the spores being ellipsoidal and not large enough to cause swelling of the vegetative cells.

Biochemical reactions typical of *B. cereus* which may be tested for include:

- anaerobic dissimilation of glucose (Hugh and Leifson's medium, page 73)
- *inability* to attack mannitol or xylose (pages 72–73)
- hydrolysis of gelatin (page 66)
- reduction of nitrate (page 71)
- ability to grow on Knisely's chloral hydrate agar.

In addition *most* strains can hydrolyse starch (page 72), and *most* are Voges-Proskauer positive (page 74). (See also Part III, page 262).

N.B. Lecithinase-negative strains of *B. cereus* are known to occur. Their significance in food poisoning is still in doubt so that in epidemiological work associated with outbreaks it may be felt desirable to search for lecithinase-negative strains in addition to the more typical lecithinase-positive strains.

H. *Vibrio parahaemolyticus*

This organism has been involved in outbreaks of food poisoning caused by fish products, mainly in Japan. There is at present not a great deal known about its frequency of occurrence or its public health significance in other countries. Although reported in fish caught off the coasts of America, and in Mediterranean and Baltic fish, it appears to be rare in fish landed in Britain (Shewan, 1971) but it has been reported in shellfish and coastal waters off S.W. England (Barrow and Miller, 1972). There have been outbreaks of *V. parahaemolyticus* food poisoning reported in the U.K. caused by imported fish products (e.g. Anon, 1975). If it is desired to examine foods for the presence of *V. parahaemolyticus*, we recommend that the procedures described by the IAMS Committee (Thatcher and Clark, 1968) or by Fishbein and Wentz (1975) be followed.

I. *Aspergillus flavus*

Certain strains of *Aspergillus flavus* are able to produce metabolites (aflatoxins) which are known to be extremely potent toxins and carcinogens for a number of animals, and which are suspected of offering a similar hazard to man. Aflatoxin production is substrate-dependent. Amongst foods which support the growth of *A. flavus*, some encourage aflatoxin production, others do not. Procedures have been developed for the detection of aflatoxins in foods. Extreme care must be taken whilst following such procedures, because of the potential carcinogenic hazard. Details of analytical procedures are given by AOAC (1970), Jones (1972), and Krogh (1973); see also Jarvis and Moss (1973).

THE MICROBIOLOGICAL EXAMINATION OF SPECIFIC FOODS

As already mentioned (see page 119 ff.) microbiological examinations of foods *may* assist in the assessment of hygienic precautions during production, and of the efficacy of a preservation process, and *may* allow predictions of the potential shelf life, and also the identification of potential health hazards by the use of suitable indicators or by direct detection of pathogens. However, it has already been pointed out that extreme care must be taken in the interpretation of results and in the conclusions drawn from them.

In this section we provide outline procedures for the examination of food components and products on a commodity basis.

A. Meat and Meat Products

1. Fresh, refrigerated and frozen meat

The muscle tissue of healthy animals contains few if any bacteria, but cut and exposed surfaces become easily contaminated after slaughter and during and after butchering. Bacteria can readily multiply on the cut surfaces, although the bacterial count of the interior of the meat usually remains much lower. Bacteria which may be present include micrococci, *Acinetobacter*, *Moraxella*, *Pseudomonas*, *Microbacterium*, *Staphylococcus*, *Lactobacillus*, *Streptococcus*, and *Salmonella*. Mould spoilage of the cut surfaces of refrigerated or frozen meat can occur at temperatures down to about −5°C. At the lower temperature mycelial growth may occur without spore production, and this will give rise to a white fluffy appearance caused by, e.g. *Mucor*, *Rhizopus*. Other types of mould spoilage which may be obvious on inspection are "White spot"—caused by, e.g. *Sporotrichum*; "Black spot"—caused by, e.g. *Cladosporium*; and green patches—caused by, e.g. *Penicillium*. In addition, certain yeasts may cause low temperature spoilage.

Surface slime on meat is usually caused by *Pseudomonas* and *Acinetobacter*, but *Streptococcus*, *Lactobacillus*, *Leuconostoc*, and *Micrococcus* may also be responsible. *Pseudomonas* and *Clostridium* may cause putrefaction. The lactic-acid bacteria on the other hand can cause souring due to the production of organic acids.

The bacterial condition of the meat is determined by taking both superficial and deep tissue samples. The microbiological condition of the surface can be assessed most rapidly by microscopic examination of contact slides stained by Gram's method but detailed cultural examinations are best achieved by taking *superficial* samples as very thin slices using sterile scalpels and forceps (see page 125). These samples are homogenised in diluent to give a 10^{-1} initial dilution. Deep tissue samples should also be taken separately.

Procedure

Prepare decimal dilutions to 10^{-5}. Carry out:

(*a*) General viable counts on plate count agar, incubated at 5, 25 (or 20°) and 37 (or 35°)C for 7, 3 and 2 days respectively.

(*b*) Coliform counts either by a pour plate procedure using violet red bile agar, or (if small numbers of coliforms are expected) by a multiple tube count.

(*c*) An examination for the presence of *Salmonella* (see page 150 ff).

(*d*) Viable counts of anaerobic bacteria using a dilution series prepared with reinforced clostridial medium as a diluent. Use plate count agar, and blood agar; incubate anaerobically at 37°C for 2 days.

(*e*) Examination for the presence of *Clostridium perfringens*.

Recommended standards

ICMSF (1974) recommend that the general viable count at 35°C (or at 20°C in the case of chilled meats) should be less than 10^7 per gram, and that *Salmonella* should be detected in not more than one of five 25 g samples.

2. Raw sausages, hamburgers and similar meat products

Meat products in this group usually contain cereal or rusk, salt and spices in addition to minced meat. In Great Britain they are usually partially preserved by the addition of metabisulphite. This has a temporary preservative action, and the general viable count and coliform count will be found usually to diminish during the first day after production. Thereafter the viable count will rise again as the free sulphur dioxide content falls. After storage *Microbacterium thermosphactum*, yeasts, *Lactobacillus* and *Micrococcus* are found as the major components of the microbial populations (Dowdell and Board, 1968, 1971); faecal streptococci and coliform organisms may be found in numbers up to about 10^4 per gram. *Pseudomonas*, *Acinetobacter*, and *Kurthia zopfii* may occur as minor components of the microflora. In certain varieties of sausage *Pediococcus* or *Leuconostoc* can be found. Gardner (1966) has described a selective medium for the detection of *Microbacterium thermosphactum*.

Procedure

Surface counts of sausages may be performed, but more usually homogenates of decimal composite samples are used for the preparation of serial decimal dilutions up to 10^{-6}. Counts should be made on:

(*a*) Plate count agar for general viable counts, incubated at 5, 25 and 37°C for 7, 3 and 2 days respectively.

(*b*) Violet-red bile agar for coliform counts (or use a multiple tube technique if small numbers are expected).

(*c*) Acetate agar (or Rogosa agar) with Acti-dione (to inhibit microfungi), as double layered pour plates for *Lactobacillus*, incubated at 30°C for 5 days.

(*d*) Selective media for faecal streptococci.

(*e*) Davis's yeast salt agar, acidified to pH 3·5, for counts of yeasts and moulds, incubated at 25°C for 5 days.

(*f*) Examine for such pathogens as appropriate.

3. Meats and meat products preserved by curing and pickling

Meat preserved by the use of sodium chloride and sodium nitrate and/or nitrite may contain salt-tolerant micro-organisms which can be detected by growing in nutrient media containing 15 per cent sodium chloride (see also page 133). The bacterial flora of such products frequently includes staphylococci or micrococci and lactic-acid bacteria. Potentially pathogenic staphylococci can be detected as described on page 158. *Lactobacillus* spp. may be detected upon acetate agar base (Varnam and Grainger, 1972, 1973). Plate count agar may be used for general viable counts. Spoilage of frankfurters, sliced ham, etc., packed in moisture-proof laminated plastic films may occur due to surface growth of yeasts and cocci promoted by the accumulation of moisture between the surface of the meat and the packaging material. Therefore surface counts as well as counts of blended or homogenised samples should be made.

Procedure

To determine the surface count of frankfurters and similar products, follow the procedure on page 125 using 1 or 2 sausages. To determine the surface counts of sliced meats, follow the same procedure, using 10 g of the product. To determine the surface flora of unsliced meat, homogenise superficial slices in diluent to give a 10^{-1} dilution as described for fresh and frozen meat.

In all cases prepare seriai dilutions for testing. Carry out counts on:

(*a*) Plate count agar for general viable counts (pour plates), incubated at 5, 25 and 37°C for 7, 3 and 2 days, respectively (but see also page 137).

(*b*) Plate count agar +15 per cent NaCl for counts of salt-tolerant organisms (pour plates), incubated at 5 and 25°C for 7 and 3 days, respectively. The diluent used for salt-tolerant counts should be 15 per cent NaCl solution.

(*c*) Baird-Parker's medium for *Staphylococcus* (surface counts) at 37°C.

(*d*) Acetate agar base (*with the acetate buffer omitted*) for *Lactobacillus* prepared as surface drop count, incubated at 30°C for 5 days, in an anaerobic atmosphere containing 5 per cent carbon dioxide (see page 63) (Varnam and Grainger, 1972, 1973).

(*e*) Davis's yeast salt agar for yeasts and moulds, incubated at 5 and 25°C for 7 and 3 days, respectively. If the presence of moulds is suspected, also prepare surface spread plates from the lowest dilutions.

4. Cooked meat pies, etc.

The contents of meat pies should have low viable counts, any surviving organisms being spores of *Bacillus* and *Clostridium*, and examination therefore for both aerobic and anaerobic bacteria should be carried out. In the case of pies which have a jelly or stock added after cooking, the stock and meat should be sampled separately as well as together in a combined sample. The usual method of production involves the injection of the cooked pie with stock from a holding reservoir in which it is kept molten at a temperature just below boiling point. If high counts of organisms are found in the stock, the holding temperature should be checked. Too low a holding temperature results in multiplication of mesophilic or thermophilic bacteria in the reservoir. Aerobic plate counts on plate count agar at 25, 37 and 55°C and anaerobic plate counts at 37°C should be performed. The examination of frozen meat pies, which often consist of raw pastry with a pre-cooked filling and which are intended for partial cooking only by the consumer, is discussed in Section S.

5. Sliced cooked meats, etc.

Cooking will destroy a very large proportion of the microflora of the raw meat. Even thermoduric organisms such as *Clostridium perfringens*, usually will be reduced to very small numbers, although improper storage after cooking can allow proliferation of such survivors. During slicing and serving operations, contamination of the food can occur from hands, slicing machines, etc. Inadequate attention to hygiene can lead to the meat being contaminated by a range of commensal and pathogenic organisms including enterobacteria and staphylococci.

Procedure

Prepare a series of decimal dilutions to 10^{-3} and perform general viable counts, coliform counts, and counts of *Staphylococcus aureus* and *Clostridium perfringens* (if appropriate) as described in sections 2 and 3 above.

6. Fresh, refrigerated and frozen poultry

The microbial flora of table poultry is largely confined to the skin surface or visceral cavity, and it follows therefore that the methods most appropriate for the examination of table poultry are surface-sampling techniques. Some indication of the kinds of micro-organisms and their relative numbers and disposition on the skin surface can be obtained by the use of contact slides or adhesive tape. These techniques are particularly useful in the examination of non-frozen specimens following keeping quality tests.

For quantitative studies, swab procedures can be used with sterile templates to define the area to be swabbed, or the whole or part carcass may be rinsed in a known volume (300 ml) of sterile diluent. Neck skin also provides convenient sampling material particularly during in-line sampling, skin homogenates being prepared in the laboratory (see also Barnes, Impey and Parry, 1973). Dilution series can be prepared from swabs, rinses and skin homogenates, and cultural studies made as described for other raw meat products (see sections 1 and 2 above).

Recommended standard

ICMSF (1974) recommend that frozen poultry when examined by rinsing should give a count at 20°C of less than 10^7 per ml of the rinsing solution and that *Salmonella* should be detected in not more than one of five 25 g samples of the poultry meat.

B. Fish and Shellfish

1. Salt-water fish

In general, fish is more prone to microbial spoilage than is meat because the latter has a lower pH and is less moist, bacterial growth being to some extent inhibited. Due to the low ambient temperatures in marine and fresh waters, the bacterial flora associated with fish includes a greater percentage of psychrotrophs and fewer bacteria with an optimum growth temperature of 37°C compared with the usual flora of meat. The predominant types of bacteria are species of *Moraxella*, *Pseudomonas*, *Arthrobacter*, *Flavobacterium-Cytophaga*, *Acinetobacter* and *Micrococcus* (Shewan, 1971). Not only does the bacterial flora of the fish include a large proportion of psychrotrophs, but very many of the bacteria are halophilic or at least salt-tolerant. However, many marine bacteria have a requirement for *low* concentrations of salt and will be inhibited on media containing 5 per cent NaCl (MacLeod, 1965). Large numbers of bacteria are frequently found in the slime on the skin surface, which can support the growth of many bacteria which contaminate fish after catching. Coliform organisms and *Staphylococcus aureus* are good indicators of the standard of hygiene during handling. Shewan (1970) suggested that the following standards be adopted:

a general viable count at 35–37°C of not more than 10^5 per gram;
a coliform count of less than 200 per gram (*E. coli* less than 100 per g);
Staph. aureus fewer than 100 per gram.

If agar liquefiers are encountered frequently, silica gel may be used as the solidifying agent.

Procedure

Carry out surface counts and tissue counts in a manner similar to that described for meat, using:

(*a*) Plate count agar for general viable counts, incubated at 5, 25 and 37°C for 7, 3 and 2 days respectively.

(*b*) Plate count agar +15 per cent NaCl for counts of extreme halophiles and salt-tolerant organisms, incubated at 5, 25 and 37°C for 7, 3 and 2 days respectively. In this case the diluent should be 15 per cent NaCl solution.

(*c*) Coliform counts by the multiple tube count method.

(*d*) Baird-Parker's medium for *Staph. aureus* (surface counts) at 37°C.

2. Fresh-water fish

Many types of Gram-negative bacteria are commonly found on fresh-water fish; Trust (1975) found that *Pseudomonas* and *Cytophaga* predominated on salmonid fish. Coryneform organisms are the most common Gram-positive types. Many of the organisms will be psychrotrophic but not salt-tolerant or halophilic. The florae of healthy and diseased fish have been surveyed by Collins (1970).

Procedure

As described for salt-water fish but omitting counts of salt-tolerant organisms.

3. Smoked fish

Traditionally smoked fish are much less prone to microbial spoilage than are raw fish, and moulds become more important as spoilage organisms, due to the partial dehydration which occurs during smoking (Shewan and Hobbs, 1967). Some modern smoking methods, however, do not decrease the susceptibility of the fish to microbial spoilage.

Procedure

As for raw fish, with the addition of a qualitative examination for moulds by streaking the 10^{-1} dilution (or a higher concentration) across the surface of Davis's yeast salt agar or Czapek-Dox agar.

Recommended standard

ICMSF (1974) recommend that cold-smoked fish should have a general viable count of less than 10^6 per gram, a faecal coliform count of less than 400 per gram, and a *Staphylococcus aureus* count of less than 2000 per gram.

4. Shellfish

In general, the bacteria responsible for spoilage are the same as those responsible for the spoilage of salt-water fish. Sewage pollution of the estuarine habitat of many shellfish may result in concentration by the animal of human intestinal organisms which may include pathogens such as *Salmonella*. Therefore, in addition to total and general viable counts, the presence or absence of *Escherichia coli* (and possibly also enterococci and *Clostridium perfringens*) should always be determined as an indication of faecal pollution. Information should be sought on whether the samples have been taken before or after keeping the shellfish in chlorinated water to allow self-cleansing to take place. Pasteurised and/or pickled shellfish products should be free from all enterobacteria.

Procedure (See Knott, 1951; American Public Health Association, 1970)

Samples should be examined within 24 h of collection. They should be transported and stored dry at between 5 and 10°C until examined.

Scrub the shells of several shellfish (100 ml of body material will be required) thoroughly under running water. Finally wash in sterile water and then place one in the middle of a wad of 6 sterile, large filter papers. In the case of oysters hold the oyster in the filter paper with the flat side uppermost and open with a sterile oyster knife by cutting through the muscle which holds the two valves of the shell together. Remove the flat shell, taking care not to spill the liquid in the concave shell. Other shellfish can be opened in a similar way by cutting through the adductor muscles. Remove the liquid to a sterile wide-mouthed glass bottle using a sterile pipette, and transfer the body of the shellfish to the bottle using sterile forceps. Cut the body into small pieces with sterile scissors and repeat the process with more shellfish until there is about 100 ml of mixed liquor and finely chopped pieces. Transfer aseptically exactly 50 ml to a sterile glass bottle with ground glass stopper, containing a small number of glass beads. Shake vigorously, add 50 ml of sterile quarter-strength Ringer's solution, and shake again. Add 20 ml of this 1 in 2 dilution to 80 ml of diluent to give a 10^{-1} dilution. Prepare further dilutions in the usual way.

Carry out the following tests:

(*a*) Presumptive coliform counts by the multiple tube technique, inoculating each of 3 or 5 tubes with 2 ml of the 1 : 2 dilution (giving a 10^0), 3 or

5 tubes with 1 ml of the 10^{-1} dilution, and 3 or 5 tubes with 1 ml of the 10^{-2} dilution. Confirm presumptive coliform organisms as *Escherichia coli* by the Eijkman test.

(*b*) Viable counts on plate count agar at 25 and 37°C.

(*c*) Counts of salt-tolerant organisms on plate count agar +15 per cent NaCl at 25 and 37°C, having used 15 per cent NaCl solution as the diluent.

(*d*) A count of faecal streptococci by the multiple tube method using 3 or 5 tubes with 2 ml of the 1 : 2 dilution, 3 or 5 tubes with 1 ml of the 10^{-1} dilution, and 3 or 5 tubes with 1 ml of the 10^{-2} dilution.

(*e*) Examine for *Clostridium perfringens*, by adding 20 ml of the 1 : 2 dilution to 100 ml of litmus milk. Heat carefully to 80°C in a water bath and hold at that temperature for 10 minutes. Cool, cover the surface of the litmus milk with melted "vaspar" and incubate at 37°C for 48 h. Stormy fermentation is presumptive evidence of *Cl. perfringens*. Record the presence or absence of *Cl. perfringens* in 10 ml of blended shellfish. An alternative procedure is to use differential reinforced clostridial medium by the multiple tube technique (see page 156). If pasteurised or heated shellfish products are being examined, the laboratory pasteurisation stage should be omitted.

C. Eggs

1. Shell eggs

Although *Salmonella* has been shown to be capable of infecting hen eggs as well as duck eggs *in utero*, the contents of the great majority of new-laid hen eggs are sterile. The outside of the shell normally carries a large number of bacteria as a result of the contamination of the egg-shell with faeces, dust, etc. These bacteria include Gram-positive bacteria of the genera *Micrococcus*, *Staphylococcus*, *Arthrobacter* and *Bacillus*, and Gram-negative bacteria of the genera *Pseudomonas*, *Escherichia*, *Enterobacter*, *Proteus*, *Serratia*, *Aeromonas*, and members of the *Acinetobacter–Alcaligenes–Flavobacterium* group (Board, 1966). Of these, it is the Gram-negative organisms which are most frequently isolated from egg contents. Rough handling or washing of the eggs facilitates penetration of the shell and membranes by the surface contaminants, whereas dry cleaning does not so readily assist penetration.

Examination of the contents of shell eggs

Scrub the eggs with warm soapy water and a stiff brush, then rinse well and drain. Immerse in alcohol for 10 min, then allow to drain well and flame quickly. Cut a hole in the end opposite the air sac (which is located at the blunt end) using a small carborundum disc on an electric drill, or using a sterile scalpel. Remove the contents aseptically and homogenise using a blender or the Colworth Stomacher. Counts can be carried out on single eggs or on the bulked contents of a number of eggs. Prepare serial decimal dilutions in the usual way. Carry out general viable counts on plate count agar at 25 and 37°C, presumptive coliform counts and an examination for the presence of *Salmonella*.

2. Frozen whole egg

Frozen whole egg is normally packed in metal containers with press-on or screw-on lids. The product may contain many contaminating bacteria derived from the processes involved in its preparation. ICMSF (1974) recommended that pasteurised whole egg should have a direct microscopic count of less than 5×10^6 per gram, a general viable count of less than 10^6 per gram, and that *Salmonella* should be absent from 125 grams. They also suggested that international trade in unpasteurised egg products should be discouraged.

Procedure

Sample the product while still in the frozen state. Clean the lid and top of the tin, swab with alcohol, flame and then remove the lid. With a sterile auger or other suitable instrument, remove two cores, one from the centre of the can and one at the edge, extending from the top surface to as deep a level as possible with the instrument used. Transfer these to a sterile container and examine as soon as possible. Allow the frozen samples to soften slightly, and while still very slightly frozen, blend thoroughly. Prepare serial decimal dilutions to 10^{-5}. Carry out:

(*a*) General viable counts on plate count agar at 25 and 37°C.

(*b*) Presumptive coliform counts using either violet red bile agar (pour plates) or the multiple tube technique.

(*c*) A direct microscopic count by the Breed's smear method, on the 10^{-2} dilution.

(*d*) Examination for the presence of *Salmonella*.

3. Dried egg

Dried egg may have a viable count ranging from a few hundred bacteria to several hundred millions of bacteria per gram. Consequently the dilutions used for viable counts should be chosen to cover this range. The procedure is similar to that used for frozen egg except that sampling is simplified because there is no question of thawing, and the use of a homogeniser is not required.

4. Frozen, dried and flake albumen

The methods used are similar to those for frozen or dried whole egg, but some samples of dried or flake albumen will be found to have high total (microscopic) counts due to the use of a bacterial fermentation stage in the manufacturing process. The subsequent drying, in the case of dried albumen, may cause a great reduction in the viable count.

D. Liquid Milk

Procedures include those which give general information as to the numbers and activities of micro-organisms present, those which attempt to enumerate only particular groups or kinds of micro-organisms, and methods for investigating various abnormalities. (See also British Standard, 1968; Bulletin, 1968; International Dairy Federation, 1966a, 1966b; Joint Committee, 1972).

The microbial counts to be expected or desired in samples of milk are obviously dependent on the extent of processing. Chilled raw milk sampled at the farm should have a general viable count at 30°C of less than 10^4 per ml, and bulked raw milk sampled at the factory or creamery should have a general viable count at 30°C of less than 10^5 per ml. After pasteurisation, the general viable count at 30°C should be not more than 3×10^4 per ml, with a coliform count of less than 10 per ml (coliform counts of less than 1 per ml should be readily attainable). Milk subjected to an ultra-high temperature (UHT) sterilising process can be expected to have a general viable count of less than 1 per ml immediately after the heat treatment, but since such milk is then aseptically packed under conditions which may allow some ingress of micro-organisms, a typical standard may require that the general viable count at 30°C (e.g. on yeast extract milk agar, or on plate count agar) be less than 10^3 per ml.

1. Determination of total count

Total (i.e. both living and dead) numbers of micro-organisms may be determined by means of the Breed's smear technique, as previously described on page 40. In the examination of whole milk, it is necessary to defat the smear and for this purpose Newman's stain either in its original form or in the modified form proposed by Charlett (1954) is very convenient. Both forms of the stain contain methylene blue with tetrachlorethane as defatting agent, but in Charlett's recipe basic fuchsin is also present, thereby giving a differential stain in which the milk casein forms a pink background to the cells and micro-organisms which are stained blue. Since the Breed's smear technique is relatively insensitive, for routine purposes associated with liquid milk samples the technique is most often used for assessing numbers of animal cells in mastitis milk.

Method of staining

Stain with Newman's stain for 2 min or with Charlett's improved form of the stain for 10–12 sec. *Take all possible precautions to avoid inhaling the harmful vapour* (use a fume cupboard if available). Pour off the stain and in the case of Newman's stain allow the slide to dry in air before washing gently in a beaker of water. With Charlett's recipe, drying before washing is not necessary. Remove excess water by gentle blotting and allow to dry in air.

2. General viable counts

Media suitable for this purpose are plate count agar and yeast extract milk agar. For official tests on milk samples in England and Wales and in Scotland, yeast extract milk agar is the approved medium and the prescribed incubation period is 3 days at 30°C. Psychrotrophic counts and thermophilic counts can be determined by incubating sets of plates at 5 and 55°C for 7 and 2 days respectively. If low numbers of thermophiles are expected, they may be counted by the multiple tube technique either by using tryptone glucose yeast extract broth, with microscopic examination after incubation, or (for milk-spoiling thermophiles only) by detecting changes in the milk or in inoculated litmus milks. With the increasing emphasis on refrigerated cooling of milk supplies, the use of refrigerated bulk tanks, and alternate day collection, the psychrotrophic count becomes of great significance (Orr, *et al.*, 1964). For a useful review of assessment methods for psychrotrophs see Thomas (1969). Lück (1972) has made a critical appraisal of the tests which may be applied to bulk-cooled milk.

3. Presumptive coliform counts

Although, as already explained on page 139, coliforms can proliferate in a food such as milk and can establish a secondary habitat on poorly cleansed equipment, tests for coliforms or Enterobacteriaceae are useful as an index of careless handling and production methods in the case of raw milk, and of post-heating contamination (or less frequently of inadequate processing) in the case of heated milks. Methods to be used are as described on pages 139–144, unless specified differently for statutory control purposes (see also International Dairy Federation 1966a, 1966b).

4. Dye reduction tests

Standard quantities of dye and milk, usually 1 ml of dye solution and 10 ml of milk, are mixed in sterile rubber-stoppered test-tubes and incubated at 37·5°C or other temperature as required for the purposes of the test. The dyes most widely used for milk testing are methylene blue and resazurin, and to a lesser extent 2,3,5-triphenyltetrazolium chloride (TTC). In the case of TTC, reduction of the colourless compound yields a red-coloured compound, formazan. It is possible for leucocyte activity to contribute to dye reduction. This is particularly likely to occur when mastitis milk is examined and large numbers of leucocytes derived from the udder are present in the milk. Resazurin may be partially reduced in this way though methylene blue is not so affected.

Extensive experimental work by Wilson (1935) established that raw milk contains a natural reducing system operating independently of bacterial activity but that it is incapable of bringing about reduction of methylene blue under normal test conditions and for practical purposes may be disregarded. Pasteurisation of milk largely destroys this natural reducing system which is generated by enzyme activity, but in autoclaved milk an inherent reducing capacity reappears, though of a different character from that in raw milk. It follows therefore that any experimental work on the reducing capacity of pure cultures of bacteria in milk must be carried out on either "sterile" raw milk (aseptically drawn milk from healthy udders) as used by Wilson (1935) or on laboratory pasteurised milk of good quality as used by Garvie and Rowlands (1952), or UHT-sterilised milk.

(*a*) *Methylene blue test*

The method described is based on the work of Wilson (1935) and follows the specifications for the Milk (Special Designation) Regulations, 1963.

1. Prepare a stock solution of methylene blue. Add one standard tablet (B.D.H. Ltd.) to 200 ml of cold, sterile, glass-distilled water in a sterile flask, shake until completely dissolved, and make up to 800 ml with more distilled water. This solution can be stored in a cool dark place, preferably a refrigerator, for up to two months. Transfer aseptically each day's requirement into a sterile container and discard at the end of the day.
2. Thoroughly mix the sample to be tested and pour the milk aseptically into a sterile test-tube up to the 10 ml mark.
3. Add 1 ml of methylene blue solution and, after a lapse of 3 sec, blow out the remaining drops. The same pipette may be used for a series of tubes provided that it does not contact the milk or the wetted side of the tube.
4. Close the test-tube with a sterile rubber stopper and invert the tube slowly twice to mix the contents.
5. Within 5 min, place the tube in a water bath at 37·5 $\pm$ 0·5°C and note the time. The level of the water in the bath should be above that of the milk in the tubes, and the bath should be fitted with a lid to exclude light.
6. Set up a control tube with each batch of tubes similar in colour and

fat content to the milk under test. Pour 10 ml milk into a sterile stoppered test-tube, add 1 ml of tap water and place in boiling water for 3 min, then cool and place in the water bath. The control tube will help to determine when decolorisation is complete.

7. Examine the tubes after half an hour. The milk is regarded as decolorised when the whole column of milk is completely decolorised or decolorised to within 5 mm of the surface. A trace of colour at the bottom of the tube may be ignored provided that it does not extend upwards for more than 5 mm.

8. When the test is to proceed beyond the half-hour period, the tubes should be examined at half-hourly intervals for the duration of the test. Tubes which have decolorised should be removed from the water bath; tubes in which decolorisation has begun should remain in the bath without inversion until decolorisation is complete. All other tubes in the water bath should be inverted once to redistribute surface cream within the milk and replaced.

Excessive inversion should be avoided since it results in reoxidation of the methylene blue and consequently will invalidate the test result.

Applications of the test

The half-hour methylene blue test was the official test prescribed by the Milk (Special Designation) Regulations, 1963, for both untreated milk and pasteurised milk in England and Wales, and indicates potential keeping quality. Prior to testing the milk is aged by storing at atmospheric shade temperature from the time of arrival in the laboratory until testing at 0930 hours on the following day. During the winter months (1 November to 30 April) storage overnight (1700 to 0930 hours) is at 65 $\pm$ 2°F (18·3 $\pm$ 1·1°C). Samples are regarded as satisfactory provided the methylene blue is not decolorised after 30 min at 37–38°C.

The methylene blue may also be used as a reference method when ambiguous results have been obtained from the resazurin test, since methylene blue is not affected as is resazurin by cellular activity present in early and late lactation milk and in mastitis milk.

(*b*) *Resazurin test*

The reduction of resazurin takes place in two stages, first irreversibly into resorufin through shades of blue and mauve to pink, and then reversibly from resorufin into the colourless dihydroresorufin. The colours produced can be matched in a Lovibond comparator (Tintometer Sales Ltd.) and given a number ranging from 0 (colourless) through 1 (pink) to 6 (blue).

1. Prepare the resazurin solution by adding one standard tablet (B.D.H. Ltd.) to 50 ml of cold sterile glass-distilled water to give a 0·005 per cent standard resazurin solution. When not in use the solution should be kept in a

cool dark place, preferably a refrigerator, and should be discarded when 8 h old.

2. Thoroughly mix the sample to be examined and add 10 ml of milk to a test-tube.

3. Add 1 ml of resazurin solution, and after a lapse of 3 sec, blow out the remaining drops, without contacting the milk or wetted side of the tube.

4. Close the test-tube with a sterile rubber stopper and invert the tube slowly twice (in 4 sec) to mix the dye and the milk.

5. Transfer the tube to a water bath at 37·5 $\pm$ 0·5°C and note the time. When large numbers of samples are being set up, resazurin should not be added to more than 10 tubes before placing in the water bath and, in the case of the 10 min rejection test (see below), not more than five tubes. Light must be excluded by using a lid on the water bath.

6. The length of the incubation period depends on the particular form of the test being carried out. At the end of the prescribed time, remove and examine each tube. Any tube showing complete reduction, appearing white, is recorded as 0. Any tube showing an extremely pale pink, or pink and white mottling, is recorded as $\frac{1}{2}$. Other tubes are inverted and immediately matched in the comparator as follows:

Place a "blank" tube of mixed milk without dye in the left section of the comparator and the incubated tube in the right section. The comparator must face a good source of daylight (preferably a north window) or a standard source of artificial daylight. Place the comparator on a bench and look down on the two apertures. Revolve the disc until the colour of the incubated tube is matched, then note the disc reading. When the colour falls between two disc numbers record as the half value, e.g. record a reading between 3 and 4 as $3\frac{1}{2}$.

Applications of the test

(*a*) *Ten minute platform rejection test:* since its introduction in 1942, this test has been widely used to detect and segregate churn milk of unsatisfactory bacteriological quality arriving at creamery platforms. Results are obtained after a ten-minute incubation period at 37 $\pm$ 0·5°C. Standards are 4–6, accepted; $3\frac{1}{2}$ or less, rejected.

(*b*) *Temperature compensated test:* this test at one time formed the basis of milk testing schemes in England and Wales. Prior to testing, milk samples are aged at atmospheric shade temperature for a specified period and the length of the incubation period in the water bath at 37·5°C adjusted according to the mean atmospheric shade temperature for this period. Incubation periods range from 15 min for temperatures of 16·1°C and above, to 2 h for temperatures of 4·4°C and under.

(*c*) *S.M.M.B. modified resazurin test:* this is an extension and modification of the temperature compensated test described above, as introduced by the Scottish Milk Marketing Board. Tests are begun at 1430 hours on the day of sampling and tubes are then placed in a water bath at 37°C. The length of the incubation period is adjusted from 15 min to 2 h according to the mean atmospheric shade temperature calculated from the minimum temperature at 0900 hours and the maximum temperature at 1430 hours on the day of testing. At the end of the appropriate incubation period, the tubes are transferred to a water bath at 20°C and held at this temperature until 0900 hours on the following day. The tubes are then inverted and a reading made in the Lovibond comparator. Samples giving a disc reading during November to April (inclusive) of $2\frac{1}{2}$ or less, and during May to October (inclusive) of 0, are regarded as failing the test (for the N. hemisphere).

(*d*) *Hygiene test:* this test is at present used by the Milk Marketing Board in England and Wales (see Joint Committee, 1972). Samples of milk are maintained at or below 7°C (churn milk) or 10°C (bulk milk) until testing. At the laboratory sub-samples are kept overnight at 0–5°C until between 0800 and 1000 hours on the day following collection of milk from the farm. After preparing tubes with resazurin as described above, the tubes are then incubated at $37 \pm 0{\cdot}5$°C for $2\frac{1}{2}$ h with examination and inversion every half hour. Those showing complete reduction, i.e. samples with disc readings of 0 or $\frac{1}{2}$, are regarded as having failed the test.

5. Laboratory pasteurisation test

This test consists of pasteurising a small sample of milk under laboratory conditions. The time-temperature combinations most conveniently carried out are equivalent to those of batch (holder) pasteurisation. Any micro-organisms surviving such heat treatment are regarded as (presumptively) thermoduric and may be subsequently enumerated and isolated by plate counts. The thermoduric nature of the isolates may be confirmed by D-value determinations (see page 96). Isolation of mesophilic thermoduric bacteria is normally carried out at 30°C as certain thermoduric bacteria, e.g. microbacteria, grow sparsely or not at all at 37°C. The significance of numbers of thermoduric bacteria in milk as determined by colony counts has been discussed by Cuthbert (1964).

The reduction in numbers following heat treatment may be calculated as a percentage of the original numbers (the percentage reduction). A milk sample with few heat resistant micro-organisms will show a high percentage reduction, while samples containing large numbers of thermoduric micro-organisms will show a low percentage reduction. The method described is based on the work of Egdell *et al.* (1950).

Procedure

(*a*) Mix the sample and place 10 ml into a sterile test-tube. Close firmly with a sterile rubber bung.

(*b*) Invert and completely immerse the tube to rest on its bung in a stirred water bath at 63·5 $\pm$ 0·5°C for 35 min. This time period includes an allowance of 5 min for the temperature of the milk to reach 63·5°C (the ability of the water bath to return rapidly to 63·5°C should be checked). Alternatively, the tube may be held in the water bath for 30 min after the pasteurising temperature has been reached in a control tube.

(*c*) Remove the tube from the water bath and cool rapidly in iced water.

(*d*) Invert the tube 3 times to mix and prepare decimal dilutions up to 10^{-3} or as required.

(*e*) Plate out on yeast extract milk agar or plate count agar, and incubate at 30°C for 3–4 days. If required the thermoduric isolates can be identified using the identification schemes in Part III, and the heat resistance can be determined by the methods described on pages 96–99.

6. Enumeration and isolation of *Bacillus*

Aerobic spore-formers can be isolated amongst other heat resistant bacteria by means of the laboratory pasteurisation test. Since, however, *Bacillus* spp. possess relatively heat resistant spores, it is possible by subjecting milk to a more severe heat treatment to arrive at a more selective method of enumeration and isolation. A "total" spore count may be determined by heating for 10–15 min at 70–80°C. Some *Bacillus* spp. produce particularly heat-resistant spores, and these can be determined after heating for 30 min at 100°C (Franklin, Williams and Clegg, 1956). When low numbers of survivors are expected a multiple tube counting method should be used. *Bacillus cereus* can be enumerated by the method given on page 158.

Procedure

(*a*) Pour 5–10 ml of well-mixed milk sample into a sterile test-tube. Heat in a water bath at 80°C for 10 min after this temperature has been reached in a control tube. Cool, and plate out on starch milk agar; the starch in the medium encourages the germination of spores by absorbing inhibitors such as unsaturated fatty acids which may be present in the medium. Incubate at 30 or 37°C for 3 days for mesophiles, and at 55°C for thermophiles. When considerable numbers of microbacteria are present in a sample, it is possible for some of these to survive this heat treatment, so colonies should be confirmed as *Bacillus*.

(*b*) Alternatively, counts may be determined by the multiple tube technique, using sterile litmus milk. A multiple tube technique needs to be used for "resistant" spore counts determined by heating at 100°C. When using this counting technique the milk can be distributed in the dilution series *before* the heat treatment, which is then given to the tubes of inoculated medium before incubation. Estimate numbers by reference to probability tables (see Appendix 2).

(*c*) Enumerate *Bacillus cereus* using MEPP agar (see page 158).

7. Investigation of faults and taints in milk

Microbial faults and taints may develop in milk when particular micro-organisms grow to such an extent that their metabolic products become discernible and distasteful. The nature of the fault or taint itself provides considerable information about the causative organism. The term "fault" is usually applied when there is a change in the physical condition of the milk, whereas the term "taint" is used when the physical condition is normal but the flavour or smell is objectionable.

(*a*) *Isolation of causative organisms*

Examine the milk sample and pour a few ml into a Petri-dish to examine the physical condition; determine the pH reaction using indicator papers. Examine microscopically, staining the smears either with Charlett's improved Newman's stain, or by Gram's method (in which case first defat the smear with xylene, which should be drained from the smear and then dried in air before staining).

Streak a loopful of the milk on to the surface of a non-selective medium such as yeast extract milk agar and on to any suitable seledtive or differential medium appropriate to the suspected organism as indicated in the examples given below. Incubate at temperatures over the range 5–63°C, including the temperature at which the milk sample had been stored prior to the detection of the fault or taint. After incubation select colonies for further study; restreak to purify the cultures, and identify them using the identification schemes in Part III. Check the pure cultures for their role as causative organism by subculturing into whole sterile milk and incubating: the causative organism has been isolated when the original fault or taint is reproduced in the subculture.

(*b*) *Examples of faults in milk*

1. *Sour milk:* milk which has coagulated as a result of lactic-acid production by lactose-fermenting bacteria (lactic streptococci and the coli-aerogenes group). In freshly soured raw milk, *Strep. lactis* may comprise about 90 per cent of the total flora. To isolate *Strep. lactis*, streak on to yeast extract milk agar, yeast glucose lemco agar or neutral red chalk lactose agar,

It is an advantage if thallium acetate is incorporated in the medium prior to pouring, at a final concentration of 1 : 2000 to inhibit Gram-negative bacteria.

2. *Gassiness or frothiness:* gas bubbles are produced on the surface of the milk and may be trapped in the cream. The most common causative organisms are coli-aerogenes bacteria, especially *Enterobacter aerogenes,* and lactose-fermenting yeasts. Violet red bile agar can be used to isolate the Enterobacteriaceae and malt extract agar (pH 3·5) to isolate yeasts.

3. *Sweet clotting or sweet curdling:* milk which has coagulated at an approximately neutral pH because of the activity of proteoloytic enzymes. The organisms responsible are predominantly *Bacillus* spp.; the fault frequently develops in heat-treated milks perhaps partly due to the removal of competing bacteria, and partly due to heat activation of the spores. For isolation methods see above.

4. *Ropiness or sliminess:* milk which is viscous and can be drawn out into threads with a wire loop. The fault is caused by the growth of capsulate organisms, frequently developing in refrigerated milk. Common causative organisms are capsulate strains of coli-aerogenes bacteria (e.g. *Klebsiella aerogenes*), *Alcaligenes viscolactis,* capsulate strains of *Micrococcus* and *Bacillus subtilis.* Colonies will also show a viscous consistency.

5. "*Broken*" *cream or bitty cream:* a condition in which the cream breaks up into separate particles on the milk surface and does not re-emulsify. It is particularly evident when poured into hot tea or coffee. The fault is caused by lecithinase produced by *Bacillus cereus* and *B. cereus* var. *mycoides.* The bacteria may be isolated from unheated or heated samples (see page 176 above), using egg yolk agar (page 78), or MEPP agar (page 158). If small numbers of the organisms are anticipated they may be enumerated by a multiple tube count enrichment technique, in which replicate quantities or dilutions are set up in litmus milks, incubated at 25°C, followed by streaking loopfuls from each tube on to egg yolk agar plates to detect lecithinase-producers (Billing and Cuthbert, 1958).

(*c*) *Taints in milk*

These may be accentuated and so more easily recognised if the milk is warmed slightly. The propagation test can be used to determine if the taint is microbial in origin:

Transfer about 5 ml of tainted milk to 50 ml of sterile whole milk in a glass-stoppered bottle, and incubate at 15–25°C or other appropriate temperature, and examine twice daily for the development of the original taint. This will only occur, i.e. can only be propagated, when it is microbial in origin.

1. *Malty or caramel taint:* the causative organism is *Streptococcus lactis*

var. *maltigenes*, which differs from "non-malty" strains of *Strep. lactis* only in its ability to produce 3-methylbutanal from the leucine component of casein.

2. *Carbolic or phenolic taint:* this occurs most commonly in bottles of commercially sterilised milk, and is caused by certain phenol-producing strains of *Bacillus circulans*, the spores of which may survive the heating process.

8. Examination of mastitis milk

Many of the procedures used in the laboratory for the diagnosis of bovine mastitis (inflammation of the udder) are methods for detecting changes in the character of the milk which follow microbial infection of the udder. Isolation by cultural methods of the specific causative organism may also be attempted—this procedure is also a necessary step in the evaluation by sensitivity tests of the appropriate antibiotic treatment. It is important that milk samples for bacteriological examination should be taken only after careful washing and disinfection both of the hands of the milker and of the udder. The sample is obtained, after discarding first milk into a strip cup, by milking directly into a sterile sample bottle. For preliminary investigations and herd tests, mixed quarter samples from single cows may be suitable, but for detailed investigations individual quarter samples from each suspect animal are required. Laboratory investigation is particularly necessary in cases of subclinical mastitis when the cow and the milk show no abnormality detectable by eye.

(*a*) *Cell count, using Breed's smear technique*

One of the most useful methods for the diagnosis of mastitis in the laboratory is a determination of the total cell count using the Breed's smear technique and Newman's or similar stain. Bovine milk even from healthy animals contains some cells, mainly epithelial cells and small numbers of leucocytes, but in mastitis milk the numbers of leucocytes are greatly increased as a result of udder inflammation. An increase in cell count also occurs in early- and late-lactation milk but in this case the increase is due to epithelial cells. These cells can be differentiated by means of a special staining technique as described by Blackburn and Macadam (1954) but, as subsequently shown by Blackburn, Laing and Malcolm (1955), differential cell counts have no particular advantage over total cell counts in the diagnosis of mastitis except for late lactation milk. When quarter samples are being examined it is convenient to prepare the four smears from each animal on the same slide.

Standards suggested by Laing and Malcolm (1956) are as follows.

Individual quarter samples:	
Below 250,000 cells per ml	Negative
250,000–500,000 cells per ml	Suspect
More than 500,000 cells per ml	Positive
Mixed quarter/single cow samples:	
More than 200,000 cells per ml	Suspect

(*b*) *Whiteside test*

This is a simple, rapid test for the diagnosis of mastitis and is basically an indirect method of assessing numbers of cells present in the milk. When these are sufficiently numerous as in mastitis milk, the nucleic acid set free from the cells on titration of the milk with alkali is sufficient to produce pronounced viscosity. Variations of the Whiteside test include the California mastitis test (CMT) in which a solution of surface active agent is substituted for the alkali of the original method (see Blackburn, 1965).

Mix 1 part of N sodium hydroxide with 5 parts of udder secretion on a glass plate or dish and stir with a glass rod for 15 sec.

Record the extent of the viscosity developed, if any. An increase in viscosity occurs with an increase in the number of cells. Absence of increased viscosity is recorded as negative and indicates a normal milk.

(*c*) *Isolation of causative organisms*

Isolations of mastitis organisms are normally carried out from individual quarter samples using either streak or pour plate techniques. The media used should include one capable of supporting the growth of most known mastitis organisms, though selective media for particular groups may be used if required.

The most satisfactory non-selective medium is a 5 per cent blood agar. It must be noted that the kind of blood used will affect the haemolysis produced (pages 81, 265, 272).

Edward's aesculin crystal violet blood agar (Edwards, 1933) is a selective medium for the isolation of mastitis streptococci. The selective agent, crystal violet, is at such a concentration, 1 in 500,000, as to permit growth of streptococci but to inhibit that of any staphylococci. Aesculin-fermenting organisms, including *Strep. uberis*, produce black colonies, while non-aesculin-fermenting organisms, e.g. *Strep. agalactiae* and *Strep. dysgalactiae* produce colourless colonies.

Use a 4 mm loop (see page 41, Part I) to smear 0·01 ml milk on to the surface of a previously prepared and well-dried blood agar plate. When 4 quarter samples from one animal are being examined these may conveniently be streaked in separate sections of one plate. Incubate at 37°C for 48 h, examining after 24 and 48 h. Since standard quantities have been used for inoculations, a comparison can be made of the numbers developing from individual quarter samples. The presence of any haemolytic colonies should also be recorded.

The predominant organisms developing on the isolation plates are of most significance and are presumed to be the causative organisms. Record the size and shape of representative colonies, haemolysis if any, pigmentation, Gram reaction and morphology. These observations should be sufficient to indicate the probable genus of the organism, but further confirmatory tests can be carried out if required. It is convenient at this stage to carry out a slide coagulase test (see page 79) on suspected staphylococci; a positive reaction indicates the presence of *Staphylococcus aureus*. In the case of suspected streptococci, in addition to physiological tests, the precipitin test (see page 104) may be used to determine the Lancefield group of the isolate (see page 271).

(*d*) *Sensitivity tests*

It is important in the treatment of mastitis to determine which antibiotics are most likely to be effective in treatment, by carrying out sensitivity tests on a pure culture of the isolate from the mastitis milk.

Subculture the isolate into a suitable broth medium (nutrient broth or yeast glucose lemco broth) and incubate at 37°C for 24 h. Pour 0·1 ml of culture on to the surface of a previously poured and well-dried agar plate (nutrient agar for staphylococci and yeast glucose lemco agar for strepto-cocci). Spread the culture over the plate and allow to dry. Place an Oxoid Multodisk (Oxoid Ltd.) in the centre of the plate on the surface of the agar, using flamed forceps with aseptic precautions. Alternatively individual antibiotic discs may be used in conjunction with a suitable dispenser (e.g. Difco Ltd.). Incubate at 37°C for 24 h and record the presence of zones of inhibition around the tips of the Multodisk. Inhibition indicates sensitivity to a particular antibiotic, which is therefore potentially valuable for mastitis treatment. Note, however, that a large zone around one antibiotic disc does not necessarily mean that it is more effective than another antibiotic which produces a small zone, since many factors are involved in determining the zone size (Cooper, 1955; Linton, 1961).

9. The detection of antibiotics in milk

Antibiotics excreted in milk following treatment for bovine mastitis are undesirable for public health reasons; for example, certain individuals may show allergic reactions, and transfer of drug resistance may be encouraged. In addition, residual antibiotics may cause interference with lactic acid fermentation in the manufacture of dairy products dependent on this process; for example, fermented milks, cheese, and ripened cream for butter making. The problem largely arises from a failure to withhold milk for the specified time after the last treatment of the cow, usually 48–72 h, the concentration of antibiotic excreted gradually diminishing during this period.

Several tests sufficiently sensitive to detect traces of antibiotics in milk have been developed. Most are based on the inhibitory effect of the residual antibiotics on the growth or activity of a chosen organism. Except in the case of penicillin—where penicillinase is available—it is not generally practicable to determine the nature of the inhibitory substance which may be an antibiotic, a detergent or a disinfectant. The possibility of phage or other heat-sensitive natural inhibitor producing a false positive result can be eliminated by the heat treatment of presumptive positive samples before carrying out confirmatory tests. In all cases where comparative results are required, it is essential that the procedure is carried out under standardised conditions using a specified strain of test organism and standardised conditions of cultivation. The sensitivity of the tests and the extent to which traces of antibiotic can be detected depend on the sensitivity of the selected organism under the particular test conditions. Two methods extensively used for routine assay of antibiotics in milk are disc assay and dye reduction (see also British Standard 1968; International Dairy Federation 1970b). A third method, the "Intertest" (Intervet Laboratories Ltd.), used by the Milk Marketing Board of England and Wales, employs bromcresol purple to indicate the amount of acid produced in the milk sample by *Streptococcus thermophilus* (Joint Committee, 1972).

(*a*) *Disc assay*

In this method an agar medium heavily seeded with the test organism is poured into a Petri-dish and a small filter paper disc soaked in milk is then placed on the surface of agar. Incubation until growth appears is at a temperature appropriate to the test organism. Clear zones occur around the discs where antibiotic or other inhibitor has diffused into the medium. *By comparison with known controls*, provided that conditions are carefully standardised, the diameter of the zone of inhibition can be used to give a quantitative estimation of a known antibiotic present in the milk (see Cooper, 1955; Linton, 1961).

Any sensitive organism may be used in this test, e.g. *Micrococcus luteus*, *Bacillus subtilis* or *B. stearothermophilus* (*B. calidolactis*). An advantage of the thermophile *B. stearothermophilus* as used by Galesloot and Hassing (1962), in the modification by Crawford and Galloway (1964), International Dairy Federation (1970b) and in the British Standard (1968) is that results may be obtained comparatively rapidly, after incubation at 55°C.

The following procedure using penicillin serves to demonstrate the principle of the disc assay, but for assays for any particular authority, reference should be made to the appropriate standard procedure.

1. Prepare penicillin control milk at the required concentration by adding penicillin solution to antibiotic-free milk. The concentration of penicillin in the control milk should be equivalent to the level of tolerance for the test – this in turn will depend on the sensitivity of the test organism. For *B. subtilis* 0·05 units of pencillin per ml provides a suitable standard, but with *B. stearothermophilus* (*B. calidolactis*) a more sensitive test is obtained and a pencillin control milk containing 0·02 units of penicillin per ml can be used.

2. Inoculate 1 ml of an exponential-phase culture of the test organism in nutrient broth into 5 ml of molten nutrient agar at 45–50°C and pour into a Petri-dish. Allow to set, on a level surface, and mark off the base of the plate as required.

3. Using clean dry forceps, pick up one of the antibiotic assay discs (6 mm diameter, Whatman) and dip into the well-mixed milk sample. Drain off the excess liquid and then place the disc in the appropriate sector of the plate. Rinse and dry the forceps before testing the control and any other samples.

4. Incubate the plate at the optimum temperature for the test organism (i.e. 30°C for *B. subtilis*, or 55°C for *B. stearothermophilus* until growth appears, or for the statutorily prescribed period.

5. Examine the plate and compare the diameter of the zone of inhibition of the control with that of the test sample. Samples of milk are regarded as satisfactory when the zone of inhibition produced is less than that of the control. When zones of inhibition are equivalent to, or greater than, that of the control, the milk sample should be retested after first heating the milk to destroy any natural inhibitory substances. (Alternatively unheated and heated samples may be examined concurrently.) A milk sample in which the presence of inhibitory substances was confirmed in a concentration equivalent to, or greater than, that of the control would fail the test.

(*b*) *Dye reduction*

In this method, developed by Neal and Calbert (1955) and Wright and Tramer (1961), the dye used is triphenyltetrazolium chloride (TTC) and the assay organism *Streptococcus thermophilus*. When TTC dye is added to antibiotic-free milk previously inoculated with an active culture of *Strep. thermophilus*, it is rapidly reduced to the red-coloured formazan, as can readily be observed in the milk sample. This reduction process is irreversible, and the formazan form of the dye is therefore not reoxidised by molecular oxygen as is the case with methylene blue. When the activity of the bacterial cells is completely inhibited by the presence of more than a certain minimum concentration of residual antibiotics or other inhibitory substances, conversion of the TTC dye into formazan does not occur and the milk therefore remains white. Partial inhibition of the test organism is indicated when intermediate shades of red colour are produced. Since, as pointed out by Neal and Calbert (1955), the sensitivity of the test is affected by the age of culture, the size of inoculum, and the duration of the incubation period before the addition of TTC, it is important that carefully standardised conditions be maintained when comparative results are required. The main features of the TTC test are demonstrated in the following simplified version of the standard method (see also Crawford and Galloway, 1964), but for assays for any particular authority, reference should be made to the appropriate standard procedure (e.g. British Standard, 1968).

1. Pipette 5 ml of well-mixed milk sample into each of two sterile, marked test-tubes with rubber bungs and make up to 10 ml with sterile antibiotic-free milk.

2. To one tube add 0·2 ml of a solution of 1000 I.U. of penicillinase (Calbiochem) per ml.

3. Add 1 ml of *Strep. thermophilus* culture (prepared from equal quantities of an 18-hour culture and sterile antibiotic-free milk to facilitate pipetting) to each tube.

4. Invert to mix and incubate in a water bath at $44 \pm 0{\cdot}5$°C for $1\frac{1}{2}$ h.

5. Add 1 ml of 1 per cent solution of TTC in sterile distilled water to each tube, invert the tubes to mix, and reincubate at 44°C for a further 1 h.

6. Examine the sample for inhibition by comparison with the control tubes. A control tube set up in antibiotic-free milk should be a deep pink in colour. A second control tube in which distilled water has replaced TTC indicates the colour to be found when complete inhibition of the test culture has occurred. A change in colour in the tube containing penicillinase indicates the presence of penicillin.

7. For comparative purposes and to check the sensitivity of the culture, tests may also be carried out using milk with added penicillin at known concentrations. (The *Strep. thermophilus* culture is sufficiently sensitive when a pink colour is obtained in milk containing 0·01 units of pencillin per ml, but complete inhibition, i.e. no pink colour, occurs in the presence of 0·02 units of penicillin per ml.) Amounts of adventitious penicillin may be determined quantitatively by using a dilution series of the milk samples, and comparing inhibitions obtained against inhibitions in a series of tubes with known amounts of penicillin added to antibiotic-free milk.

8. All presumptive positive samples should be retested within 36 h of the time of sampling. Milk samples which cause complete inhibition of the culture in confirmatory tests are regarded as having failed. A similar test series using a sample which has been heated to 95°C will differentiate antibiotics from heat-sensitive inhibitors, as explained on page 182.

E. Milk Powder

Important factors influencing the microflora of milk powder are the heat treatment given the milk prior to the drying process and the method of drying the milk. Where comparatively severe exposure to heat occurs (for example in roller-drying) the resultant powder shows a more restricted flora than does one in which the temperatures involved are less extreme (for example, spray-dried powder). Further factors influencing the microflora of the powder are the extent of contamination from the milk-plant and the extent to which microbial multiplication can occur prior to the drying process. It is particularly important at this stage that numbers of coagulase-positive *Staphylococcus aureus* do not reach levels at which enterotoxin production creates a health hazard in the subsequent milk powder. The packaging process may also allow the introduction of contaminants, particularly atmospheric contaminants such as yeasts and moulds. On storage, numbers of micro-organisms gradually decline, and in year-old powders, spore-formers may comprise the dominant flora. When powdered milks are reconstituted, any surviving micro-organisms are capable of growth, and milks should therefore not be kept for longer than fresh milk once reconstituted.

1. Sampling

A standard procedure for sampling milk powder, requires that samples should be taken with a dry, sterile, metal spatula or spoon after mixing the top 150 mm (6 inches) of the contents and then transferring not less than 115 g (4 oz) to a sterile sample jar of sufficient size to allow mixing by shaking. It is more informative if surface and sub-surface contents of the packaged powder are taken as separate samples (see also International Dairy Federation, 1969 and British Standard, 1974).

2. Preparation of dilutions

Use the method of Higginbottom (1945). Weigh out aseptically 10 g of milk powder and transfer to 90 ml of sterile distilled water at 50°C in a wide-mouthed bottle. Shake the bottle 25 times in 12 sec with an excursion of 30 cm. Place the reconstituted milk in a water bath at 50°C for 15 min, then invert the bottle several times and examine immediately. Prepare further dilutions in 9 ml amounts of quarter-strength Ringer's solution or 0·1 per cent peptone water as required, or as indicated by examination of the Breed's smear preparation (see 3 below).

Note that the International Diary Federation Methods (1970a, 1971a, 1971c) specify the use of quarter-strength Ringer's solution for reconstitution.

Unless it is required to follow the IDF standard method, we recommend the use of distilled water for reconstitution (see page 132).

3. Microscopic examination

Set up a Breed's smear preparation, using Charlett's improved Newman's stain. The total count gives an indication of the extent to which micro-organisms may have proliferated in the milk prior to drying.

4. Cultural examinations

Plate out 1 ml of each dilution on suitable media for particular groups of organisms as follows:

(*a*) *General viable counts:* use plate count agar or yeast extract milk agar, incubated at 30°, 37° and 55°C for 5, 3 and 3 days respectively (see also International Dairy Federation, 1970a).

(*b*) *Coli-aerogenes count or "total Enterobacteriaceae" count:* use a multiple tube method, incubating at 30°C; for the "total Enterobacteriaceae" count use similar media containing glucose. The International Dairy Federation (1971c) method employs brilliant green lactose bile broth in a multiple tube method to test 100 ml, 10 ml and 1 ml amounts of reconstituted milk for the presence of coli-aerogenes bacteria.

(*c*) *Faecal streptococci:* see pages 144 ff.

(*d*) *Yeasts and moulds:* use malt extract agar or Davis's yeast salt agar. either acidified to pH 3·5 or with added antibiotics (see page 106), and incubate at 20–25°C for 5–3 days.

(*e*) *Staphylococcus aureus:* use Baird-Parker's medium (surface counts), incubated at 37°C for 24 h. To detect low concentrations of staphylococci, use a liquid enrichment technique preceding plating on Baird-Parker's medium (see page 157). Note that the International Dairy Federation (1971a) method recommends preliminary enrichment on Giolitti and Cantoni's medium for milk dried more than 15 days previously, and a salt enrichment medium (also containing lactose and phenol red) for milk dried less than 15 days previously. Note that the absence of viable staphylococci implies no guarantee of the absence of enterotoxin (page 157).

(*f*) *Bacillus spores, and Bacillus cereus:* use the procedures described on pages 158, 176.

(*g*) *Clostridium spores:* use plate count agar or blood agar (the latter as surface counts) incubated at 30°, 37° and 55°C anaerobically for 3–5 days. *Clostridium perfringens* can be enumerated by a multiple tube method as described on page 156.

(*h*) *Salmonella:* follow the isolation procedures described on pages 150–155, including a non-selective resuscitation stage, in which the reconstituted milk

(10^{-1} dilution) is incubated at 37°C for 24 h. North (1961b) has suggested the introduction of some selectivity into this stage by the addition of sterile solutions of either brilliant green or crystal violet to give final concentrations of 0·002 per cent and 0·004 per cent respectively.

5. Results and suggested standards

Report results per gram of milk powder. Davis (1968) suggested that freshly manufactured spray-dried milk powder should have a direct microscopic count of less than 10^6 per gram, a general viable count of less than 10^4 per gram, and counts of coli-aerogenes bacteria, yeasts and moulds each of less than 10 per gram. Galesloot and Standhouders (1968) proposed that the count of *Staphylococcus aureus* should be less than 10 per gram, and that *Salmonella* should be absent from 100 g. For freshly manufactured roller-dried powder Davis (1968) suggested that the general viable count should be less than 10^3 per gram, and that counts of coli-aerogenes bacteria, and yeasts and moulds should each be less than 10 per gram.

F. Canned, Concentrated Milk

Concentrated milks are commonly available either in the unsweetened form as evaporated milk, or with added sucrose as sweetened condensed milk. The keeping quality of evaporated milk depends on the efficiency of sterilisation of the final product and on the prevention of post-heating contamination. This contamination is most likely to occur through the seams developing temporary leaks during the cooling stage, but access of potential pathogens and spoilage organisms can be prevented by chlorination of the cooling water as described by Bettes (1965) Observations on the microbiological examination of canned foods in general are given on page 222.

Sweetened condensed milk is not a sterile product and depends for its keeping quality on the preservation factors operating following the addition of a sufficient concentration of sucrose to produce an almost saturated solution of sucrose at ambient temperatures. The low available water which results is sufficient to restrict growth of most constituent micro-organisms, though occasionally spoilage may result from growth of osmophilic and sugar-tolerant yeasts or moulds, and activity of lipolytic micro-organisms. (See also British Standard, 1970.)

1. Evaporated milk (unsweetened condensed milk)

The product should be sterile. The procedures described are therefore tests for sterility.

(a) Pre-incubation of cans

Incubate representative cans at 55°C for 7 days, 35–37°C for 14 days and 25–27°C for one month as recommended by Davis (1963). Examine the cans after the stipulated incubation period and report on their appearance. Cans containing viable gas-producing micro-organisms may become swollen, i.e. blown, and are easily recognised.

(b) Investigation of the bacteriological condition of the milk

Prepare the can carefully for opening aseptically as follows. Shake the can and its contents thoroughly. Remove paper labels and wash the outside of the can with warm water. Wipe the can dry with a clean paper towel, sponge with a suitable disinfectant and dry. Swab the surface to be punctured with ethanol and flame. Make the opening with a sterile can opener (taking great care for blown cans) and cover the opened can with a sterile Petri-dish lid.

To carry out a microscopic examination, prepare a smear of the milk on a slide and stain by Gram's method, defatting first with xylol if necessary.

To carry out a cultural examination, inoculate 1 ml quantities of the milk into Crossley's milk peptone medium, and into 10 ml of molten yeast glucose lemco agar for a shake tube. Also inoculate 1 ml into a Petri-dish and pour with yeast extract milk agar or plate count agar. Incubate at the appropriate temperature (i.e. that used for the pre-incubation of the can).

Cans which show milk in a normal condition and no organisms in 1 ml may be assumed to have been sterile.

2. Sweetened condensed milk

(a) Preparation of the can

The product may be viscous and it is therefore advisable to warm the contents of the can in a water-bath at 45°C for not more than 15 min before opening, in order to reduce viscosity. The can should then be opened aseptically as described above for evaporated milk.

(b) Cultural examination

Using a sterile 10-ml pipette with a large orifice, weigh out 10 g of milk into a sterile sample jar. Add 90 ml of diluent at 37°C and shake 25 times. Prepare further dilutions up to 10^{-3} in the usual way. Plate out 1 ml of each dilution on to suitable media as follows:

1. *General viable count:* use yeast extract milk agar or plate count agar and incubate at 30°C for 3 days.

2. *Lipolytic count:* use tributyrin agar or victoria blue butterfat (or margarine) agar and incubate at 30°C for 3 days.

3. *Coli-aerogenes count:* use violet red bile agar and incubate at 30°C for 24 h; alternatively if small numbers are expected, use a multiple tube counting technique. Presumptive positive results must be confirmed by subculture into fresh tubes of MacConkey's broth or streaking on MacConkey's agar.

4. *Yeast and mould count:* use malt extract agar or Davis's yeast salt agar acidified to pH 3·5 or with added antibiotics (see page 106) and incubate at 25°C for 3–5 days. Low levels of contamination may be detected using Davis's yeast salt broth in a multiple tube count.

(*c*) *Results and recommended standards*

Report the results per gram of condensed milk. Davis (1968) suggests that sweetened condensed milk should have a general viable count of less than 100 per gram, and a lipolytic count of less than 10 per gram, and counts of yeasts, moulds and coliform bacteria should each be less than 1 per gram.

G. Cream

The microbiology of cream is similar to that of milk in that micro-organisms present in the original milk may also be present in the cream, and survivors of any subsequent heat treatment, e.g. sterilisation or pasteurisation, together with post-heating contaminants may have an adverse effect on the keeping quality, particularly if subsequent storage temperatures are insufficiently low to inhibit microbial growth. Sterilised cream in cans, bottles or cartons should contain few if any viable micro-organisms and appropriate methods of examination would therefore resemble the sterility tests as described for evaporated milk rather than tests for fresh cream described below.

An indication of the hygienic quality of fresh cream can be obtained using a dye reduction test devised by the Public Health Laboratory Service (Report, 1958b) which is similar to that for ice-cream, or by the grading test of Crossley (1948). More detailed information can be obtained by a cultural examination carried out on a gravimetric basis to determine general viable counts and the extent of any coliform contamination. The sampling methods described below and the procedure for the methylene blue dye reduction test are based on the Report (1958b) of the Public Health Laboratory Service.

1. *Sampling*

Samples may consist either of individual cartons or packages or of a sample from bulk taken with a sterile dipper sufficient to fill a sterile sample jar. When samples cannot be delivered to the laboratory within 2 h, they should be packed in ice and delivered to the laboratory before 1700 hours on the day of sampling. On arrival at the laboratory, samples should be stored in the refrigerator until testing begins.

2. *Methylene blue dye reduction test*

This may be carried out using either a volumetric basis of testing as described below, or it may be based on a 1 : 5 dilution prepared gravimetrically. The test should be set up at 1700 hours on the day on which the sample is taken. Any other convenient time may be chosen provided a continuous incubation period of 17 h is maintained.

Procedure

(*a*) Deliver 7 ml of quarter-strength Ringer's solution into a sterile tube with a 10-ml mark.

(*b*) Add 1 ml of standard methylene blue solution (as used for dye reduction tests on milk).

(*c*) Using a wide-tipped pipette, deliver cream to the 10-ml mark. Pipetting of thick samples is facilitated by first warming the sample for a few minutes at 37°C.

(*d*) Insert a sterile rubber bung and invert the tube to mix.

(*e*) A control tube is prepared by transferring 8 ml of quarter-strength Ringer's solution to a sterile tube and adding cream to the 10 ml mark. Insert a sterile rubber bung.

(*f*) Incubate the tubes in a water bath at 20 $\pm$ 0·5°C for 17 h.

(*g*) Transfer the tubes to a water bath at 37 $\pm$ 0·5°C. Examine and invert once every 30 min for 4 h or until decolorisation is complete as compared with the control tube. Record the time taken for decolorisation of the methylene blue (see table on page 192). If the methylene blue is decolorised at the time of removal from the 20°C water bath, the time is recorded as 0 h.

3. Crossley's bromcresol purple milk test

This is a simple procedure originally devised by Crossley (1948) for the rapid arbitrary grading of bulk cream at the creamery but it was considered by the Public Health Laboratory Service (Report, 1958b) to be too "sensitive" for the examination of retail samples.

Procedure

(*a*) Inoculate 1 ml of cream into 10 ml of sterile bromcresol purple milk (separated milk +0·01 per cent bromcresol purple).

(*b*) Insert a sterile rubber bung and invert to mix.

(*c*) Incubate at 30 $\pm$ 1°C. Examine the tubes and report appearance after 16–17 h and again after 24–25 h.

Reporting results

Appearance of tubes	Results
No visible change	–
Slight acidity, indicated by faint yellow mottling	+
Definite acidity, yellow colour but no clot	++
Acid and clot	+++

4. Cultural examinations

Prepare decimal dilutions in 9 ml amounts of general-purpose diluent up to 10^{-4} or as required. Carry out counts on:

(*a*) Yeast extract milk agar or plate count agar for general viable counts incubated at 30°C for 3 days.

(*b*) Violet red bile agar, incubated at 30°C for 24 h (or use a multiple tube technique), for coli-aerogenes organisms.

(*c*) Baird-Parker's medium for *Staphylococcus aureus* (surface counts) at 37°C.

(*d*) Tributyrin agar incubated at 5°C and 30°C for 7 days and 3 days respectively for lipolytic psychrotrophs and for a general lipolytic count.

5. Results and recommended standards

Report the counts per gram of cream. Davis (1963) suggested that retail cream should have a general viable count of not more than 10,000 per gram and a coli-aerogenes count of not more than 10 per gram. Other standards based on Crossley's bromcresol purple milk test and the methylene blue dye reduction test are indicated below. Although the microbiological quality of cream has attracted much attention, in the U.K. at present only in Northern Ireland are there detailed specifications for cream based on cultural tests.

Suggested grades for cream (Crossley, 1948)

Result of bromcresol purple milk test		
At 16 hours	At 24 hours	Grade
– or +	– or +	1
– or +	++	2
– or +	+++	3
++ or +++	+++	4

Suggested dye reduction test standards for cream (Report, 1958b)

Time taken to reduce methylene blue	Interpretation
Fails to decolorise in 4 hours	Satisfactory
Decolorised in $1\frac{1}{2}$–4 hours	Significance doubtful
Decolorised in 0 hours	Unsatisfactory

Taylor (1975) has proposed that since the dye reduction test as applied to cream appeared to be unsatisfactory if the bacteria are not actively growing at the time of sampling, it would be preferable to use a very simple colony count procedure in which the bacteria are allowed to grow in a film of diluted cream adsorbed on to the surface of a non-nutrient agar. However, a statistical analysis of the results of methylene blue test, general viable count, and *E. coli* count on a large number of cream samples (Report, 1971) suggested that the methylene blue test was an excellent screening or advisory test provided that the history of the sampled cream was known. Nevertheless, the present authors believe that the extra trouble and expense entailed in the cultural tests outlined in (4) above are justified by the extent of the information obtained by the quality control microbiologist.

H. Ice-Cream

The Public Health Laboratory Service (Report, 1947) devised a methylene blue reduction test which was thought to be suitable for routine grading purposes and for providing a simple method of detecting ice-cream of poor hygienic quality (but see page 38). More detailed information can be obtained by cultural examination, which should be carried out on a gravimetric basis, and numbers of particular groups of micro-organisms determined as required by the use of appropriate media. In the United Kingdom at the present time, control of microbiological quality is achieved by statutory regulations relating to processing operations, in particular to the heat treatment of the mix, its subsequent cooling and the maintenance of low temperatures prior to sale.

1. Collection of samples

(See also British Standard, 1974)

(*a*) *Hardened ice-cream*

For packages and tubs, one or more unopened packages constitute the sample which should be delivered intact to the laboratory in a sterile container. For multi-layered ice-cream, the sample should contain the same proportions of each layer as in the original ice-cream. In the case of ice-cream in bulk containers, first remove the surface layer with a sterile spatula or spoon and with a second sterile spatula take a sample of not less than 60 g (2 oz) into a sterile jar. For information on bulk ice-cream as served to the consumer, the sample is taken from the surface layer with the retailer's own server.

(*b*) *Soft ice-cream*

This is freshly frozen ice-cream sold direct from the freezer. In this case, fill the sample jars directly from the freezer outlet—the sample should be a minimum of 60 g.

(*c*) *Transport of samples*

Samples should be transported to the laboratory in a refrigerated container and maintained at not more than −18°C until examined in the laboratory. A more convenient procedure is to transport samples to the laboratory in an insulated container for delivery within 2 h of the time of sampling, but when or where this time limit is not practicable, samples should be packed in ice and should arrive at the laboratory within 6 h of sampling.

2. Treatment of samples

(*a*) If the sample is in the original retail package, transfer aseptically the whole, or a representative portion, to a sterile container of not less than 60 ml capacity.

(*b*) Frozen samples should be left at room temperature for a maximum period of 1 h until melted. Alternatively, the sample may be liquified by holding the sample jar in a water-bath at 42–45°C for no more than 15 min as recommended in Memorandum (1948).

(*c*) If the sample is unfrozen, examine immediately.

3. Cultural examination

The analysis should be carried out on a gravimetric basis since, as shown by Patton (1950), the weight of 10 ml of ice-cream may range from 4·5 to 10·5 grams.

Procedure

Invert the sample bottle three times to mix the sample. Using a sterile 10 ml pipette, weigh out 10 g of melted ice-cream into a sterile container. Add 90 ml of sterile diluent and invert three times. This constitutes the

10^{-1} dilution. Prepare further dilutions up to 10^{-4} or as required. Carry out counts on:

(*a*) Plate count agar or yeast extract milk agar incubated at 5 and 30°C for 7 and 3 days respectively for psychrotrophs and mesophiles.

(*b*) Violet-red bile agar incubated at 30°C for 24 h for coli-aerogenes organisms (this conforms to the British Standard, 1970). Alternatively use a multiple tube technique for the detection of smaller numbers (the International Dairy Federation, 1971b, specified the use of brilliant green lactose bile broth in this case). Positive cultures can be confirmed as coliforms or as *Escherichia coli* as described on page 143.

(*c*) Davis's yeast salt agar (pH 3·5) for yeasts and moulds. If moulds are suspected (e.g. in frozen confections containing fruit puree), prepare surface inoculated plates. Incubate at 25°C for 5 days.

(*d*) Baird-Parker's medium (surface counts) incubated at 37° for 24 h, for *Staphylococcus aureus*.

(*e*) Thermoduric bacteria can be counted using the 10^{-1} dilution in a laboratory pasteurisation test as described for milk (see page 175).

4. Methylene blue dye reduction test

The test (Report, 1947) should be set up at 1700 hours on the day on which the sample is taken. (Any other convenient time may be chosen provided that the test is commenced on the day the sample is taken and that a continuous incubation period of 17 h is maintained.)

Procedure

(*a*) Deliver 7 ml of quarter-strength Ringer's solution into a sterile tube with a 10 ml mark.

(*b*) Add 1 ml of standard methylene blue solution (as used for dye reduction tests on milk).

(*c*) Using a wide-tipped pipette, deliver ice-cream up to the 10 ml mark, thus adding 2 ml.

(*d*) Insert a sterile rubber bung and invert the tube once. If the ice-cream contains much air, the meniscus will fall as the air is freed and if this occurs, fill up with further ice-cream to the 10 ml mark.

(*e*) Set up controls as follows.

1. *Ice-cream colour control:* with a 10 ml graduated pipette, deliver 8 ml of quarter-strength Ringer's solution into a sterile 10 ml-mark tube. With a wide-tipped pipette, add ice-cream to the 10 ml mark as in the test. Insert a sterile rubber bung, invert the tube and fill up with ice-cream to the 10 ml mark if required.

2. *Methylene blue control:* about 5 ml of ice-cream should be heated in a boiling water bath for 15 min, cooled and used as in the test proper.

3. For coloured ice-cream, a separate colour control and methylene blue control should be set up for each sample.

(*f*) Incubate the tubes in a water bath at 20 ± 0·5°C, until 1000 hours on the following morning, i.e. a 17 h incubation period.

(*g*) Transfer the tubes to a water bath at 37 ± 0·5°C, and invert once every half hour until decolorisation is complete as compared with the control tubes.

Recording results

Record the time taken for decolorisation of the methylene blue. If the methylene blue is decolorised at the time of removal from the 20°C water bath, the time is recorded as 0 h.

5. Results and recommended standards

Report counts per gram of ice-cream. In the survey by Lloyd (1969) it is apparent that while general viable counts and coli-aerogenes or coliform counts are most often used to define standards, levels acceptable in different countries range from general viable counts of less than 25,000 to less than 300,000 per gram with coli-aerogenes or coliform organisms required to be absent from quantities ranging from 0·1 to 10 grams. In the U.K. the methylene blue test is generally used for routine screening purposes, suggested standards being given below.

Suggested grades for ice-cream in the methylene blue test (Report, 1947, 1950)

Time taken to reduce methylene blue	Provisional grade
Fails to reduce in 4 h	1
$2\frac{1}{2}$–4 h	2
$\frac{1}{2}$–2 h	3
0	4

I. Dairy Starter Cultures

The term "starter culture" as used in the dairy industry refers to a culture of lactic-acid bacteria in milk which is used to induce a lactic-acid fermentation in those dairy products in which fermentation is an essential part of the manufacturing process. The particular lactic-acid bacteria required in any given starter culture depend on the purpose for which it is to be used. In starter cultures for cheese making, an active production of lactic acid is an essential requirement and cultures may consist of single strains of *Streptococcus lactis* or *Strep. cremoris*, or of combinations of both, with or without aroma-forming bacteria. An activity test can be used to confirm that the rate of lactic-acid production is satisfactory. For products in which development of aroma is a special requirement, e.g. in ripened cream for butter making, cultures usually consist of citrate-fermenting organisms capable of producing diacetyl, e.g. *Strep. lactis* subsp. *diacetylactis* and/or *Leuconostoc cremoris* (*L. citrovorum*) in addition to *Strep. lactis* and/or *Step. cremoris*. The ability of a culture to produce acetoin (acetylmethylcarbinol) and diacetyl can be checked by application of the Voges-Proskauer test in its creatine modification.

1. Microscopic examination

Examine smears stained by Gram's method. Defatting with xylene is not required for the examination of skim milk cultures. Note any differences in relative numbers of pairs of cocci and short or long chains of cocci. In starter cultures long chains of cocci are indicative of *Strep. cremoris*.

2. Biochemical tests

The general appearance of a starter culture is a good indication of its condition, but the application of simple biochemical tests can confirm that its condition is satisfactory.

(*a*) *Catalase test* (see page 78)

Add 1 ml of 10 vol. hydrogen peroxide to about 5 ml of starter culture in a test-tube. Since lactic-acid bacteria are negative in the test, a positive reaction indicates gross contamination.

(*b*) *Voges-Proskauer test* (*O'Meara's modification*) (see page 74)

To 2·5 ml of starter culture add an equal volume of 40 per cent sodium hydroxide. Add a knife point of creatine and shake the tube well. A pink coloration developing within 30 min constitutes a positive reaction.

(*c*) *Activity test*

Add 1 ml of starter culture to 100 ml of sterile skim milk and incubate at 30°C in a water bath for 6 h. After incubation determine the acidity that has developed by adding 1 ml of 0·5 per cent phenolphthalein solution as indicator to 10 ml of the culture and titrating with N/9 sodium hydroxide until a faint pink colour is obtained. Divide the number of ml of N/9 sodium hydroxide required by 10 to obtain the titratable acidity of the culture.

(*d*) *Modified activity test for the detection of phage*

Inoculate 1 ml of starter into each of two flasks containing 100 ml of sterile separated milk. To one of the flasks add 0·1 per cent (1 ml of 1 : 10 dilution) of whey; to the other add 1·0 ml of diluent to serve as a control. Incubate in a water bath at 30°C for 6 h.

Determine the titratable acidity for each culture as previously described. If the activity of the culture with added whey is more than 10 per cent below that of the control, it may be assumed that the whey contains phage specifically affecting that particular starter culture.

3. Cultural examination

(*a*) *Enumeration of viable starter bacteria*

Prepare dilutions of the starter culture in diluent up to 10^{-8} or as indicated from the Breed's smear examination. Plate out 1 ml of each of the last three dilutions and pour with media as follows:

1. General viable numbers: use yeast glucose lemco agar and incubate at 30°C for 3 days.
2. Citrate-fermenting organisms (*Leuconostoc*, *Strep. lactis* subsp. *diacetylactis*): use the tomato juice lactate agar of Skean and Overcast (1962). This is a simplified form of the medium described by Galesloot, Hassing and Stadhouders (1961), in which the whey agar base of the original medium is replaced by tomato juice agar (Difco). A suspension of calcium citrate incorporated in the medium enables citrate-fermenting organisms to be recognised by the formation of clear zones around the colonies, although according to Waes (1968) certain lactic-acid-producing organisms may produce clear zones even though not citrate-fermenting. Growth of *Strep. lactis* and *Strep. cremoris* is inhibited by the incorporation of 0·5 per cent calcium lactate in the medium. Incubate at 21°C for 4 days.
3. *Differentiation of Strep. lactis and Strep. cremoris:* use arginine tetrazolium agar (Turner, Sandine, Elliker and Day, 1963). Incubate at 30°C for 24–48 h. Colonies of *Strep. lactis* are red (arginine +), whereas colonies of *Strep. cremoris* are white (arginine –).

(*b*) *Detection of contaminants*

In most cases sufficient monitoring of contaminating organisms can be achieved by inoculating suitable selective media with 0·1 ml of starter culture. If quantitative estimations are required, use 1 ml quantities of each of the decimal dilutions already prepared.

1. To detect yeasts and moulds use Davis's yeast salt agar (pH 3·5), incubated at 22–25°C for 3–5 days.

2. To detect the presence of bacteria other than lactic-acid bacteria use nutrient agar plates incubated aerobically and anaerobically at 30°C for 3 days.

3. To detect coli-aerogenes organisms use a MacConkey's broth. Incubate at 30°C for 3 days.

4. To detect citrate-fermenting organisms in cultures of *Streptococcus lactis* and/or *Strep. cremoris* use semi-solid citrate milk agar prepared as below, in which fermentation of citrate is indicated by gas production. Inoculate citrated milk (prepared by adding 0·5 ml of 10 per cent sodium citrate solution to 10 ml of milk) with 1 per cent (0·1 ml) of starter culture and add to 4 ml of molten 2 per cent agar at 48°C. Mix by inversion and incubate at 30°C for 3 days. In certain circumstances, as shown by Crawford (1962), the presence of these organisms in starter cultures for cheese making may prove deleterious in that gas holes are produced in the cheese as a result of the carbon dioxide evolved during the fermentation of citrate.

(*c*) *Isolation and maintenance of pure cultures of starter bacteria*

Obtain isolated colonies of starter bacteria either by using the colony count method described above or, when qualitative information only is required, by streaking on to the surface of the particular agar medium.

Select suitable isolated colonies and examine Gram-stained smears. Subculture into yeast glucose lemco broth or litmus milk and incubate at 30°C for 3 days. For the subculture of *Leuconostoc* spp. the broth medium should be used, since these organisms produce little or no acid in litmus milk and successful subculture in this medium is not immediately apparent. The culture obtained should be re-examined and restreaked if required. Once the culture has been established, single colony isolation should *not* be used during routine maintenance subculturing.

Maintenance of the pure culture can be achieved by inoculation into yeast glucose chalk litmus milk and incubating at 30°C for 24 h or until a *slight* acidity develops. As an alternative, Robertson's cooked meat medium may be used. Store the culture at room temperature and subculture every 3–6 months.

4. Demonstration of the effect of phage on starter bacteria

(*a*) *Liquid culture method*

Inoculate a loopful of an 18-h culture of lactic streptococci into each of two tubes of litmus milk and two tubes of yeast glucose lemco broth. Similarly inoculate the pure culture isolate obtained from the starter.

Add two drops of the phage preparation to one of the litmus milk tubes and to one of the broths for each culture. Label the tubes containing added phage. The uninoculated tubes serve as controls. Incubate at 30°C overnight. Report the appearance of cultures with added phage and compare with that of the controls. In the presence of the specific bacteriophage the broth cultures will show lysis and the litmus milk cultures will fail to produce acid.

If incubation of the cultures is prolonged there may be a subsequent secondary growth following the development of phage-resistant mutants, shown by the development of turbidity in the previously lysed broths and of acid in the previously inactive litmus milks.

(*b*) *Agar plate method*

Spread a few drops of an 18-h broth culture of lactic streptococci over the surface of a previously dried plate of yeast glucose lemco agar, using a sterile, bent glass rod. Allow the inoculum to dry.

Prepare decimal dilutions of the phage preparation in quarter-strength Ringer's solution. Using a 4 mm standard loop (page 41), spot inoculate on to previously marked sectors of the plate a loopful from each of the phage dilutions and from the undiluted preparations. Incubate at 30°C overnight.

Examine the plates for the presence of plaques, *viz.* areas of clearing where bacterial lysis has occurred.

5. Demonstration of the effect of penicillin on starter cultures

This may be demonstrated by inoculating the test culture into a series of litmus milks containing added penicillin at a range of concentrations. The appearance of the litmus milk cultures following incubation indicates whether or not inhibition has occurred.

Procedure

(*a*) Prepare a stock solution of penicillin at 100 units per ml by adding one tablet containing 10,000 units of benzyl penicillin to 100 ml of sterile distilled water, allowing to stand for 15 min and then shaking to dissolve.

Prepare working solutions of 10 units, 1 unit and 0·1 unit of penicillin per ml by serial decimal dilution of the stock solution.

(*b*) Use the prepared penicillin solutions at 100, 10, 1 and 0·1 units per ml, to inoculate a series of 9 ml amounts of sterile litmus milks to give penicillin concentrations ranging from 10 units per ml to 0·001 units per ml as indicated below:

Tube no.:	1	2	3	4	5	6	7	8
Litmus milk, ml	9	9	9	9	9	9	9	9
Ml of penicillin solution in units/ml	1 of 100	1 of 10	0·5 of 10	1 of 1	0·5 of 1	0·1 of 1	0·1 of 0·1	0
Sterile distilled water, ml	0	0	0·5	0	0·5	0·9	0·9	1
Final concentration of penicillin in units/ml	10	1	0·5	0·1	0·05	0·01	0·001	Control

(*c*) Inoculate a loopful of the culture to be tested into each tube of the series. Incubate overnight at a suitable temperature: 30°C for lactic streptococci and starter cultures, and 37°C for *Strep. thermophilus* and yoghurt cultures. Also incubate an uninoculated tube of litmus milk for comparative purposes.

(*d*) Report the results for the cultures tested. When the control tubes show acid production, the absence of acidity indicates inhibition by the penicillin at the concentration tested. *Strep. thermophilus* is one of the most sensitive organisms to penicillin and may be used therefore in penicillin detection tests. As it is a constituent of yoghurt starters, cultures for this fermented milk product are very sensitive to residual antibiotics in milk.

(*e*) Examine Gram-stained smears prepared from the control tube and from tubes showing some inhibition of acid production. Note any differences in morphology between the control and the inhibited cultures. There is a tendency for elongate bacillary forms to be produced by the cocci in the presence of penicillin. Report the concentration at which such abnormalities are observed.

J. Fermented Milks

Fermented milks are produced by the growth in milk of sufficient numbers of lactic-acid bacteria to produce curdling or thickening of the milk and to give it a typical sour flavour. This can occur by the development of bacteria already present in the milk as in the natural souring of raw milk when lactic streptococci make up the predominant flora, but for satisfactory large scale production, fermentation is usually induced by added cultures.

When lactic-acid bacteria are the predominant organisms, an acid fermented milk is obtained as for example in cultured buttermilk, acidophilus milk and yoghurt. An association of lactic-acid bacteria with yeasts results in a product which is both sour and weakly alcoholic, and when fermentation is carried out in closed containers the fermented milk will be effervescent, as, for example, in kefir. The ability of yeasts to thrive in fermented milks may present problems in the manufacture and distribution of yoghurt and buttermilk, as frothiness and/or off-flavours may develop from low initial numbers of contaminating yeasts.

1. Microscopic examination

Prepare smears of the fermented milk in a drop of water on a slide and allow to dry. If whole milk products are being examined, first defat the smear by flooding with xylene, draining and drying before staining. Stain the smears by Gram's method. Report the kinds and relative proportions of micro-organisms present in each type of product being examined. For example, in yoghurt cocci and rods should be present in approximately equal numbers. Contaminants will only be detected when they are present in considerable numbers. To detect small numbers of Enterobacteriaceae or other contaminants, use the appropriate selective techniques already described, if necessary employing a multiple tube technique.

2. Isolation of constituent flora

Although not normally required for routine testing purposes, these procedures also include methods to indicate whether particular culture organisms still remain viable. For quantitative investigations prepare decimal dilutions and inoculate the specified media with 1 ml quantities in the usual way. For detection or isolation, streak or spread plates may be used.

(*a*) *Streptococcus* spp.

Use neutral red chalk lactose agar. Incubate as necessary—for example at 30°C to isolate lactic streptococci from sour milk and from cultured buttermilk, and at 37°C to isolate *Strep. thermophilus* from yoghurt. The differential medium may also be made selective for Gram-positive bacteria by the inclusion of thallium acetate at a final concentration of 1 : 2000. This is useful when suppression of coliforms is required, as for example in the isolation of streptococci from sour raw milk.

On neutral red chalk lactose agar, acid-producing colonies of streptococci are small, deep red and surrounded by a a clear zone where the acidity developed has dissolved the chalk present in the medium.

Select isolated presumptive streptococci for examination, subculturing with a straight wire into litmus milk. Also prepare a smear for microscopic examination and stain by Gram's method to confirm the presence of Gram-positive cocci. Incubate the litmus milk at 30 or 37°C until acid is produced, then re-examine. Obtain a pure culture by restreaking on a solid medium if required and confirm identity as indicated on page 272.

(*b*) *Lactobacillus* spp.

Use a suitable selective medium for lactobacilli, for example Rogosa agar, and then cover with a further 5 ml of the molten agar medium to form a layer plate. The temperature of incubation depends on the particular species of *Lactobacillus* to be isolated. Streptobacteria are lactobacilli (e.g. *L. casei*) having a low optimum temperature, and require incubation at 30°C. Thermobacteria are lactobacilli (e.g. *L. acidophilus* and *L. bulgaricus*) with a higher optimum temperature and incubation at 37°C is necessary.

Select isolated colonies and examine microscopically. Subculture from colonies of Gram-positive rods—presumptive positive lactobacilli—into litmus milk and incubate as previously. Re-examine the culture when growth is obtained. If further identification (page 274) is required, purification of the cultures on solid media is necessary.

(*c*) *Yeasts*

Use a selective medium for yeasts which is sufficiently acid to inhibit the growth of bacteria, e.g. malt extract agar, at pH 3·5, or Davis's yeast salt agar at pH 3·5. Incubate at 22 or 25°C for 3–5 days. Isolated yeast colonies may be picked off to malt extract agar slopes for further study. In particular, it may be of interest to investigate the fermentative capacities of the isolates by inoculating into a series of fermentation media containing, e.g. glucose, galactose and lactose. See page 277 for details of procedures for identification of the cultures.

(*d*) *Detection of contaminants*

Methods for detecting contaminants in starter cultures combining streak plates and selective media can be applied as appropriate, and purified isolates identified using the diagnostic keys in Part III. (See also International Dairy Federation 1971d, 1971e, 1971 f.)

It should be noted that, as pointed out by Davis (1970), the coliform test is only indicative of the standard of plant hygiene when performed on samples within 24 h of manufacture, since the coliform bacteria will die on storage. Yeasts may be of significance as potential spoilage organisms, and small numbers may be detected by a multiple tube count with Davis's yeast salt broth at pH 3·5 as the selective medium (yeasts must be confirmed microscopically in tubes showing growth with or without gas production).

K. Cheese

Fermented foods such as cheese may be expected to contain large numbers of lactic-acid bacteria, the micro-organisms responsible for the fermentation process. For a short period after manufacture, the predominant lactic-acid bacteria are those derived from the starter culture, the species composition of which largely determines the microbial flora of unripened soft cheese, e.g. cottage cheese and immature cheese of other varieties. In cheese of more prolonged ripening periods, the starter-derived organisms are gradually displaced by a population of more acid-tolerant bacteria, the lactobacilli, and in many hard pressed varieties of cheese, e.g. Cheddar, these may be largely responsible for the satisfactory ripening of the cheese. In other varieties of cheese, the particular character of the final product may also depend on the growth and metabolic activity of other micro-organisms, e.g. *Penicillium* spp. in mould-ripened cheese, and propionibacteria in Swiss cheese.

The procedures described below are designed to give general information concerning the numbers and kinds of micro-organisms developing in a cheese and may be adapted for the isolation of any particular group of micro-organisms by the use of appropriate selective media. (See also British Standard, 1970.)

1. Sampling

Sampling of cheese for bacteriological purposes should be carried out using sterile sampling equipment and the sample transferred aseptically to a sterile sample jar. Precise details of sampling technique, which will vary according to the type of cheese to be examined, have been suggested by the International Dairy Federation (1958, 1969), and form the basis of the methods described below. Similar procedures are prescribed in the British Standard (1974). The weight of the sample should not be less than 50 g.

For small soft cheese and small packets of wrapped cheese, an entire cheese or packet may be taken as a sample. For other soft cheese and semi-hard cheese where use of a trier is not practicable, a sample may consist of a wedge of cheese taken with a sterile knife by making 2 cuts radiating from the centre of the cheese. Any inedible surface layer should be removed before transfer to the sample jar.

For hard cheese of large size, the sample is most conveniently obtained by the use of a sterile cheese trier. This should be inserted obliquely towards the centre of the cheese on one of the flat surfaces, not less than 10–20 cm from the edge. The outer 2 cm of cheese containing the rind is cut off and replaced in the cheese.

2. Preparation of dilutions

Several methods are available for the preparation of the initial 10^{-1} cheese emulsion. Earlier methods requiring sterile sand are now seldom used. The diluent may be either quarter-strength Ringer's solution or 2 per cent sodium citrate solution at 45°C (Naylor and Sharpe, 1958a). The latter diluent facilitates dispersal of the curd and consequent release of micro-organisms. Emulsification may be achieved by shaking by hand or, more efficiently, by use of a macerator or the Colworth Stomacher. If shaking by hand is used, first grate or mince the sample aseptically, using a sterile mincer or grater (ordinary domestic apparatus is satisfactory, and may be sterilised in the autoclave).

Further dilutions can be prepared from the 10^{-1} dilution in 0·1 per cent peptone water in the usual way. Since the counts obtained will be influenced by the method of preparing the initial emulsion, the method selected should be adhered to throughout any series of experiments.

3. Microscopic examination

(a) Qualitative

Press a clean slide firmly over a freshly cut level surface of the cheese or cut a section of cheese and press between two slides, then separate and remove excess cheese with the edge of a slide. Defat with xylene for one minute, then pour off and dry in air. Stain by Gram's method and examine. Report the kinds and *relative* numbers of micro-organisms present. A number of fields should be scanned since in a mature cheese the bacteria are often localized in colonies rather than scattered uniformly throughout the cheese.

(b) Quantitative

Examine dilutions of the cheese sample by Breed's smear preparations stained with Charlett's improved Newman's stain. The information so obtained may be used as a guide in the preparation of dilutions for cultural examinations.

4. General cultural examinations

(*a*) *General viable count.* Use yeast glucose lemco agar. This non-selective medium will support the growth of most lactic-acid bacteria, but will not reveal the presence of lactobacilli when these are greatly outnumbered by streptococci. Incubate at 30°C for 3 days.

(*b*) *Lactobacilli.* Use acetate agar, Rogosa agar or the modified Rogosa agar of Mabbitt and Zielinska (1956), and pour layer plates. Any of these similar media will partially select for lactobacilli, suppressing growth of streptococci, though some growth of leuconostocs and pediococci may occur. This type of medium is therefore particularly useful for the detection and enumeration of lactobacilli in the early stages of cheese-ripening when streptococci are the predominant organisms. Incubate at 30°C for 3–5 days for the isolation of streptobacteria from Cheddar and similar cheeses.

(*c*) *Coli-aerogenes bacteria.* Use either violet red bile agar or a multiple tube technique, incubating at 30°C.

(*d*) *Yeasts and moulds.* Use malt extract agar (pH 3·5) or Davis's yeast salt agar (pH 3·5), incubated at 22–25°C for 3–5 days.

(*e*) *Staphylococcus aureus.* Use Baird-Parker's medium, surface inoculated, incubated at 37°C for 24 h.

(*f*) *Clostridium* spp. Use Angelotti's sulphite-polymyxin-sulphadiazine agar in pour plates incubated anaerobically. Alternatively, differential reinforced clostridial medium may be used in a multiple tube technique.

5. Isolation and maintenance of pure cultures of *Lactobacillus* spp.

Select isolated colonies from Rogosa agar or modified Rogosa agar. Examine Gram-stained smears and pick off presumptive *Lactobacillus* colonies into a suitable broth medium, e.g. the MRS broth of de Man, Rogosa and Sharpe (1960). Incubate broths at 30°C for 2–3 days.

The resultant culture should be re-examined by Gram's method and restreaked on to a suitable agar medium, e.g. acetate agar, Rogosa agar or the agar medium (MRS agar) of de Man *et al.* (1960). Confirm that the isolate is a *Lactobacillus* sp. and identify as far as possible by methods described on page 274.

Pure cultures of lactobacilli may be maintained in a similar manner to streptococci (see pages 19–20).

6. Isolation of propionibacteria from Swiss cheese

Propionibacteria may be isolated from Swiss cheese by means of a yeast extract lactate medium devised by van Niel (1928). This medium, which may be termed elective (page 55) since it satisfies the minimum nutritional requirements of the propionibacteria but of few others, can thus also be used for enumeration of propionibacteria if required. Fermentation of lactate in the medium results in the evolution of carbon dioxide and if the original broth medium of van Niel is converted to the semi-solid form by the addition of agar, disruption of the medium is presumptive evidence of propionibacteria. Isolation of propionibacteria should, if possible, be carried out under anaerobic conditions in an atmosphere enriched with 5 per cent carbon dioxide as provided by the GasPak system (see page 63).

Procedure

(*a*) Prepare dilutions as in (2) above and inoculate into molten yeast extract lactate semi-solid medium at 45°C. Mix well by rotating the tubes between the hands and allow to cool. Seal the surface with molten 2 per cent agar at 45°C. Incubate at 30°C, preferably anaerobically in an atmosphere enriched with 5 per cent carbon dioxide, for 7–10 days.

(*b*) Tubes of medium showing gas fissures are presumed positive and can be confirmed by microscopic examination and subculture. Propionibacteria frequently appear as small Gram-positive coccobacilli which are non-motile and can thus readily be differentiated from coli-aerogenes bacteria and clostridia which may also be present in the cultures. Report the highest dilution at which gas production is observed and hence estimate numbers of propionibacteria per gram of cheese.

(*c*) Streak from one of the presumptive positive tubes, in effect an enrichment culture on to yeast extract lactate agar. Incubate at 30°C for 5–7 days, either anaerobically, in an atmosphere enriched with 5 per cent carbon dioxide or aerobically, using layer plates, in a carbon dioxide-enriched atmosphere. Anaerobic incubation is to be preferred.

(*d*) Select isolated colonies for examination and stain by Gram's method. Subculture presumptive *Propionibacterium* colonies into yeast extract lactate broth or yeast glucose lemco broth and incubate at 30°C, restreaking if required.

(*e*) Pure cultures of isolates can be maintained by stab inoculation into yeast glucose lemco agar, incubating until growth is established and then covering with a layer of sterile liquid paraffin.

7. Isolation of moulds from mould ripened cheese

Mould-ripened cheeses include semi-hard cheese—e.g. Stilton, Roquefort—in which the ripening agent, *Penicillium roqueforti*, grows in the interior of the cheese, and also soft cheese of comparatively small size or shallow depth—e.g. Camembert or Brie—in which the mould-ripening agent *P. camemberti* or more usually *P. caseicolum* (*P. candidum*) grows on the surface of the cheese. Isolation of a particular mould in pure culture from a mould ripened cheese can be achieved by plating a fragment of cheese showing mould growth on appropriate selective media.

Procedure

(*a*) Remove a fragment of mould growth from the required cheese with a sterile loop in about 2 ml of sterile quarter-strength Ringer's solution.

(*b*) Use a dilution technique to obtain isolated colonies. Inoculate one loopful of the prepared suspension into 10 ml of molten malt extract agar (pH 3·5) or Davis's yeast salt agar (pH 3·5) at 45°C. Mix by rotation between the hands and transfer a loopful of the agar suspension mixture to a second tube of molten medium. The first tube of medium is then poured into a Petri-dish. The second tube is mixed by rotation and used to inoculate a third tube before itself being poured into a Petri-dish. The process is repeated until 4 plates have been prepared. Incubate at 25°C for 3–5 days.

(*c*) Examine the plates for isolated colonies typical of the required mould species and record details of colonial appearance. The identity of the isolate can be confirmed by microscopic examination as described on page 286.

(*d*) Subculture from a suitable isolated colony using a sterile straight wire on to a malt extract agar slope, and incubate at 25°C for 3–5 days. Pure cultures can be maintained on this medium or on potato dextrose agar and should be subcultured at intervals of 3–6 months.

(*e*) Proteolytic activity of the isolates can be determined by inoculating on to 10 per cent milk agar as described on page 67. Incubate at 25°C for up to 14 days. Proteolysis is indicated by clear zones surrounding the growth.

(*f*) The lipolytic activity of the isolates can be examined by subculture on to tributyrin agar, butter-fat agar or victoria blue butter-fat agar as described on page 76. Incubate at 25°C for up to 14 days.

8. Isolation of *Brevibacterium linens* from bacterial-ripened soft cheese

A bacterial-ripened soft cheese is one in which a surface smear or coat consisting of the ripening agent *Brevibacterium linens* develops on the surface of the cheese and imparts a characteristic butyrous texture and orange or orange-brown colour. In addition to occurring on typical bacterial-ripened soft cheeses, such as Limburger, *Bbm. linens* may also be isolated from orange or orange-brown spots on the surface of mature mould-ripened soft cheese.

Procedure

(*a*) Prepare a suspension of the surface smear in about 2 ml of quarter-strength Ringer's solution and streak on to the surface of nutrient agar containing 5 per cent sodium chloride. Alternatively, the cheese agar of Albert, Long and Hammer (1944) may be used to enhance orange pigment production by *Bbm. linens*. Incubate at 25°C for 5 days.

(*b*) Examine orange pigmented colonies by Gram's method. Subculture from isolated presumptive *Bbm. linens* colonies into nutrient broth. Incubate at 25°C and restreak on to nutrient agar. Confirm the identity of the isolate as indicated on pages 267 ff.

(*c*) Pure cultures may be maintained on nutrient agar slopes by subcultures at intervals of 3–6 months.

L. Butter

Butter may be manufactured either from unripened cream to make sweet cream butter or from cream ripened by the addition of a starter containing citrate-fermenting organisms to make ripened cream butter. Many of the organisms present in the cream are removed in the buttermilk during the process of manufacture but some will survive to the finished product. Butter may be either unsalted or salted, the concentrations of sodium chloride varying from 0·5–2·5 per cent. The moisture content of butter is usually restricted by legislation—in the United Kingdom it should not exceed 16 per cent. The effective concentration of salt in the aqueous phase is considerably higher than the overall concentration and for a 2 per cent salt and a 16 per cent moisture content, the salt concentration in the aqueous phase is 12·5 per cent. This is a considerable deterrent to the growth of many micro-organisms, and growth is further restricted when the water droplets are of small size and evenly distributed throughout the butter. Refrigeration and storage of the butter at low temperatures also retards microbial growth. Spoilage of butter is therefore most likely to arise from the activity of micro-organisms capable of growing at low temperatures, particularly those capable of lipolysis, proteolysis or loss of flavour, or those causing discoloration.

1. Sampling

The sampling of butter for bacteriological purposes requires the use of sterile equipment and the aseptic transfer of samples to sterile sample jars. The butter should not come into contact with paper or any absorbent surface. In a a standard procedure described in the British Standard (1974) for the sampling of butter in bulk containers, the samples are taken with a sterile trier inserted vertically at the centre of the block and near two diagonally opposite corners of the opened end. The final sample, weighing not less than 60 g (2 oz), is made up of the top third of the first core, the middle third of the second core, and the bottom third of the third core. A similar sampling procedure by the International Dairy Federation (1958, 1969) requires a final sample of not less than 200 g total weight.

To obtain comparable results with smaller units of butter, care should be taken that the surface area included in the sample remains constant. This may be achieved by sampling with a trier as in the standard method of the American Public Health Association (1972). When the surface is to be examined separately, remove the surface butter to the required depth with a sterile knife or spatula and transfer to a sterile sample jar. Samples should be kept cool during transport to the laboratory and examined as soon as possible.

Several standard methods have been described for the bacteriological examination of butter. Analysis may be on either a volumetric or a gravimetric basis, and in the case of the method described by the British Standard (1970) is based on a shaken melted sample which thus includes both the fat and aqueous fractions. Alternatively, as in the method described by the International Dairy Federation (1964b) the butterfat may be allowed to separate from the aqueous fraction, in which most of the organisms are present, and this aqueous fraction only is examined.

2. Procedure for cultural examinations

Melt the sample quickly at a temperature not exceeding 45°C by immersing the sample bottle for a short time in a water bath at 45°C. Mix the melted sample by shaking 50 times with an excursion of 30 cm in 1 min.

For analysis on a volumetric (v/v) basis, transfer 10 ml of melted sample to 90 ml of diluent in a bottle of 200 ml capacity. The diluent should consist of quarter-strength Ringer's solution with 0·1 per cent agar to stabilise the emulsion. The pipettes and diluent should be at a temperature of 45°C. For analysis on a gravimetric (w/v) basis, the butter may be transferred to the diluent with a pipette but using a balance to weigh 10 g of sample to which are added 90 ml of diluent. Shake the diluent and butter 25 times with an excursion of 30 cm to give a homogeneous suspension, the 10^{-1} dilution.

Prepare subsequent decimal dilutions up to 10^{-4} or as required.

Carry out counts on:

(*a*) Plate count agar for general viable counts incubated at 5 and 30°C for 14 and 3 days respectively.

(*b*) Davis's yeast salt agar (pH 3·5) incubated at 22–25°C for 3–5 days for yeasts and moulds (equivalent to method of International Dairy Federation, 1964).

(*c*) 30 per cent milk agar of Smith, Gordon and Clark (1952) for caseolytic micro-organisms, as suggested by Druce and Thomas (1959); incubated at 30°C for 5 days. Caseolytic colonies are those surrounded by clear zones and may be confirmed by flooding the plates with a protein precipitant before counting the colonies (see also page 67).

(*d*) Lipolytic test media: a number of different media are available (see page 76). The disadvantage of the simplest medium, tributyrin agar, is that many organisms clearing this medium are incapable of hydrolysing more complex fats, e.g. butter-fat or margarine. For this reason, media using complex fats may be preferred or used in addition to tributyrin agar. An alternative procedure is to use one medium as the sole isolation medium, but subsequently to test for lipolysis on other media. This may be done by the separate subculture of individual colonies, or more speedily, by use of the replicator technique (page 49). Incubate at 22°C for 5 days or 30°C for

3 days. The International Dairy Federation (1966c) and British Standard (1968, 1970) recommend the use of Victoria blue butter fat agar (see page 77).

(*e*) Violet-red bile agar for coli-aerogenes bacteria, incubated at 30°C for 24 h; alternatively use a multiple tube technique.

(*f*) Halophilic/salt-tolerant counts may be performed using 15 per cent sodium chloride to prepare a dilution series, and media containing 15 per cent sodium chloride.

3. Results and recommended standards

Record results per ml or gram of butter tested according to method of analysis used. Davis (1968) suggested that butter manufactured from unripened cream should have a general viable count of less than 10,000 per gram, counts of proteolytic, lipolytic and psychrotrophic micro-organsisms each of less than 1000 per gram, of yeasts and moulds less than 100 per gram and coli-aerogenes bacteria of less than 10 per gram.

4. The isolation and further study of lipolytic micro-organisms

Select isolated colonies showing lipolysis. Stain smears by Gram's method and subculture into semi-solid yeast glucose lemco agar. Incubate at 22 or 30°C until growth is established, then re-examine and restreak on solid media, to obtain pure cultures.

Determine the lipolytic activity of each isolate by single streak inoculations on to the surfaces of a range of test media (page 76 ff.). Several cultures may be tested on one plate and media not used for counting purposes may be examined in this way, e.g. Tween agar.

To determine the salt tolerance of isolates, use the semi-solid nutrient agar cultures above to prepare 24-h nutrient broth cultures of the lipolytic isolates. Inoculate a standard loopful of each broth culture into each of a series of tubes of nutrient broth containing sodium chloride at a range of concentrations, e.g. 5, 10, 15 and 20 per cent. Include also nutrient broth (0·5 per cent NaCl). Incubate at 22 or 30°C for 3–5 days and record the relative amount of growth in each tube. The salt tolerance of isolates obtained in "halophilic" counts may be determined similarly.

M. Fruit and Vegetables

Fresh fruit and vegetables normally carry a surface flora of micro-organisms of soil saprophytes and some plant parasites. Some of these micro-organisms will play an important part in any subsequent spoilage. Microbiological analysis of these foods is not usually carried out, but in the case of fruit and vegetables to be eaten raw (particularly when considerable handling is involved, e.g. prepared salads or when the source is dubious), an examination for the presence of food poisoning pathogens, or for indicators of faecal contamination, may be advisable. Fruit and vegetables grown on land which has been fertilised with organic manures (or with "night soil") should be examined similarly. ICMSF (1974) recommend that vegetables to be eaten raw should have an *Escherichia coli* count of less than 10^3 per gram, and that *Salmonella* should be absent from ten 25-gram samples. Frozen fruit and vegetables have a microflora similar to that of fresh products, although the proportions of types and their absolute numbers will be somewhat different due to effects of general hygienic measures in the freezing plant, blanching procedures, storage temperatures before and after blanching, etc. Freeze-dried products may also show a similar flora to the original, with the same provisos concerning proportions and absolute numbers.

1. Fungal spoilage of fruit

Examination of fruit for incipient or potential spoilage, or of already spoiled samples may be required but the types of organisms sought depend on the type of fruit and on the method of storage. The fungi can be placed in two main groups: firstly those which are true parasites and which have been able to invade the host tissue through the lenticels (e.g. *Glœosporium*), and secondly those which invade only or mainly through wounds (e.g. *Monilia fructigena*), although some of this latter group can attack senescent fruit through the lenticels.

The main economically important fungi causing spoilage of pome fruits are *Botrytis cinerea*, *Glœosporium* (especially *G. album*), *Monilia fructigena* and *Penicillium expansum*. The most important fungi causing spoilage of citrus fruits are *Penicillium* species, especially *P. italicum* and *P. digitatum*, *Alternaria* and *Colletotrichum*.

A clue to the fungal type is given by the form of spoilage: for example *Penicillium* will usually cause blue and green mould rots (although *P. expansum* causes brown rots of pome fruits, and the blue-green spots of conidia production only develop in the advanced stages of rotting), *Botrytis* causes fluffy grey mould growth, *Glœosporium* and *Colletotrichum* cause anthracnose with small sunken spots being evident on the surface of the fruit. A comprehensive review of fungal spoilage of pome fruits is given by Fidler, Wilkinson, Edney and Sharples (1973). Many moulds can be identified microscopically by reference to Part III. *Colletotrichum* and *Glœosporium*, when growing on the host plant, produce clumps of conidiophores which emerge from breaks in the plant cuticle as cushion- or disc-shaped structures known as acervuli. The differentiation of *Colletotrichum* and *Glœosporium* requires the detection and microscopic examination of these acervuli.

Colletotrichum typically produces dark-pigmented pointed spines or setæ amongst the conidiophores, whereas *Glœosporium* does not have these setæ. The conidia of both genera are colourless, sickle- (crescent-) shaped, ovoid or cylindrical. The macroconidia of *Gl. album* tend to be curved with rounded ends, whereas those of *Gl. perennans* are straight and pointed at one end; *Gl. perennans* also frequently produces many spherical microconidia.

2. Fruit pastes, purees, and comminuted fruit

Frequently these products are used in food manufacture, and the specific methods of examination will depend on the nature of the food or the end product. For example: the examination for coli-aerogenes organisms in fruit preparations being used in recipes for ice cream and frozen dairy confections; the examination for thermoduric organisms, flat-sour spoilage organisms, etc., in fruit preparations being used in food products to be canned (especially when the product as a whole will not be very acidic). The types of organisms sought and the methods of examination chosen depend on the food product or production method involved (see the appropriate section).

A direct microscopic examination of the fruit preparation for fungal hyphae can give a very rough indication of the extent of fungal spoilage of the original fruit, and arbitrary limits may be set for acceptance purposes. An example of this procecure is the Howard mould count for tomato products (see Association of Official Analytical Chemists, 1975; Williams, 1968; Aldred, Evans and Husbands, 1971) which can be modified for other fruits. The standard Howard mould count makes use of a special Howard counting chamber or cell, but any suitable haemocytometer or counting chamber could be used in an agreed modified procedure, assessing the number of hyphae in a given volume and expressing the result as the number of hyphae per ml. In microscopic examinations, phase contrast microscopy, dark field microscopy or appropriate stains (e.g. erythrosin solution, ethanolic solutions of methylene blue or basic fuchsin) may be used to visualise the fungal hyphae, depending on the nature of the background formed by the suspension of the product.

Outline procedure for Howard mould count on tomato products

(*a*) Mix the sample. Add water to give a total solids content which, at 20°C, results in a refractometer reading of 45·0–48·7 or a refractive index of 1·3447–1·3460. Mix well.

(*b*) Add 4–5 loopfuls to the Howard cell, carefully place the coverslip on the cell, lowering from one side to exclude air bubbles.

(*c*) Adjust the drawtube of the microscope to give a field of view with a diameter of 1·382 mm using the ×10 objective. The quantity of liquid in each field of view will then be 0·15 mm^3 when a Howard cell is used. If a microscope is being used which is incapable of such adjustment, determine the field diameter and calculate the required correction factor.

(*d*) Examine at least 25 randomly selected fields on each of two or more slide preparations, using a mechanical stage to facilitate the choice of fields. The field of view should be selected whilst *not* looking down the microscope!

(*e*) A "positive" field is recorded when the aggregate length of not more than 3 mould hyphae exceeds $\frac{1}{6}$ of the field diameter. Record the number of "positive" and "negative" fields. The most difficult task in performing a Howard mould count is the differentiation of fungal hyphae from plant tissue.

(*f*) Various government and legislative control agencies have set maximum limits in the case of tomato products. For example the Food and Drugs Administration of the U.S.A. have set a maximum limit of 40 per cent "positive" fields for tomato paste, and 20 per cent "positive" fields for tomato juice. It is important to remember that the test procedure is artificial and arbitrary, that the definition of a "positive" field is arbitrary, and also that there is not necessarily any good correlation with the amount of spoilage of the product as judged chemically or organoleptically (since the amount of spoilage caused by a given amount of mould growth will depend on the fungal species involved and its particular metabolic activity and physiological capability).

3. Vegetables

These frequently can be affected by bacterial soft rots caused particularly by Gram-negative organisms of the genera *Erwinia* and *Pseudomonas*. The development of such soft rots is encouraged by prepacking vegetables in insufficiently ventilated water-impermeable wraps, and vegetables showing spoilage of this type can be observed on shelves of supermarkets which have inadequate stock control procedures.

Fungi, particularly *Botrytis*, *Fusarium*, *Rhizopus*, and the parasitic fungus *Peronospora* (which causes downy mildews), may also cause spoilage on stored vegetables.

Cultural examinations may be made using the usual media for general viable counts, yeasts and mould counts etc. Pectinolytic organisms may be counted using polypectate gel medium, which is inoculated as spread plates (American Public Health Association, 1966). After incubation at 25–30°C a count is made of the colonies forming depressions in the surface of the gel. The medium may be made partially selective for bacteria by the addition of Acti-dione (Upjohn Ltd.), and for Gram-negative bacteria by the addition of Acti-dione and crystal violet.

N. Fruit Juices and Squashes

Bottled fruit juices and squashes frequently will have been pasteurised or may contain preservative, and in consequence the microflora will be considerably modified from the original microflora of the raw fruit. At ready-to-drink concentrations fruit squashes and fruit drinks usually have a limited shelf life with high bacterial counts often developing. The growth of acetic acid bacteria can cause off flavours and the loss of oxygen from the headspace can result in partial collapse of plastic containers.

Frozen fruit juice concentrates normally show very low microbial counts (usually below 10^3 per ml), but after storage of the reconstituted ready-to-drink product to the end of shelf-life very high bacterial counts may be found (up to 10^7 per ml). These may be identified by the procedures listed in Part III. Bacterial spoilage of citrus juices caused by diacetyl-producing species of lactic-acid bacteria can be assessed by a quantitative version of the Voges-Proskauer test (see Murdock, 1968; Hill and Wenzel, 1957); this test may be used in control situations in processing factories.

Viable counts should be carried out on an agar medium containing 2 or 20 per cent sucrose or glucose. The lower concentrations of sugar are used for a general viable count of micro-organisms; the higher sugar concentrations for viable counts of osmophiles (alternatively osmophilic agar may be used). Acetic acid bacteria can be identified by the procedures listed in Part III.

Procedure

(*a*) Prepare serial dilutions and carry out:

1. A general viable count of yeasts and moulds on orange serum agar, or Davis's yeast salt agar (at ph 3·5 or with antibiotis—see page 107).
2. A general viable count of yeasts, moulds and bacteria on malt extract agar, pH 5·4.
3. An osmophilic count on osmophilic agar or on orange serum agar containing 20 per cent sucrose adjusted to pH 5·4 or pH 3·5. Plates should be incubated at 25–30°C (for 7 days) and examined after 3 and 7 days.

(*b*) *Spore count.* Transfer aseptically 10 ml of the sample to each of two sterile, plugged test-tubes and into one insert a thermometer through the cotton-wool plug, so that the thermometer bulb is completely immersed in the sample. Place both tubes in a water bath at 80°C and allow the tubes to remain in the bath for 15 min after the temperature in the control tube attains a maximum (usually just below the temperature of the water-bath). An alternative time-temperature combination which may select a slightly different thermoduric population is obtained by holding at 100°C for 5 min.

Remove the tubes and cool quickly in cold water. Carry out pour plate counts or multiple tube counts from the sample and its dilutions and incubate duplicate sets at 30 and 55°C, for the mesophilic spore count and the thermophilic spore count respectively, using the media specified.

(*c*) *Microscopic count.* Carry out a direct microscopic count for yeasts by placing a known volume (1/100 ml) of the sample or a low dilution of the sample on a slide, adding one drop of 0·02 per cent erythrosin solution and mixing well. Spread the mixture over 1 cm^2 of the slide and allow to dry. Alternatively the sample-stain mixture may be examined as a wet preparation using a counting chamber or haemocytometer slide. The yeasts and mould fragments can be seen and counted but due to the high solids content of most fruit juices and squashes, bacteria cannot usually be distinguished. A modification of a Howard mould count may also be adopted using wet mounts with a suitable counting chamber or haemocytometer slide.

O. Sugars and Sugar Syrups

Sugar syrup is used as a constituent in the preparation of many food products and the nature of the microbiological examination will depend on the type of end product. For example, if the sugar is to be used in the preparation of a heat-treated canned food, samples should be examined for the presence of organisms able to cause flat-souring (*Bacillus stearothermophilus*), hydrogen sulphide-producing clostridia using shake tube counts in iron sulphite agar, and non-H_2S-producing clostridia using liver broth (see Section T). In general, plate counts for bacteria, yeasts and moulds should be carried out using orange serum agar, Davis's yeast salt agar, and osmophilic agar or orange serum agar with 20 per cent sucrose. Incubate at 25–30°C for up to one week (see Section N). The genera most frequently isolated are *Bacillus*, *Lactobacillus*, *Leuconostoc*, *Hansenula*, *Pichia*, *Saccharomyces* and *Torulopsis*. Mesophilic and thermophilic spore counts and microscopic counts can be carried out as described for fruit juices and squashes. For methods of examination specifically related to the type of end product, refer to the appropriate section. In general, viable counts should not exceed 10 per gram (Muller, 1972).

P. Salted, Pickled and Fermented Vegetables

The microbial flora of salted and pickled vegetables may be very different from fermented vegetables. A microbiological examination of such products can be carried out by sampling the brine or liquor. In the case of the finished products, it should be noted that frequently pickled and fermented vegetables are pasteurised to increase the shelf-life. Such pasteurised products are bottled or canned and the presence of heat-sensitive organisms indicates either faulty pasteurisation or post-pasteurisation contamination (e.g. due to jar or can leakage).

Non-pasteurised fermented vegetables must be expected to contain many organisms, usually lactic-acid bacteria, particularly *Lactobacillus* and *Leuconostoc*, but also *Pediococcus*. Yeasts and moulds, members of the Enterobacteriaceae, *Bacillus*, and *Clostridium* may all cause spoilage of fermented products. However, some fermented vegetable products are produced as the result of action by mixtures of lactic acid bacteria, yeasts and/or moulds.

Procedure

Prepare serial dilutions in the appropriate diluent from the liquor or brine. In an actively fermenting brine the following population levels per ml may be present: 10^7 to 10^9 lactic acid bacteria, 10^6 to 10^8 yeasts, 10^6 to 10^9 salt-tolerant or halophilic organisms, and up to 10^7 coliform organisms. In the finished, pasteurised product the expected counts will be low, requiring plating of 10^0 to 10^{-2} dilutions. Carry out:

1. A direct microscopic count of the undiluted liquor or brine by a Breed's smear using Gram's staining method. It is advisable to heat fix slightly more thoroughly than usual to help prevent the organisms being washed from the slide when the salt or sugar is dissolved.

An examination for yeasts or moulds may also be carried out using erythrosin as described in Section N.

2. A viable count of yeasts and moulds on Davis's yeast salt agar at pH 3·5 (or with antibiotics, see p. 107), incubated at 25–30°C for up to 1 week.

3. A viable count of yeasts, moulds and bacteria on malt extract agar at pH 5·4, incubated at 25–30°C for up to 1 week.

4. A count of lactic-acid bacteria on Rogosa agar or acetate agar, prepared as layer plates, and incubated at 30°C for 5 days.

5. A count of *Staphylococcus aureus* on Baird-Parker's medium, surface-inoculated and incubated at 37°C for 24 h.

6. Counts of salt-tolerant and halophilic organisms, in the case of brine-cured and salted products. Use 15 per cent sodium chloride as diluent for these counts, and media incorporating 15 per cent sodium chloride.

7. Coliform counts using violet red bile agar plate counts or a multiple tube count for smaller numbers.

Q. Alcoholic Beverages

The microbiology of the *production* of beers, ciders and wines is outside the scope of this book (interested readers are referred to Findlay, 1971 and to Reed and Peppler, 1973) but the food microbiologist may occasionally be called on to examine reputedly spoiled samples.

Wild yeasts (i.e. yeasts other than the *Saccharomyces cerevisiae*, *S. carlsbergensis*, and *S. ellipsoideus* strains involved in the primary fermentation) frequently prove to be more resistant to the presence of cycloheximide (Acti-dione, obtainable from Upjohn Ltd.) in a medium than the species named. Thus the incorporation of Acti-dione to a final concentration of 10 parts per million will enable yeasts such as *Pichia*, *Brettanomyces* and *Torulopsis* to be detected. However, it should be remembered that the flor yeasts of sherry and the initiating yeasts of Bordeaux wines are types which may grow on media containing Acti-dione, or on the lysine agar described below. Obviously in such cases yeasts so detected may or may not be spoilage yeasts. Acti-dione-containing media will also enable bacterial contaminants to be detected, since bacteria are also resistant to Acti-dione.

A suitable medium is plate count agar with the glucose content increased to 20 g per litre to which is added 3 g of malt extract per litre, one set of plates being prepared at pH 7·0 and one set of plates at pH 4·0. Acti-dione may be added to the medium before sterilisation or it may be added as a sterile solution just before pouring the plates. If the latter procedure is adopted a 0·1 per cent stock solution of Acti-dione can be prepared, and sterilised by filtration or by autoclaving; 1 ml of this solution is added to every 100 ml bottle of molten medium and mixed immediately before pouring the plates. Such a medium should provide for the nutritional requirements of most bacteria and yeasts likely to be found in these samples. However, some lactobacilli found as contaminants in breweries or distilleries may not grow in media such as MRS medium unless the medium is supplemented with either hopped beer (Kirsop and Dolezil, 1975) or filter-sterilised malt extract and yeast autolysate (Bryan-Jones, 1975).

Sets of plates should be incubated aerobically and anaerobically at 30°C. Possible spoilage bacteria include lactobacilli, pediococci, acetic acid bacteria, and *Zymomonas anaerobia*. *Zymomonas anaerobia*, an important spoilage organism in the brewing industry, causing turbidity and off-flavours, will grow only on the anaerobically incubated plates. The commonest microbial spoilage of cider is acetification caused by acetic acid bacteria.

If it is required to assess low levels of contamination, the liquid version of such media can be used as an enrichment stage with a multiple tube counting technique to obtain counts of the organisms found. Alternatively, membrane filtration can be used to remove the micro-organisms from a relatively large amount of sample. Membrane filtration will however be practicable only with filtered or clarified products, but in such situations it may be possible to attempt detection of contaminants at concentrations as low as one cell in 10 litres.

Another medium suitable for the detection of many wild yeasts is lysine agar, a defined medium (Morris and Eddy, 1957). *Saccharomyces cerevisiae* and *S. carlsbergensis* are unable to utilise lysine as a sole nitrogen source, whereas many other yeasts can utilise this amino acid: lysine agar is thus an elective medium. It is available in dehydrated form from Oxoid Ltd. When lysine agar is used to detect wild yeasts it is important to wash and centrifuge the yeast suspension at least three times using sterile distilled water to ensure that there is no adventitious extracellular source of nitrogen added to the medium. The pellet is finally resuspended to a known volume using sterile distilled water. In consequence this medium is more suitable for the detection of wild yeasts in samples such as yeast concentrates. It is also important with this medium to use levels of inoculum on the plates which provide 10^4 to 10^6 yeast cells per plate.

R. Bread, Cakes and Bakery Goods

These food products usually do not support the growth of bacteria, but cream and similar fillings of cakes are highly favourable for bacterial multiplication. Very rarely, species of *Bacillus* may cause a defect in bread known as "ropiness" due to the production of capsular material. The methods used in the microbiological examination of cake fillings depend on the nature of the constituents and are given in the sections appropriate to the constituents. Microbiological control of bakery products other than fillings is concerned mainly with the possibility of spoilage due to the growth of moulds. Since the baking temperature is sufficient to kill fungal spores, subsequent spoilage is usually caused by mould contamination from the atmosphere or from wrapping material.

1. Examination for moulds

If an item is suspected of being spoiled by the growth of moulds, make a visual examination for the presence of mould mycelium by teasing out with needles small samples of the food in a Petri-dish placed on a black background with oblique lighting illuminating the sample. Follow this by a microscopic examination. Tease out small portions (including any which appear to contain fungal hyphae) on a microscope slide in a drop of lactophenol-picric acid or lactophenol-cotton blue, and cover with a coverslip. Identify the mould by reference to Part III if the sporing stages are present. If only vegetative mycelium is present it would be necessary to culture the mould on a range of media in order to obtain spore formation.

2. Examination of compressed bakers' yeast

This may be examined for viability, the presence and numbers of bacterial contaminants and the numbers of wild yeasts (see also Reed and Peppler, 1973).

(*a*) Weigh out a 10 g sample and prepare serial dilutions to 10^{-9}.

(*b*) Perform microscopic (total) counts either by haemocytometer or by Gram-stained Breed's smear, using an appropriate dilution to give not more than 30 organisms per field.

(*c*) Set up a viable count of the yeast on the high dilutions using Davis's yeast salt agar, incubated at 30°C for 3–5 days. From the microscopic count and the viable count the percentage viability can be calculated.

(*d*) Set up viable counts for bacterial contaminants on a wide range of dilutions (10^{-1} to 10^{-7}) using plate count agar with Acti-dione (Upjohn Ltd.) added to a final concentration of 10 p.p.m. (add 1 ml of 0·1 per cent sterile stock solution of Acti-dione to each 100 ml of molten medium immediately prior to pouring the plates). Incubate at 30°C for 5 days.

(*e*) Set up viable counts for lactobacilli similarly, but using Rogosa agar or acetate agar, containing Acti-dione, as layer plates and incubating at 30°C for 5 days.

(*f*) Set up a count for wild yeasts on the low dilutions (10^{-1} to 10^{-3}) using Davis's yeast salt agar with 10 ppm Acti-dione. Incubate at 30°C for 5 days.

3. Examination of stored cereal grains

The examination of whole cereal grain does not often come within the purview of the food microbiologist. The microflora of cereal grains is extremely varied, with bacteria (including actinomycetes and streptomycetes) outnumbering the fungi. Usually counts performed at 20–25°C will exceed counts obtained at 37°C, but when grain is malted not only will bacterial counts increase, but also the ratio of the count at 37°C to the count at 20–25°C often will be found to increase.

The main concern of the microbiologist, however, is to assess the extent of fungal growth. In recent years it has become apparent that in addition to gross spoilage rendering grain unacceptable to the consumer, fungal growth frequently may involve the production of potentially hazardous metabolites. Of course the toxicity of *Claviceps purpurea* is well known, since ergotism is a disease well described in many ancient accounts. Chemical methods of assessing ergot in rye have been described (see American Association of Cereal Chemists, 1962 *et seq.*). But the full scale of the possible toxicity of metabolites of *Aspergillus flavus* and other *Aspergillus* spp., *Penicillium* spp., *Stachybotrys*, *Fusarium* spp., *Alternaria* and other fungi has been realised only relatively recently.

It is obvious that in this context a viable mould count on cereal grains or on flour provides little useful information on the possible amount of metabolites – the mould colonies on the plates may have developed from spores, or from small or large hyphal fragments as already explained (page 138). When a *specific* toxin is being sought, as in the case of aflatoxin, chemical methods of detection and assessment may be used. Any attempt at assessing a potential hazard from *unspecified* toxic metabolites must surely be based on a method of estimating the amount of mycelial growth present (e.g. by a modification of the Howard mould count—see page 212). Any acceptability standards based on such microscopic counts will be arbitrary and empirical. It may become possible eventually to correlate microscopic findings with concentrations of metabolites determined by quantitative chemical assay techniques. However, it must be emphasised that the presence of a species known to be capable of producing toxin is not evidence *per se* of the presence of toxin in the foodstuff. Firstly, toxigenicity varies from strain to strain within a species, and secondly toxin production is substrate dependent and can also be affected by environmental factors such as temperature, pH, etc. Nevertheless, until much more is known about the toxicity of cereals and flours (and other foods) which have been subject to fungal attack, we suggest that the presence of any fungal hyphae at a level detectable microscopically as the equivalent of 20 per cent "positive" fields determined in the Howard mould count must be regarded with suspicion.

4. Examination of flour

The indigenous flora of grain includes coli-aerogenes organisms, therefore coliform counts on flours may be advisable when these are being incorporated into food products on which coliform counts are normally conducted, although usually the flour incorporated into products will receive a heat treatment sufficient to kill these organisms.

Since flour is usually to be subjected to a heat treatment, the most significant micro-organisms to be sought are species of *Bacillus* and *Clostridium*. As already mentioned certain *Bacillus* spp. (especially *B. subtilis*) may cause ropiness in bakery products; the presence of *Clostridium* is of special significance if the flour is being used as a thickener in meat-containing products because good conditions for development will be provided.

Procedure

1. Prepare decimal dilutions in 0·1 per cent peptone water for aerobic counts, and in reinforced clostridial medium for anaerobic counts. (See pages 61,132). Use a Colworth "Stomacher" or other homogeniser to ensure adequate mixing.
2. Carry out the following counts:

(*a*) General and thermophilic viable counts on plate count agar incubated aerobically and anaerobically at 25, 37 and 55°C for 3 days.

(*b*) A coliform count at 30°C using violet red bile agar, or by a multiple tube technique.

3. Using the appropriate 10^{-1} dilutions perform both aerobic and anaerobic spore counts on plate count agar at 30 and 55°C (see Section N for method).

Recommended standard: Flour to be used in soups and meat products should contain not more than 15 thermophilic spores per g; not more than 10 flat-sour spores per g, and less than 1 *Clostridium* spore per gram (see also Amos, 1968).

S. Frozen Foods

Generally the methods will be those used for the similar product in the unfrozen state, as has already been mentioned in some of the previous sections. In addition, keeping quality tests are advisable in which samples of the foods in their original packets are stored under the conditions advised on the packets. Following this storage, carry out viable counts, coliform counts and direct microscopic counts.

1. Frozen raw foods

The procedure for the microbiological examination of such products corresponds to that already given in the appropriate section for the type of product in question.

2. Frozen pre-cooked foods

Examples of frozen pre-cooked foods are stews, sliced meat in gravy, meat pies with cooked fillings and raw pastry, pies, soups, and also complete meals which require heating to serving temperature or, less frequently, partial cooking.

Procedure

1. While the food is frozen, remove it from its pack, with full aseptic precautions, into a sterile container. Allow the frozen sample to soften slightly and weigh 10 g aseptically. Homogenise the weighed sample with 90 ml of sterile diluent to give the initial 10^{-1} dilution, preferably using a Colworth "Stomacher". Prepare serial decimal dilutions as required. Carry out the following tests:

(*a*) Plate counts on plate count agar, incubated at 25 and 37°C for 3 and 2 days respectively.

(*b*) A presumptive coliform count using either violet red bile agar layer plates incubated at 35°C for 24 h, or a multiple tube technique. Confirm presumptive coliforms as *Escherichia coli* by the Eijkman test (see page 143).

(*c*) A direct microscopic count on the 10^{-1} dilution or, if that is not possible, the 10^{-2} dilution, staining the Breed's smear by Gram's method or with Loeffler's methylene blue as appropriate.

(*d*) An anaerobic viable count on pre-cooked meat products, and an examination for the presence of *Clostridium perfringens*, using for these examinations a dilution series prepared in reinforced clostridial medium.

(*e*) A count of *Staphylococcus aureus* on Baird-Parker's medium (surface counts), incubated at 37°C for 24 h.

2. Store further packets of the frozen food at the maximum storage temperature recommended by the manufacturer (for example, if it is recommended that the frozen food can be stored for 3 days in a refrigerator or for 12 h at room temperature, carry out the keeping test by storing for 12 h at room temperature), and repeat the tests after storage.

T. Canned Foods

Although most canned foods are processed for sterility, some are not. For example, foods with a pH below 4·5 may be given a heat treatment sufficient to kill yeasts, moulds and their spores, and the vegetative forms of bacteria, without killing bacterial spores since bacterial spores cannot usually germinate and grow in an acid food. Canned food products with a pH above 4·5 are, with few exceptions, given sufficient heat treatment to destroy the spores of *Clostridium botulinum*. The chief exceptions are some canned cured meats. These products would suffer from a significant deterioration of quality if given the usual fairly rigorous heat treatment. However, they can be given a less severe heat treatment, since the presence of nitrate, nitrite and sodium chloride combined with refrigerated storage and a limited shelf life prevents the growth of surviving organisms. Thus, canned hams and similar products may be found to contain a variety of spore-bearing bacteria, or even *Streptococcus*. However, Gram-negative bacteria should be absent, as they would be indicative of grossly inadequate processing or of post-processing contamination.

Since the spores of some of the non-pathogenic spoilage organisms commonly associated with canned foods (particularly "flat-souring" organisms such as *Bacillus stearothermophilus*) are more heat resistant than *Cl. botulinum* spores, the examination of the food for the presence of non-pathogenic thermoduric bacteria can suffice in the determination of the efficiency of the heating process. Canned foods showing evidence of flat-sour *spoilage* in all probability have been insufficiently cooled after processing.

After heat treatment cans are usually cooled in water. Temporary leaks through the seams of the cans may occur as a result of the stresses introduced by sudden cooling. When such leaks occur, small amounts of cooling water will enter the can. Consequently, water used for cooling must be of better than potable quality and contain fewer than 100 bacteria per ml, to reduce to a minimum the possibility of post-processing contamination. Post-processing contamination may involve micro-organisms of many types including *Flavobacterium* (Bean and Everton, 1969), cocci and other non-sporing bacteria.

The methods of examination should include those used for the similar product when uncanned, *with the addition* of sterility tests both before and after storage. Pathogens—*Clostridium botulinum*, *Cl. perfringens*, *Staphylococcus aureus*, *Salmonella* and *Shigella*—should be completely absent. A large microscopic count coupled with the absence of viable organisms indicates the possibility of spoilage of the product before canning.

Procedure

Select a representative number of cans from each batch (see page 128). Examine the cans for physical defects which include faulty side or end seams, perforations, rust or other corrosion, dents, and bulging ends. Bulging at one or both ends may be due to bad denting or to the multiplication in the food of micro-organisms which ferment sugars with the production of gas. If any of the cans selected show such defects, examine these separately and remove further cans from the batch to make up the number of normal cans examined. Since most cans are partially evacuated, opening the can provides an opportunity for contamination of the contents with air-borne organisms. Therefore, cans should be opened in an inoculating chamber which should preferably ensure a sterile atmosphere by the use of forced ventilation of sterile air, ultraviolet light sterilisation or other means, or at the very least provide completely draught-free conditions.

(*a*) Examine half the normal cans selected as follows:

Swab the top of a can with alcohol and then flame it. Open the can with a sterile can opener (sterilise this also by swabbing with alcohol and flaming). If the food product is liquid, remove a sample with a sterile pipette to a sterile container. In the case of a solid sample, use a sterile cork borer to remove a core of the food from the centre of the can. Also remove a core from the immediate vicinity of the side seam.

Carry out a microscopic examination of a smear of the food, stained by Gram's method if possible.

Inoculate five tubes each of tryptone soya broth, and glucose tryptone broth with portions of the food, to examine for the presence of viable bacteria, and "flat-sour" organisms respectively. Inoculate five tubes of *freshly prepared* liver broth (Oxoid) and incubate anaerobically either by sealing with 2 per cent agar or by placing in an anaerobic jar. Incubate these media at 25°C for 3 days. Similarly inoculate two further sets of media and incubate at 37 and 55°C for 3 days. After incubation, examine the tubes for the growth of micro-organisms and prepare Gram-stained smears from tubes which show growth. In the case of the glucose tryptone broth "flat-sour" organisms will produce acid from the glucose and cause the medium to change colour from purple to yellow.

(*b*) Incubate the remainder of the normal cans for one week at 37°C and then examine the cans for evidence of blowing. Sample and examine the contents of the cans as described above.

(*c*) Examine blown cans by opening the cans aseptically with precautions to prevent the high pressure in the cans causing the contents to be scattered. For example, after sterilising the top of the can with alcohol and by flaming, invert a sterilised funnel over the top of the can. Insert a sterile metal punch through the hole in the funnel and puncture the top of the can. When the pressure has been released, open the can with a sterile can opener and examine the contents. As well as the tests indicated above, inoculate media as follows:

1. Lauryl tryptose broth or violet-red bile agar	to detect coliform bacteria
2. Robertson's cooked meat medium or anaerobically incubated liver broth (Oxoid)	to detect *Clostridium* (mesophilic or thermophilic)
3. Tryptone soya broth	to detect *Bacillus*

Direct determination of F-value

If the heating and cooling curve has been obtained by the use of thermocouples inserted into cans of food (radiotelemetry being required in the case of hydrostatic cookers), then the microbiological effect can be determined mathematically. This can be achieved either by calculation, using the methods described by Stumbo (1973), provided that some D-values and the *z*-value of the reference organism are known, or the thermocouples can be linked directly to an automatic F_0 computer (e.g. as made by Ellab A/S, which assumes a *z*-value of 10°C).

Alternatively, the total effect of a thermal process can be determined experimentally as follows.

(*a*) The thermal destruction curve for the reference organism (usually *Bacillus stearothermophilus*) at 121°C is determined in the laboratory as described on page 96 ff. a graph of log survivors: time being plotted in the usual way.

(*b*) Some more ampoules of the spore suspension prepared at the same time are placed without undue delay in cans of the food product to be processed. Each ampoule is located in place within a can at the "heating centre" using a suitable cradle (e.g. of thin but rigid wire).

(*c*) After processing, the ampoules are recovered and the counts of surviving spores determined, and the average count calculated.

(*d*) The average count obtained in (*c*) is matched on the curve obtained in (*a*) to find the time of heating at 121°C which is *equivalent* to the entire heating and cooling of the cans during the heat processing under investigation.

It should be noted that the accuracy of this determination depends on the ampoule cradle providing an insignificant change in the heat transfer characteristics. In addition since the effect of heat on bacteria may be modified by other environmental parameters (e.g. pH, concentration of protein, fat or sugar) the suspending liquid used in the ampoules should be carefully chosen to match as far as is practicable the foodstuff being canned. One of the advantages of this technique over inoculated pack studies is that the bacterial suspensions being examined in the laboratory after the two types of heat treatment are identical in all respects except for the heat treatments.

U. Water

1. Introduction

Most microbiological examinations of water samples are carried out on water supplies or proposed water supplies to test the potability of the water. The bacteria found in water are mainly of three types: natural aquatic bacteria, soil dwelling organisms, and organisms that normally inhabit the intestines of man and other animals. Most of the bacteria which are normally aquatic are Gram negative (including *Pseudomonas*, *Flavobacterium*, *Cytophaga*, *Acinetobacter* and *Chromobacterium*) although a few Gram positive bacteria (coryneform bacteria, *Micrococcus* and *Bacillus*) may be found. Although some aquatic bacteria are extremely difficult to cultivate, most will be capable of growing on very dilute media, for example CPS medium (Collins *et al.*, 1973) or pond water agar. Many of these organisms will not grow on standard nutrient agar or plate count agar. Such bacteria usually have an optimum temperature for growth of 25°C or less, and plates should be incubated for 14 days. Counts on nutrient agar or plate count agar will usually be only one tenth of counts on CPS medium. The soil dwelling bacteria which may be washed by rain into streams, ponds, etc., include species of *Bacillus*, *Streptomyces* and saprophytic members of the Enterobacteriaceae such as *Enterobacter*. Most of these organisms will have an optimum temperature for growth around 25°C and will be capable of growing on nutrient agar or plate count agar.

The expected viable counts on CPS medium depend upon the nature of the water sample. Unpolluted rivers may show comparatively low counts of up to around 100 per ml, although after heavy rain the run-off entering the river will contain large numbers of soil dwelling organisms, but counts will also depend on the trophic state of the river. Downstream of towns counts will be much larger, partly because of the discharge of sewage and industrial effluents into the rivers. Water taken from the river at or immediately downstream of the town may contain more than one million bacteria per ml, but more usual counts for polluted river water are between about 10,000 and 200,000 per ml. Lakes and reservoirs may exhibit counts of 100 per ml or less when unpolluted although this depends on the trophic state. Unpolluted waters from deep wells in good condition normally have counts of 100 per ml or less, but if contamination is able to occur (e.g. by seepage through a cracked well-casing) the counts will be much higher although still usually below 10,000 per ml. Generally counts on plate count agar will be about one tenth of those given above, although not such a great difference in counts may be seen in highly polluted waters. Although high viable counts are usually indicative of both contamination of the water and the presence of bacteria other than aquatic bacteria, they do not necessarily indicate pollution by faeces, untreated sewage, etc., and the bacteria may be mostly soil saprophytes. Thus, a high viable count alone is not evidence that a water supply is potentially dangerous due to the possible presence of intestinal pathogens, but water supplies with high viable counts are nevertheless undesirable since they may contribute to food spoilage problems.

Faecal pollution of water gives rise to the presence of organisms derived from the intestine, including *Escherichia coli*, *Streptococcus faecalis*, *Clostridium perfringens*, and possibly intestinal pathogens such as *Salmonella* and *Vibrio cholerae*. In certain circumstances enteric pathogens may survive in water for comparatively long periods (see Geldreich, 1972).

For the purpose of determining the potability of a water supply, it is necessary to establish that the water is not contaminated with pathogenic micro-organisms such as those named. Since intestinal pathogens, if present, would be greatly outnumbered by normal intestinal inhabitants such as *Escherichia coli*, *Streptococcus faecalis* and *Clostridium perfringens*, it is more satisfactory to examine the water for the presence of the latter organisms, which, if found, indicate that pathogens could be present. The variety of *E. coli* which is primarily intestinal is *E. coli* type 1, and the presence of this bacterium in water in temperate climates indicates that recent faecal pollution has occurred. *Strep. faecalis* is not as efficient an indicator as *E. coli* since it is usually present in the intestine in smaller numbers than *E. coli* and it dies as quickly as *E. coli* in water. Nevertheless, occasionally *Strep. faecalis* may be predominant, since in a few animals *Strep. faecalis* may outnumber

E. coli. *Clostridium perfringens* is capable of surviving in water for longer than *Strep. faecalis* and *E. coli* and, in the absence of these two organisms, serves as an indicator of remote faecal pollution.

In addition to examining for the presence of these intestinal organisms, the standard method for the bacteriological examination of water supplies (Report, 1969) includes an estimation of general viable counts at 22°C and 37°C as an indication of the level of general contamination, e.g. by surface waters or from soil.

The procedure described below is based upon that described in the above report of the Public Health Laboratory Service Standing Committee on the Bacteriological Examination of Water Supplies, for the determination of the potability of a water supply. It should be realised, however, that other procedures may be required if the water is to be used for purposes other than drinking. For example, in the case of water being used by a food manufacturer, the numbers and types of psychrotrophic bacteria present may be of great importance.

2. Sampling procedure

Water samples are best collected in sterile wide-mouthed bottles with dust-proof ground glass stoppers. In the case of chlorinated water samples, the sample bottles should contain 0·1 ml of a 2 per cent solution of sterile sodium thiosulphate for each 100 ml of water sample to be collected.

Great care should be taken during the sampling procedure to prevent contamination of the sample. The sample bottle should be completely filled at the time of sampling. If the sample is to be taken from a tap, the outside and inside of the tap nozzle should be cleaned thoroughly, after which the tap is turned on for a few minutes. The tap should then be turned off and heat-sterilised using, for example, an alcohol lamp. The tap is cooled by allowing water to run to waste for a minute or two, after which the sample bottle is filled with water. It should be noted that this method of sampling ensures that the bacteriological quality of the water supply delivered to the tap is tested, but if the reason for the bacteriological examination is to trace the source of contaminating organisms it would be advisable in addition to take either an initial sample from the tap before the sterilisation procedure or a swab of the inside and outside of the nozzle to determine the possibility of contamination of the tap itself. The results of a single sample of a water supply are of limited value since often contamination may be intermittent.

In collecting water samples from reservoirs, ponds, wells, rivers etc., it is advisable to use some sterilised mechanical device for holding the bottle and removing the stopper, and, in the case of obtaining samples from moving water, the mouth of the bottle should be directed against the current.

The forms of apparatus suitable for collecting deep water samples have been reviewed by Collins *et al.* (1973). If such a device is unobtainable and cannot be improvised, extreme care should be taken in the manual collection of a sample so that no water is collected which may have contacted the hands.

3. Cultural examinations

Since the sample bottle will be completely full, the sample should be mixed by inverting the bottle 25 times with a rapid rotary motion, then about a quarter of the contents poured away and the bottle shaken vertically 25 times with an excursion of 30 cm. Prepare aseptically serial decimal dilutions up to 10^{-2} in sterile quarter-strength Ringer's solution.

Carry out the following counts.

(*a*) *General viable counts*

Use CPS medium and plate count agar, incubated at 25 and 37°C for 7–14 and 2 days respectively.

(*b*) *Coliform counts by the multiple tube technique*

Use 5 (or 3) bottles or tubes at each of the dilutions 10^2, 10^1, 10^0 and 10^{-1}. Quantities of water greater than 1 ml are inoculated into *double strength* medium equal in volume to the amount of water inoculated. If the water sample is suspected of being highly polluted, dilutions higher than 10^{-1} should be tested setting up 5 (or 3) tubes at each dilution. It has been suggested (Report, 1969) that MacConkey broth is not satisfactory as a medium because of the variability in the composition of bile salts and peptone and a complex chemically defined medium (formate lactose glutamate broth) was recommended. However, the present authors feel that the advantage of comparability of results obtained on water samples with those obtained on food samples (bearing in mind that low dilutions of food may modify the characteristics of the medium) is probably great enough for the food microbiologist to use identical media for these examinations and we suggest standardization on one of the media already described (see page 142) which have been recommended by the IAMS Committee (Thatcher and Clark, 1968).

Incubate at the appropriate temperature, and record the presumptive coliform count. Subculture each positive tube into a further lactose broth to be incubated at 44·5°C for an Eijkman test, and also into a peptone water also incubated at 44·5°C which is tested for indole production after incubation (see page 143). *Escherichia coli* type 1 gives gas from lactose at 44·5°C, and is capable of producing indole at 44·5°C. Thus an *E. coli* type 1 count can also be recorded.

(*c*) *Faecal streptococci by the multiple tube technique*

Use 5 (or 3) bottles or tubes at each of the dilutions 10^2, 10^1, 10^0, and 10^{-1}. Quantities of water greater than 1 ml are inoculated into double strength medium equal in volume to the amounts of water being inoculated. Incubate the inoculated glucose azide broths at 37°C for 72 h. Record those tubes which are positive (that is, those in which acid is produced), and confirm by subculturing a loopful from each positive tube into a fresh single strength glucose azide broth and incubating at 45°C (see page 144).

(*d*) *Detection or counting of* Clostridium perfringens

Since, in faeces, the ratio of *Cl. perfringens* to *Escherichia coli* is very low, the detection of *Cl. perfringens* in a water sample is an insensitive method of examination for recent contamination compared with the presumptive coliform count. The value of the tests for *Cl. perfringens* lies in the fact that the spores are able to survive in water for comparatively long periods, and therefore their presence in water free from *E. coli* can be taken as some indication of remote faecal pollution. Three methods of detection may be used.

(1) *Litmus milk method:* Add 200 ml of water sample to 400 ml of sterile litmus milk (heated to 100°C in a steamer to drive off dissolved air and then cooled, immediately before inoculation) in a container of about 1000 ml capacity. (A large container must be used to allow for blowing without risk of contaminating the incubators.) Heat in a water bath to 80°C for 10 min. The timing of this heating is best established by placing in the water bath an exactly similar container (with the same type of closure) in which is 600 ml of water and a thermometer inserted so that the bulb of the thermometer is not in contact with the container wall. The moment when the contents attain 80°C can thus be established. Cool, cover the surface of the milk with melted sterile "vaspar", and incubate at 37°C for 5 days. Examine every 24 h for the production of acidity, clotting and gas – the reaction known as a "stormy clot"—which indicates the presence of *Cl. perfringens*.

This test can be made semi-quantitative by putting up a range of volumes of the water sample and estimating the most probable number, using probability tables. It should be noted, however, that the number of *Cl. perfringens* present will be very low so that normally *Cl. perfringens* would not be detectable in quantities of less than 100 ml. Therefore, the amounts to be tested should be, for example, 200 ml, 100 ml and 50 ml; thus specially designed probability tables may be required.

(2) *Multiple tube technique:* using differential reinforced clostridial medium, followed by confirmation (see page 156).

(3) *Wilson and Blair's sulphite medium:* heat to boiling 20 ml of Wilson and Blair's medium (dispensed in a Miller-Prickett tube) and cool to 50°C.

Add 20 ml of water sample (previously heated to 80°C for 10 minutes to destroy vegetative forms of bacteria), also at 50°C, mix well and allow to set. Incubate aerobically at 37°C for 24 h. After incubation, examine for the presence of black (i.e. sulphite-reducing) colonies. The two main disadvantages of this method are that comparatively small amounts of water are tested, and that it is difficult to remove inocula from sulphite-reducing colonies to confirm them as *Cl. perfringens*.

4. Membrane filtration methods

General viable counts, coliform counts and *Strep. faecalis* counts can be carried out by using membrane filtration. It has already been mentioned that one advantage of membrane filtration is that small numbers of organisms can be detected, since the amount of water passed through the membrane is restricted only by the amount of gross suspended matter present in the water. (See pages 32–34 in Part I, for the basic methodology.)

General viable counts can be carried out on a water sample by using membrane-type nutrient broth or membrane-type tryptone soya broth, and incubating the filters at 35–37°C for 18–24 h or at 25°C for 2 days. For counts of *E. coli* use Millipore type HC filters. Incubate the filters for 2 h at 37°C on pads saturated with membrane-type nutrient broth, and then transferring the filters to pads saturated with membrane-type MacConkey's broth and incubating for 18 h at 44°C (Taylor, Burman and Oliver, 1955). By this means, *E. coli* type 1 counts can be determined within one day. Similarly, *Strep. faecalis* counts should be determined with pre-selective incubation of the filters on nutrient broth for 2 h at 37°C, before transferring the filters to pads saturated with glucose azide broth and incubating at 45°C for 18 h.

5. The special requirements for food manufacture

It has already been mentioned that the food manufacturer may have more rigorous microbiological requirements of a water supply than that it is potable. For example the presence of saprophytic psychrotrophs and other aquatic organisms has significance for spoilage of refrigerated or other foods. In order to get sufficiently low counts, additional chlorination at the factory may be necessary, e.g. in the case of cooling water in canning factories (water chlorination in canneries is frequently obligatory since a potable water supply is not often used for cooling purposes).

Another type of micro-organism which may be occasionally of significance is the *Sphaerotilus- Leptothrix-Gallionella* groups of sheathed and appendaged organisms. Some of these, in addition to some strains of *Pseudomonas* and *Acinetobacter*, may cause trouble by blocking water distribution systems within the factory, and some may even cause spoilage troubles such as haze formation and the development of sliminess in low-pH fruit-flavoured liquid products such as those designed for domestic production of "ice lollies". For isolation and identification of these organisms refer to Bergey's Manual (Buchanan *et al.*, 1974), Collins (1964) and Mulder (1964).

V. The Examination of Food Processing Plant

A preliminary inspection of the premises, preferably during processing, is an invaluable aid in assessing the microbiological significance of general organisational procedures, particular patterns of layout and the processing methods used. An opportunity should also be taken at this stage to note the general condition of the equipment, and the presence of any residues, film or scale. These observations should be recorded so that they may be correlated later with the results of any subsequent tests.

Information on the microbiological condition of the equipment can also be obtained by comparing the results on samples (taken at suitable points) of the first product passing over, and in effect rinsing, the equipment since the last cleaning and sterilising process. This procedure is applicable to farm and creamery conditions, and to food processing equipment carrying liquids (e.g. pipelines carrying soups, sugar syrups, etc.). The samples obtained should be examined for general viable count and coliform count, and the presence of other organisms should be tested for as appropriate (see the relevant section).

When detailed information is required on a particular utensil or reasonably small piece of equipment, the rinse method is suitable, for example in the examination of churns and cans, bottles, and cartons. It may also be used for the examination of pipeline installations by drawing an appropriate volume of rinse through the system.

For the examination of defined areas of large pieces of equipment, e.g. vats, or for assessing the microbiological condition of equipment or parts of equipment where the rinse method is not applicable (e.g. conveyor belts and cutting blocks), the swab method is suitable. The principal methods of examining surfaces have already been described in the introductory section of Part II (see pages 124–127).

In all cases where chemical sterilising agents are known to have been used, the quarter-strength Ringer's solution of the rinse or swab should contain a supplement of an appropriate inactivator. For hypochlorites and iodophors, sodium thiosulphate is used and incorporated in the rinse at 0·05 per cent final concentration. For quaternary ammonium compounds a mixture of 4 per cent lecithin and 6 per cent Cirrasol ALN-WF in water is incorporated to give a final concentration of 1 per cent of mixture (Cirrasol ALN-WF is obtainable from Honeywill-Atlas Ltd.).

The methods for determining the numbers and kinds of micro-organisms collected in the rinse and swab diluents usually involve a colony count on a non-selective medium and tests for the presence of coliforms and other groups using appropriate selective media. It is also possible to determine numbers of thermoduric organisms by laboratory pasteurisation of the diluent, but in this case it is advisable to add a supplement of sterile separated milk prior to heat treatment, e.g. 5 ml of sterile milk to 5 ml of diluent as recommended by British Standard, (1968).

An alternative method of examining the rinse and swab diluents is to use membrane filtration (see pages 32–34). This has the advantage that larger volumes of diluent can be examined than by traditional methods.

Since the proportion of micro-organisms recovered from surfaces is greatly influenced by the extent of the rinsing and swabbing process, it is necessary that procedures should be carefully standardised if comparative results are to be obtained. For example, the Ministry of Agriculture, Fisheries and Food (Bulletin, 1968) and British Standard (1968) prescribed the methods to be used for examining farm dairy equipment: in both cases the non-selective medium recommended for viable counts was yeastrel milk agar (yeast extract milk agar).

1. Examination of processing plant, equipment, working surfaces, etc.

(a) By swabs

In testing any piece of equipment more than one area should be swabbed, paying particular attention to those points which are difficult of access for cleaning (e.g. valves and junctions of pipelines that have been cleaned in place).

(i) *Preparation of swab.* Cotton-wool or alginate-wool swabs may be used (see page 125), or alternatively larger swabs prepared from unmedicated ribbon gauze. These are prepared by winding a 15 cm length of 5 cm wide ribbon gauze on to a 30 cm long stainless steel wire support. The wire support is made from 35 cm of stainless steel wire of about 3 mm diameter, formed into a loop at one end and notched at the other to hold the gauze without slipping. Cotton-wool or gauze swabs should be placed in alloy or stainless steel tubes containing a known volume of quarter-strength Ringer's solution (containing an inactivator if a chemical sterilant has been used on the equipment).

(ii) *Swabbing.* Whenever possible a minimum of 100 sq cm should be swabbed. Before swabbing, press the swab against the side of its container to express excess liquid. Swab the surface as described on page 126, i.e. by rubbing firmly over the surface using parallel strokes with slow rotation of the swab, and swabbing a second time using parallel strokes at right angles to the first set. Return the swab to the tube.

(iii) *Examination of swabs*. This should be completed as soon as possible, and in any case within 6 h of sampling. Examine alginate swabs by using Calgon Ringer's solution as described on page 126. In the case of cotton-wool or gauze swabs, after not less than 5 min contact between the swab and the Ringer's solution, mix by twirling the swab vigorously in the Ringer's solution 6 times. After thorough mixing prepare any dilutions thought to be necessary. Set up general viable counts on plate count agar (or yeast extract milk agar if so specified by a standard procedure) and incubate at 30°C for 3 days. Examine also for the presence of coliforms and any other groups of significance; for example, in the case of equipment carrying hot gelatin stock for meat pies examine for the presence of thermophilic spore-bearers.

(iv) *Interpretation of results:*

General viable count per sq cm	Conclusion
Not more than 5	Satisfactory
5–25	Requires further investigation
More than 25	Highly unsatisfactory – requires immediate action

Coliform counts: equipment used for carrying, dispensing or holding heat-treated foods should bear less than 10 coliform bacteria per 100 sq cm. No coliform bacteria found on 100 sq cm can be regarded as satisfactory.

(*b*) *By rinses*

Reasonably small pieces of equipment, e.g. buckets, can be examined using 500 ml of quarter-strength Ringer's solution. When a chemical sterilant has been used, the rinse liquid should contain the correct inactivator (see page 231). When cleaned-in-place pipeline systems are examined by a rinse method, much larger quantities of sterile rinse liquid are required, which are pumped round the system and then aliquot quantities taken for examination.

In farm dairies, milking machines can be examined by either static or pulsating rinse procedures. A static rinse is carried out by passing the rinse solution twice through the teat cups and long milk tube into the original container. A pulsating rinse is carried out in a similar way but with the machine pump operating so as to pulsate the teat cup liners. The advantage of this latter method is that it is more closely related to actual milking conditions.

Rinses should be tested as soon as possible and in any case within 6 h of sampling. Transport temperatures and conditions should be regulated and recorded.

Examine rinses for general viable counts, coliform counts and any other appropriate selective counts. Counts on rinses of small pieces of equipment should be calculated per utensil. Standards to be set will obviously depend on the individual situation, type of equipment, etc., and it is recommended that standards should be determined by prior survey of counts attainable under model cleaning conditions.

(*c*) *By agar sausages and impression plates*

See page 127 for methods of use. These techniques have a relatively limited range of uses in the examination of equipment as they can only be used on flat or nearly flat surfaces (e.g. conveyor belts, inside surfaces of rectangular vats, etc.). They have the virtue of providing a record of the *position* of viable organisms relative to any observed surface defects, corrosion, etc. A further advantage is that in surveys of processing plant remote from the laboratory transport and storage problems are obviated.

2. Examination of Washed Bottles and Food Containers

Bottles should be selected for examination immediately after washing, closed with a sterile rubber bung and examined as soon as possible, and in any case within 6 h. Other food containers should be closed or covered to prevent contamination; they may be examined by a procedure similar to that for bottles described below.

(*a*) *Rinses*

Add 20 ml of sterile quarter-strength Ringer's solution containing 0·05 per cent sodium thiosulphate to the bottle and replace the bung. Hold the bottle horizontally in the hands and rotate gently 12 times in one direction so that the whole of the internal surface is thoroughly wetted. Allow the bottle to stand for not less than 15 and not more than 30 min and again gently rotate 12 times so as to wet the whole of the internal surface.

Prepare general viable counts using an appropriate medium; for example in the case of milk bottles use plate count agar or yeast extract milk agar, in the case of bottles for fruit squashes or juices use orange serum agar, and so on. Also perform a coliform count by the multiple tube technique. Incubate plates at 25–30°C for 3 days. Record results as the colony count or coliform count per bottle.

(*b*) In situ *culture of contaminants*

An alternative method of examining washed clear glass bottles for the presence of contaminants is to add molten agar medium to the bottle, in sufficient volume to form a thin (5 mm) layer over the inside surface. The bottle is rolled slowly horizontally until the agar sets on the inside surface (this can be hastened by rolling the bottle in a shallow dish of iced water).

The bottle should be incubated, bung down, in the same way as a roll tube. *In situ* culture is also very successful for demonstrating the presence of micro-organisms on cracked and chipped crockery etc. (see also Angelotti and Foter, 1958, Favero *et al.*, 1968).

(*c*) *Cans and churns*

Larger food containers such as churns and cans used for transport and storage of milk, cream, sugar syrups, fruit syrups, etc., may be examined by a rinse method employing 500 ml of sterile quarter-strength Ringer's solution.

(i) *Selection and visual examination.* Cans and churns should be examined not less than half an hour and not more than 1 h after washing. First a visual inspection should be made and the general condition recorded, in particular: (*a*) bad dents, rusting, open seams or poor lids, (*b*) presence or absence of film, scale or food or milk solids, (*c*) degree of wetness of the churn—recorded as dry, moist or wet (with an obvious pool of water at the bottom of the churn or can).

Cans and churns containing turbid water or easily removable food or milk solids, as distinct from film or scale, can be reported as unsatisfactory without testing.

(ii) *Rinsing.* Pour 500 ml of sterile quarter-strength Ringer's solution over the inside of the lid into the can. Replace the lid, lay the can on its side, and roll it to and fro through 12 complete revolutions. Allow the can to stand for 5 min, and repeat the rolling. Pour the rinse solution from the can into the lid and then into the original sterile container.

(iii) *Testing the rinses.* Rinses should be tested as soon as possible, and in any case within 6 h of sampling. Mix by inverting the container slowly 3 times. Pour 1-ml and 0·1-ml plates using plate count agar, yeast extract milk agar, or other appropriate medium, and incubate at 30°C for 3 days. Test also for the presence of coliforms by a multiple tube technique, using tubes at 10^1, 10^0 and 10^{-1} dilutions (double strength media are used for inocula exceeding 1 ml). Yeast and mould counts should also be performed on rinses of churns and cans used for sugar syrups, fruit syrups, etc., using Davis's yeast salt agar. Record the results per can.

(*d*) *Interpretation of results*

(i) *Bottles and small containers*

Viable count per bottle	Classification
Not more than 200	Satisfactory
200–1000	Cleaning procedure needs improvement
Over 1000	Unsatisfactory – take immediate action

(ii) *Churns and cans*

Viable count per churn or can	Classification
Not more 10,000	Satisfactory
10,000–100,000	Cleaning procedure needs improvement
More than 100,000	Unsatisfactory – take immediate action

N.B. A wet can is degraded to the next class below as it is unlikely to remain in a satisfactory condition.

Rice (1962) has suggested that milk containers should not show more than one colony per ml of container capacity.

3. Air

Occasionally it may be useful to examine the air in food factories, dairies, etc., for the presence of specific organisms. This is particularly so in the case of mould contamination problems, since fungal spores are very readily distributed through the air. If the source of a known and identified contaminant is being sought, it may be sufficient to use simple air exposure plates. Plates of already poured and set media are exposed in a variety of locations for 15 min. The number of colonies developing on one plate following a 15 min exposure period represent the number of particles carrying that type of organism settling on approximately 0·1 metre^2 per minute. The variation in counts of the organisms being traced is likely to indicate the focal points of the contamination. A second visit to the factory may then be sufficient to locate the source and to suggest possible remedial measures.

It is not often necessary in this type of work to require the use of a slit sampler (C. F. Casella and Co.) although such sampling devices are obviously needed to count absolute concentrations of micro-organisms in the air. Membrane filtration may also be used—air sampling apparatus is available from the manufacturers of membrane filters.

A further environmental sampling procedure which can conveniently be used following the air sampling is the sweep plate technique as described by Cruickshank (1975). Plates of culture media similar to those used for air sampling are exposed face down on the test surface (e.g. work surface) and are rubbed to and fro ten times over a distance of 30 cm. The lid is then replaced and the plate incubated with the air exposure plates. Although sweep plates obviously yield essentially qualitative information, the technique is particularly suitable for demonstrating reservoirs of organisms on infrequently cleaned surfaces and in demonstrating improvements following a satisfactory cleaning routine.

4. Detection of *Salmonella* in Processing Plant Effluents

In investigations of food processing plant or premises where a particular hazard of *Salmonella* contamination exists (e.g. poultry or meat processing plants) suitable samples can be obtained by the use of sewer swabs installed at convenient points in drains and gulleys (Harvey and Phillips, 1961; Harvey and Price, 1970, 1974; Georgala and Boothroyd, 1969). A 2 m length of string is attached to the swab (which may be folded gauze or a sanitary towel) and the whole wrapped in paper and sterilised in the autoclave. On site, the swab is suspended in the chosen drain or outlet, secured with the string and left in position for 1–7 days. Alternatively the swab may be used to wipe the drain or gulley surface and immediately removed to the laboratory. Extending the exposure time may increase the possibility of picking up salmonellae, but it will also increase the extent of contamination by other organisms, and in general the longer the exposure time, the longer the incubation time required in the liquid enrichment stage of isolation (and in the case of heavily soiled swabs a 2-stage liquid enrichment may be required before plating).

After transporting the swab to the laboratory, liquid enrichment cultures are set up using selenite broth, etc. In the method of Patterson (1969) the swabs are transferred to a sterile plastic bag, a corner of which is cut off and the absorbed fluid in the swab then expressed and collected in a sterile container. (Sterile sodium thiosulphate solution can be added to neutralise any residual chlorine.) Ten ml of the fluid are added to each of 90 ml of selenite broth, tetrathionate broth and Hajna's GN broth (see page 151). An alternative procedure recommended by Harvey and Price (1974) is to place the entire swab in a wide-mouthed jar (or to leave it in the jar which is used for transport if this is suitable) and to add 750 ml of selenite broth. Although this does not allow the use of more than one enrichment medium per swab, Harvey (1965) considered that the culture from the entire swab increased the likelihood of isolating *Salmonella*.

The liquid enrichment cultures should be incubated and then used for isolation and identification procedures as already described (pages 152 ff.).

The use of sewer swabs in this way enables premises to be screened for the possibility of the existence of a *Salmonella* problem more quickly and more easily than by the examination of many end-of-line samples. In the event of salmonellae being found a return visit can be paid and detailed samples taken in order to attempt to detect the source of the contamination.

PART III

SCHEMES FOR THE IDENTIFICATION OF MICRO-ORGANISMS

SCHEMES FOR THE IDENTIFICATION OF MICRO-ORGANISMS

In the schemes that follow, methods are given for the identification of the micro-organisms most commonly isolated by, or of importance to, microbiologists in the food and dairy industries.

The diagnostic keys are separated into: A. A scheme for the identification of Gram-negative bacteria; B. A scheme for the identification of Gram-positive bacteria; and C. A scheme for the identification of yeasts and moulds. Certain fastidious and/or anaerobic animal pathogens have not been considered; for further details of such organisms see Buchanan *et al.* (1974), Cowan (1974), Stableforth and Galloway (1959) and Graham-Jones (1968).

The methods for identification in these schemes have been put forward on the assumption that the use of selective and diagnostic media has not already enabled a tentative identification—in these cases fewer tests may be required. For example, the screening with Kohn's 2-tube media for presumptive *Salmonella* obtained on media such as desoxycholate citrate agar has already been described. Sequential identification methods have been used because they employ a minimum of tests. However, as already appreciated by all who have attempted to use a sequential key for "identifying" a flowering plant whose identity is already known, and as reiterated by Sneath (1974), such sequential keys are very sensitive to error. Therefore when the identity has been determined, the general description of the genus or the species should be checked against the observed characteristics of the isolate. Some of the organisms described in Section B require special media for their isolation and in such cases it is obvious that choice of primary isolation media may allow a restriction in the identification procedures to those employed for the group likely to be isolated, but the possible hazards of wrong identification should be fully realised.

Difficulties in identification may also be caused by the Gram-staining reaction: for example *Acinetobacter* may stain purple, whereas old cultures of *Bacillus* or young cultures of *Arthrobacter* may be Gram-variable or Gram-negative. In addition, many biochemical tests used for classification and identification show poor reproducibility within a laboratory or between different laboratories (see for example Sneath and Collins, 1974). Eventually statistical analyses of tests, such as those reported in the reference cited should lead to the selection and use of the most reliable test methods, to be used both for classification and for identification.

The symbols used in the diagnostic tables (unless otherwise indicated) are as follows:

+	Positive reaction (usually in 1 or 2 days)
—	No reaction, or no growth
V	Variable in reaction, depending on strain, biotype or variety
V(+)	Variable, most are positive
(V−)	Variable, most are negative
(+)	Slowly developing positive
×	Late and irregularly positive
±	Slightly or weakly positive
?	Test not applicable or significant, or reactions not fully determined.

DIAGNOSTIC TABLES FOR GRAM-NEGATIVE BACTERIA

SECTION 1. GRAM-NEGATIVE BACTERIA WHICH CAN GROW ON NUTRIENT AGAR OR PLATE COUNT AGAR

(After Buchanan *et al.*, 1974; Edwards and Ewing, 1972; Park and Holding, 1966;
Report, 1958a; Ewing and Edwards, 1960)

The organism to be identified should be examined for its ability to produce a pigment when grown on nutrient agar. It should also be tested to determine its mode of utilisation of glocose. When these two characters have been determined, appropriate tests can be selected to complete the identification of the organism to generic level. In the case of tests not already described in the Manual, methods are given at the end of this section in the order in which they occur in the diagnostic scheme.

Production of pigment on nutrient agar. Streak the test organism on a nutrient agar plate. Incubate for 3 days at 25–30°C and examine for evidence of pigment production.

Mode of utilisation of glucose. Inoculate the test organism, by stabbing with a straight wire, into Hugh and Leifson's medium with and without glucose (see page 73). In some instances the modified medium described by Board and Holding (1960) may be more suitable. The original method of the Hugh and Leifson test requires two tubes of media + glucose to be inoculated and one to be sealed with sterile liquid paraffin or vaspar, but the latter tube may not be necessary provided that the tube is examined frequently, e.g. daily for 7 days, especially if the double-indicator version of the medium is used (see page 342). The result obtained will be one of the following:

(*a*) Oxidation of the glucose, resulting in acid production at the surface of the medium, although, on prolonged incubation, the acid reaction may later spread downwards through the tube.

(*b*) Fermentation of the glucose, resulting in acid production uniformly throughout the medium along the entire length of the stab, with or without gas production. The reaction is first seen in the medium immediately surrounding the stab, and spreads rapidly throughout the tube. On prolonged incubation pH reversal may occur.

(*c*) Inability to utilise glucose with either no change in the colour of the medium or the production of alkali at the surface of, or throughout, the tube.

Interpretation of Results of the Two Tests and Further Tests Required

Division 1: bacteria producing an intracellular (non-diffusible) purple pigment

Chromobacterium organisms produce a pigment of this type; tryptophan is required for pigment production. The violet pigment is soluble in ethanol but insoluble in water; it becomes green in a 10 per cent ethanolic solution of sulphuric acid. *Chromobacterium* is catalase positive, indole negative, and Voges-Proskauer negative; it can reduce nitrate. There are two species recognised: *C. violaceum* grows at 37°C but not at 4°C, whereas *C. lividum* grows at 4°C but not at 37°C.

Division 2: bacteria producing an intracellular (non-diffusible) yellow or orange pigment

Cytophaga may be either oxidative or fermentative, and consists of long thin, often curved rods. Organisms from actively growing cultures may show flexing movement when examined in hanging drop preparations. On nutrient agar, many strains form spreading colonies due to gliding motility (see Weibull, 1960, and Perry, 1973). Some strains show motility better on a less rich medium such as Hayes's medium. Most strains of *Cytophaga* hydrolyse starch, which is a characteristic relatively uncommon in the other soil bacteria that produce orange and yellow pigments. Many strains of *Xanthomonas* and some strains of *Flavobacterium* may also hydrolyse starch, however. *Flavobacterium*, *Xanthomonas* and the yellow- or orange-pigmented enterobacteria will not show marked spreading on nutrient agar or Hayes's medium nor display gliding motility. They can be separated by using the following tests.

Group or genus	Mode of attack on glucose	Formation of domed mucoid colonies on nutrient agar + 2% w/v glucose	Plant pathogen	Position of flagella	Xanthomonas pigment absorption spectrum	Growth at 40°C
Flavobacterium	Oxidative or no attack	—	—	Peritrichous	—	—
Xanthomonas	Oxidative	V(+)	V(+)	Polar	+	—
Pseudomonas	Oxidative	—	V	Polar		V(+)
Enterobacteriaceae	Fermentative	For further characterisation see page 248				

(See also Symposium on Myxobacteria and Flavobacteria, 1969; McMeekin, Patterson and Murray, 1971; McMeekin, Stewart and Murray, 1972.)

For further details of the characterisation of *Xanthomonas*, refer to Hayward (1966).

Division 3: bacteria producing an intracellular (non-diffusible) red pigment

Two types of Gram-negative bacteria produce a red, intracellular pigment. They are certain *Pseudomonas* species, which do not ferment glucose and *Serratia* species, which do ferment glucose. These two genera are therefore differentiated on the type of reaction in Hugh and Leifson's medium. In addition flagella staining can be used since *Pseudomonas* possesses polar flagella, whereas *Serratia* possesses peritrichous flagella. The red-pigmented methanol-utilising pseudomonads described by Anthony and Zatman (1964) are capable of growth on nutrient agar.

Division 4: non-pigmented fermentative bacteria

Bacteria of this type can be separated by using the following tests.

Group or genus	0/129 sensitivity	Curved, S-shaped and spiral cells. Many spherical cells in 7 day cultures	Oxidase test	Motility	Polar flagella	Thornley's arginine test	Gelatin hydrolysis	Starch hydrolysis	Luminescence
Vibrio	+	V(+)	+	+	+	−	+	+	−
Aeromonas	−	−	+	+	+	+	+	+	−
	−	−	+	−	−	+	?	?	−
Plesiomonas	V(+)	−	+	+	+	V	−	−	−
*Photobacterium**	+	−	V	+	+	?	×	−	+
*Lucibacterium**	−	−	+	+	−	?	+	+	+
Enterobacteriaceae‡	−	−	−	+	−	−	V(−)	V(−)	−
	−	−	−	−	−	V(−)	?	?	−

* *Photobacterium* produces gas from glucose, *Lucibacterium* does not; see also Hendrie, Hodgkiss and Shewan (1970). Most isolations of luminous bacteria have been made from marine environments in nutrient media based on sea water. They are inserted in this section because although they may have a nutritional requirement for constituents of sea water they are not halophilic in the sense of requiring high concentrations of sodium chloride, and they may be isolated and be luminescent on plate count agar plates incorporating low dilutions (i.e. high concentrations) of samples of marine origin.

‡ For further differentiation of the Enterobacteriaceae see page 248.

Morphology. Most *Vibrio* cultures at 18–24 h on nutrient agar or in nutrient broth consist of slender organisms which are curved, S-shaped or spiral. After 7 days these cultures contain many coccal forms. For photographs illustrating the morphology of *Vibrio* see Baker and Park (1975). Neither *Aeromonas* nor enterobacteria display this characteristic morphology though some of the organisms in a culture may be slightly curved.

Thornley's arginine test gives positive results with all *Aeromonas* strains. Some *Plesiomonas* strains are positive (Eddy and Carpenter, 1964) but positive results are very rare in other fermentative Gram-negative bacteria (see page 69).

Non-motile strains which are oxidase positive are classified as *Aeromonas* while those which are oxidase negative are classified as enterobacteria.

Additional tests. The decarboxylase tests of Møller (1955) can be used for further differentiation between *Vibrio* and *Aeromonas*.

	Lysine	Ornithine	Arginine
Vibrio	+	+	−
Aeromonas	−	−	+

Additional features that can be used to distinguish between *Aeromonas* and enterobacteria are given by Eddy (1960b).

The transfer of marine vibrios including *Vibrio parahaemolyticus* to the genus *Beneckea* has been proposed (Baumann, Baumann and Mandel, 1971; Baumann, Baumann and Reichelt, 1973).

Division 5: bacteria producing a yellow, green-yellow or green pigment which diffuses into the medium and non-pigmented bacteria that do not ferment glucose

Production of yellow, green-yellow or green pigments which diffuse into the medium is a property of many, but not all, *Pseudomonas* spp.

	Glucose oxidised	3-Ketolactose produced	Inorganic N used as sole N source	Coccal morphology	Tendency to retain Gram's stain	Thornley's arginine test	Fluorescent, diffusible pigment	Motility in a hanging drop	Polar flagellation	Penicillin sensitivity	Oxidase test
Agrobacterium	+	V(+)	+	−	−	−	−	+	−	−	+
Pseudomonas	V	−	V(+)	−	−	V(+)*	V	+	+	−	V(+)
Acinetobacter	+	−	V(+)	+	±	V	−	−	−	−	−
Moraxella	−	?	V(−)	+	±	?	−	−	−	+	+
Alcaligenes	−	−	V(+)	V	−	V(+)	−	+	−‡	−	+

* Plant pathogenic pseudomonads are often arginine negative, as also are *Alteromonas* spp. (Baumann, Baumann, Mandel and Allen, 1972).

‡ *Acinetobacter* is degenerately peritrichous, with only 1–8 flagella, and will thus often appear to be polarly flagellate.

For further details of *Agrobacterium* and its interrelationships with other genera, refer to De Ley *et al.* (1966) and Graham (1964). The genus *Acinetobacter* consists of coccobacilli, some of which may retain the crystal violet-iodine complex of the Gram strain, and so may resemble Gram-positive cocci except that division only occurs in one plane so that groups of organisms do not occur, and occasional short rods can be seen in the cultures. In earlier published work, these organisms were sometimes called *Achromobacter*. The scheme that arose from the studies of Baumann, Doudoroff and

Stanier (1968) has been used as the basis for this table for reasons of simplicity. However, it should be noted that this classification and taxonomy is not in accord with that proposed by, for example, Thornley (1967). Henriksen started a recent review (Henriksen, 1973) with the observation that "The literature about *Moraxella*, *Acinetobacter* and the Mimae is very extensive and, at least to the uninitiated, highly confusing". The problem of the classification of this group has not yet been resolved (Buchanan *et al.*, 1974).

For further details of *Pseudomonas*, refer to Stanier, Palleroni and Doudoroff (1966) and Palleroni and Doudoroff (1972). Species identification within *Pseudomonas* can be achieved by tests for utilisation of single sources of carbon. Rosenthal (1974) has described a simple disc assay method for performing these tests.

Differentiation of Members of the Enterobacteriaceae

The members of this family are rod-shaped bacteria, motile with peritrichous flagella, or non-motile; they ferment glucose rapidly with or without gas production. With the exception of certain strains of *Erwinia* and *Pectobacterium* they reduce nitrate to nitrite.

The diagnostic key which follows is based upon the descriptions given by Buchanan *et al.* (1974); Edwards and Ewing (1972); Ewing and Edwards (1960). It is designed to enable further characterisation of members of the Enterobacteriaceae which have been identified as such within Divisions 2 and 4. Pink or red pigmented strains of *Serratia* have already been distinguished within Division 3. Rapid tests for the characterisation of *Salmonella* and *Shigella* isolated on selective media have already been described on page 153, but such tests are not suitable for Gram-negative rod-shaped isolates obtained from media not selective for *Salmonella* or *Shigella*.

The tests in this section for differentiating the Enterobacteriaceae—with the exception of the test for the liquefaction of gelatin which is carried out at 20°C – should be carried out at an incubation temperature of 35–37°C unless otherwise indicated.

1. Isolated from diseased plant tissue . . . . *Division 6*
Not isolated from diseased plant tissue *2*
2. Phenylalanine deaminase positive . . . . *Division 7*
Phenylalanine deaminase negative *3*
3. Growth in Simmon's citrate agar at 37°C, 22°C or both temperatures *4*
No growth in Simmon's citrate agar . . . . *Division 8*
4. Methyl red positive . . . . *Division 9*
Methyl red negative . . . . *Division 10*

Division 6: plant pathogenic bacteria

This division includes the genera *Erwinia* and *Pectobacterium*, which can be distinguished from certain other members of the Enterobacteriaceae only on the basis of their plant pathogenicity. They are motile; may or may not reduce nitrate to nitrite; may or may not liquefy gelatin; and normally do not require organic nitrogen for growth. They frequently possess a cream, yellow or orange intracellular pigment. *Erwinia* causes dry necroses, galls or wilts. *Pectobacterium* produces a pectinase and causes soft rots. *Bergey's Manual* (Buchanan *et al.*, 1974) groups both these within the single genus *Erwinia*, and subdivides the genus into three groups of species which can de differentiated as follows:

	Lactose fermentation	Yellow pigment	Gas from glucose
Amylovora group	−	−	−
Herbicola group	V(+)	+	−
Carotovora group	V	−	V(+)

For species differentiation consult Buchanan *et al.* (1974).

Division 7: bacteria positive to the phenylalanine deaminase test

Organisms in this division are also motile when grown at 20°C (but may be non-motile when grown at 37°C), methyl red positive, and ferment neither lactose nor dulcitol.

Some *Proteus* (including *Pr. vulgaris* and *Pr. mirabilis*) are capable of swarming to produce concentric zones on the surface of agar media. Swarming can be inhibited in a variety of ways (e.g. see Smith, 1975).

	Fermentation of				Indole production	Production of hydrogen sulphide	Liquefaction of gelatin	Urease production	Ornithine decarboxylase
	Xylose	Mannitol	Inositol	Maltose					
Proteus vulgaris	V	−	−	+	+	+	+	+	−
Pr. mirabilis	+	−	−	−	−	+	+	+	+
Pr. morganii	−	−	−	−	+	−	−	+	+
Pr. rettgeri	V	+	+	−	+	−	−	+	−
Pr. inconstans	−	−	V	−	+	−	−	−	−

McKell and Jones (1976) have suggested that *Proteus mirabilis* and *Pr. vulgaris* are sufficiently similar that they should be retained in the genus *Proteus*, whereas the other phenylalanine deaminase positive bacteria are sufficiently different from the former to justify their being placed in a separate genus, *Providence*.

Division 8: bacteria that are phenylalanine deaminase negative, Simmon's citrate agar negative

Apart from *Klebsiella*, whose reactions vary in these tests, organisms in this division are also methyl red positive; they do not ferment adonitol; do not produce urease; and most cannot liquefy gelatin or utilise malonate.

	Gas from glucose	Fermentation of				Production of hydrogen sulphide	Motility	Voges-Proskauer test
		Lactose	Salicin	Inositol	Sucrose			
Shigella	—[1]	—[2]	—	—	V	—	—	—
Escherichia	+[3]	+ or ×[3]	V(+)	—	V	—	+[3]	—
Salmonella[4]	V	—	—	V(−)	—[5]	V(+)[6]	+	—
Edwardsiella	+	—	—	—	—	+	+	—
Klebsiella	V	V	+	+	+	—	—	V(+)

[1] Some strains of *Shig. flexneri* type 6 produce small amounts of gas.
[2] *Shig. sonnei* ferments lactose very slowly.
[3] Certain strains of *Escherichia* – sometimes known as the Alkalescens-Dispar group – do not produce gas from glucose, ferment lactose slowly or not at all, and are non-motile.
[4] *S. typhi*, some strains of *S. paratyphi* and a few other salmonellae.
[5] Sucrose-fermenting salmonellae have been described.
[6] *S. paratyphi* and certain rare types do not produce hydrogen sulphide.

The Alkalescens-Dispar group are very similar to *Shigella* but can be differentiated with Møller's lysine decarboxylase test. *Shigella* gives a negative result whereas members of the Alkalescens-Dispar group are usually positive. For species differentiation of *Shigella*, refer to Buchanan *et al.* (1974) and Report (1958a).

Some strains of *Escherichia* may produce a yellow or brown intracellular pigment. For further information on *E. coli*, particularly concerning its role in causing infections in domestic and farm animals, see Sojka (1965).

Cultures suspected of being *Salmonella* are best confirmed by serological methods using agglutination tests (see pages 101 ff.).

Division 9: bacteria that are phenylalanine deaminase negative, Simmon's citrate positive (at 37°C or 22°C or both), and methyl red positive

	Fermentation of			Production of hydrogen sulphide	Møller's lysine decarboxylase test	Møller's ornithine decarboxylase test
	Lactose	Salicin	Inositol			
Salmonella	V(−)	−	V	+	+	+
Citrobacter	+ or ×	V	− or ×	V	−	V
Klebsiella	V	+	+	−	V	−

Cultures believed to be *Salmonella* can be confirmed by the use of agglutination techniques (see pages 101 ff.). The salmonellae can be subdivided into subgenera on the basis of the following tests:

	Salmonella subgenus			
	I	II	III[1]	IV
Dulcitol fermentation	+	+	−	−
Malonate utilisation	−	+	+	−
β-galactosidase	−	×	+	−

[1] *Salmonella arizonae* (Arizona group).

The two species of *Citrobacter* are differentiated by the production of hydrogen sulphide:
C. freundii is hydrogen sulphide +, *C. intermedius* is hydrogen sulphide −.

Division 10: bacteria that are phenylalanine deaminase negative, Simmon's citrate positive (at 37°C or 22°C or both), methyl red negative

These organisms do not produce hydrogen sulphide; they are usually indole negative; Voges-Proskauer positive; they usually ferment sucrose and mannitol; all possess β-galactosidase, and all can utilise gluconate.

Tests should be carried out at 22°C as well as 37°C.

	Motility	Fermentation of lactose	Liquefaction of gelatin	Utilisation of malonate	Møller's ornithine decarboxylase test
Klebsiella	−	V	V(×)	V	−
Enterobacter	+	+	×	+	+
Serratia	+	−	+	−	+
Hafnia	+	−	−	V	+

The species of *Enterobacter* can be differentiated as follows:

	Gas from glycerol	Møller's lysine decarboxylase test	Møller's arginine decarboxylase test
E. cloacae	−	−	+
E. aerogenes	+	+	−

Tests Used: Description of Methods (see also Sneath and Collins, 1974)

Absorption spectrum of Xanthomonas *pigment*. The absorption spectrum of *Xanthomonas* pigment is characteristic and in critical studies the spectrum should be determined by the technique of Starr and Stephens (1964).

Flagellation. Staining of flagella is not difficult provided that care is taken at every stage, following the staining technique described on page 13.

0/129 sensitivity. Sensitivity to compound 0/129 (2,4-diamino-6,7-di-iso-propylpteridine, from Calbiochem Ltd. or British Drug Houses Ltd.) can be determined by spotting one or two crystals of the compound on a plate streaked uniformly with the test organism. In practice contamination is not a problem.

Luminescence. As already explained, most luminous bacteria have been isolated from marine sources, and may have a requirement for constituents of sea water. Organisms are likely to have been isolated on seawater-containing media, or alternatively on low dilution plates (incorporating high concentrations of samples) in counts on samples of marine origin. Luminescence should be checked on the original plates, and, in the case of isolates, after 2–3 days incubation on the seawater yeast peptone agar of Hendrie *et al.* (1970).

Cultures should be examined in a dark room, after allowing 10 min for the eyes to accommodate.

Oxidase test. See page 79.

Thornley's arginine test. See page 69.

Møller's decarboxylase tests. Lightly inoculate the organism, preferably with a straight wire, into a series of four tubes of Møller's decarboxylase medium. The series consists of a control (no added amino acid), medium + lysine, medium + ornithine, and medium + arginine. Ensure that the wire penetrates beneath the layer of liquid paraffin. Incubate at 25°C and examine daily for up to 7 days. A positive result is indicated by the medium colour changing to violet after an initial change to yellow. Controls and negative reactions are yellow in colour. In the case of a positive reaction in the arginine-containing medium, if ammonia is found in the medium (tested for by the use of Nessler's reagent), and the organism does not possess a urease, the reaction is due to arginine dihydrolase.

Test for 3-ketolactose production (Bernaerts and De Ley, 1963). Inoculate the organism to be tested onto a glucose yeast chalk agar slant and incubate at 25°C for 3 days. From this culture take a loopful of bacterial growth and spot-inoculate a plate of lactose yeast extract agar so that a heavy inoculum of bacteria is concentrated in a spot about 5 mm in diameter. Incubate at 28°C for 2 days. After incubation, flood with Benedict's reagent and leave at room temperature for one hour. If 3-ketolactose has been produced a yellow ring of cuprous oxide extending up to 2–3 cm in diameter will form around the bacterial inoculum.

Test for ability to use inorganic nitrogen as sole nitrogen source (Holding, 1960). *Lightly* inoculate the organism, preferably with a straight wire, into Holding's inorganic-nitrogen medium and incubate at 25°C for up to 5 days, examining daily. An organism capable of utilising inorganic nitrogen as its sole nitrogen source will grow and produce visible turbidity. If only slight turbidity is obtained, reinoculate into another tube of the same medium to ensure that the growth was not due to a carry-over of nutrients from the original broth culture.

Detection of ability to produce a diffusible pigment which fluoresces in ultra-violet light. The production of a u.v.-fluorescent pigment, which is a characteristic of many *Pseudomonas* spp., is very dependent on the composition of the growth medium. The most satisfactory medium is medium "B" of King, Ward and Raney (1954). After incubation for 1–5 days at 20–25°C, the medium is examined in a darkened room under a u.v. lamp for fluorescence, which may be blue or green.

Test for sensitivity to pencillin. An Evans "Sentest" tablet No. 1, "Penicillin low potency", is placed on a plate seeded with the test organism. Sensitivity is shown by a zone of clearing around the tablet after incubation.

Phenylalanine deaminase test. (1) Inoculate a tube of phenylalanine malonate broth (Shaw and Clarke, 1955) with the organisms to be tested. Incubate for 24 h at 30 or 37°C. After incubation and examination for malonate utilisation (see below), add a few drops of 0·1 N HCl and a few drops of 0·5 M ferric chloride solution. A positive reaction—due to phenylpyruvic acid produced by the action of phenylalanine deaminase—is indicated by a green colour developing in the slope and in the liquid.

(2) Alternatively (Clarke and Steel, 1966), grow the culture as a "lawn" on a nutrient agar plate by spreading 0·2 ml of a broth culture over the surface of the plate and incubating for 24 h at 37°C (or 30°C). Place a phenylalanine disc on the surface of the "lawn" culture and reincubate at 37°C (or 30°C) for 2 h. Test for production of phenylpyruvic acid by adding one drop of ferric chloride (10 per cent in dilute HCl) to the disc. A deep green-blue colour developing in *c.* 1 min constitutes a positive reaction.

Motility test. Semi-solid media can be examined for spreading turbidity, and hanging drop preparations from broth cultures or moist agar slope cultures examined microscopically (see page 10).

Indole production. (see page 68).

Simmon's citrate agar. See page 75.

Methyl red test. See page 74.

Voges-Proskauer test. See page 74.

Fermentation of carbohydrates. See page 72. Nutrient broth has been recommended as the basal medium for fermentation tests (Report, 1958a). The carbohydrate should be added aseptically as a filter-sterilised solution to the sterile basal medium.

Liquefaction of gelatin. Use ferrous chloride gelatin (see page 70), stab-inoculated, and incubated at 20°C for up to 30 days. This can be used at the same time to test for the production of hydrogen sulphide.

Hydrogen sulphide production. Use ferrous chloride gelatin, see page 70. Incubate at 20°C for 7 days, examining daily.

Urease test. Use Christensen's urea agar. See page 70.

β-galactosidase production. (1) Heavily inoculate a tube of ONPG peptone water (Lowe, 1962). Incubate at 37°C. β-galactosidase activity is indicated by the development of a yellow colour, due to release of orthonitrophenol. The reaction will usually be detectable within 3 h, but the incubation time should not exceed 18 h.

(2) Alternatively (Clarke and Steel, 1966), grow the culture as a "lawn" culture on a nutrient agar plate by spreading 0·2 ml of a broth culture over the surface of the plate and incubating for 24 h at 37°C (or 30°C). Place a lactose disc on the surface of the plate culture and incubate at 37°C (or 30°C) for 2 h to induce β-galactosidase synthesis. Then apply an ONPG disc to the culture so that it overlaps the lactose disc. In the case of a positive reaction, a yellow colour will develop within about 15 min. This test can be performed on the same plate and at the same time as the phenylalanine test.

Gluconate utilisation. Inoculate a tube of gluconate broth, and incubate for 48 h at 37°C. After incubation, add 1·0 ml of Benedict's reagent, and place the tubes in boiling water for 10 min. A positive test is shown by the development of a yellow-brown precipitate. (Alternatively quick diagnostic reagents such as those produced by Ames may be employed.)

Malonate utilisation. Inoculate a tube of malonate broth or phenylalanine malonate broth and incubate for 24 h at 37°C. Utilisation of malonate is indicated by the development of an alkaline reaction.

SECTION 2. GRAM-NEGATIVE BACTERIA WHICH GROW ON GLUCOSE YEAST EXTRACT AGAR, WORT AGAR, MALT EXTRACT AGAR OR SIMILAR MEDIA

(See also Asai, 1968; Carr, 1968)

Bacteria showing these cultural characteristics which have been isolated from plant products, especially those which are acidic or alcoholic, are likely to be acetic acid bacteria of the genera *Gluconobacter* (i.e. *Acetomonas*) and *Acetobacter*. The two genera can be differentiated from each other and from *Pseudomonas* as follows. Tests should be carried out at 30°C or just below.

	Growth at pH 4·5	Ethanol oxidised to CO_2	Lactate oxidised to CO_2	Flagellation
Gluconobacter	+	−	−	Polar or none
Acetobacter	+	+	+	Peritrichous or none
Pseudomonas	−	V	+	Polar

Growth at pH 4·5. Ability to grow at pH 4·5 is readily checked in the solid media mentioned above, suitably acidified with sterile lactic acid or citric acid immediately before pouring the plates.

Oxidation of ethanol to carbon dioxide. This is tested by the use of bromcresol green ethanol yeast extract agar slopes (the ethanol must be added to the molten medium immediately before setting in the sloped position (see page 325). *Gluconobacter* and *Acetobacter* produce acid from the ethanol, effecting a colour change from bluish-green to yellow, but only *Acetobacter* continues the oxidation to give a reversion in colour back to bluish-green.

Oxidation of lactate to carbon dioxide. This test can be performed using calcium lactate-yeast extract agar plates. Spot inoculate or inoculate by a single streak. *Acetobacter* grows well on this medium and produces a precipitate of calcium carbonate in the medium, whereas *Gluconobacter* grows only poorly with no white halo in the medium.

SECTION 3. BACTERIA REQUIRING AT LEAST 12 PER CENT SODIUM CHLORIDE FOR GROWTH

Obligate extreme halophiles of the genera *Halobacterium* and *Halococcus* are found in salt lakes, solar salt, and foods such as fish which have been preserved with solar salt. They are strict aerobes, and characteristically produce red, pink or orange intracellular carotenoid pigments. Such micro-organisms could be isolated on media containing 15 per cent NaCl used for halophilic counts on foodstuffs. In cultural studies they are usually grown on complex media containing 20 per cent sodium chloride. Alternatively, the chemically defined medium of Onishi, McCance and Gibbons (1965) may be used.

Halobacterium is rod-shaped, and may be motile by polar flagella, or non-motile. *Halococcus* is coccal and non-motile.

A SIMPLE KEY FOR THE IDENTIFICATION OF GRAM-POSITIVE BACTERIA

Primary separation is based on morphological and staining characters, supplemented by observation of cultural characters and by simple biochemical tests.

1. Persistent mycelium formed of branching non-fragmenting hyphae. Aerial hyphae bear chains of exospores (conidiospores). Forms hard colonies that are partially embedded in the medium. These develop a powdery surface when the conidiospores are produced. Rarely the aerial hyphae are less well developed or absent and the spores are produced singly........................**1.** *Streptomyces* group
Not as above..*2*
2. Acid fast to 10–20 per cent sulphuric acid or to acid-alcohol **2.** *Mycobacterium* group
Not acid fast..*3*
3. Endospores produced..*4*
Endospores not produced..*5*
4. Catalase positive, aerobic........................**3.** *Bacillus*
Catalase negative, anaerobic or anaerobic-aerotolerant..**4.** *Clostridium*
5. Catalase positive..*6*
Catalase negative..*9*
6. Cells spherical, occurring in irregular masses or in packets of 4 cells or multiples of 4............**5.** *Staphylococcus-Micrococcus* group
Cells rod-shaped or coccobacillary........................*7*
Branching or non-branching filaments or a mycelium formed. Fragmentation into shorter rods or coccobacilli frequently occurs in older cultures........................**6.** *Nocardia* group
7. Ferments lactate with the production of carbon dioxide........................**7.** *Propionibacterium*
Does not ferment lactate with the production of carbon dioxide.....*8*
8. Good growth on nutrient agar or soil extract agar.............. **8.** Saprophytic and plant coryneform group
No growth (or very sparse growth) on nutrient agar or soil extract agar **9.** Animal coryneform group
9. Aerobic, microaerophilic or facultatively anaerobic..............*10*
Anaerobic..**4.** *Clostridium*
10. Cells spherical or ovoid.. **10.** *Streptococcus-Leuconostoc-Pediococcus* group

Organisms rod-shaped *11*

11. Digests Loeffler's serum . . . . *Corynebacterium pyogenes* (see **9.** Animal coryneform group)

Does not digest Loeffler's serum . . . . **11.** *Lactobacillus*

1. *Streptomyces* group

Streptomyces characteristically produces a many-branched, persistent mycelium that does not fragment as the culture ages. Aerial hyphae are formed, and these bear chains of exospores (conidiospores). The substrate mycelium gives a colony which is partially embedded in the agar and extremely hard, usually requiring more force to break up the colony than can normally be applied with a platinum wire loop.

When young, the colonies are usually spherical, dome shaped, slightly shiny and with no particular distinguishing coloration, being colourless or buff to fawn. As the cultures age and conidiospores are formed, the colonies develop powdery surfaces which may be white, grey or other colours depending on the species. The powdery surface layer of condiospores is easily scraped off using a wire loop, leaving the hard substrate mycelium as before. The colonies may become wrinkled or folded. Some species produce pigments that diffuse into the surrounding medium. In some cases the colour of the pigment varies with the pH of the medium. Many species produce an odour which is characteristic of moist soil. For species differentiation, which is based largely on morphological and cultural characteristics, refer to Buchanan *et al.* (1974).

From the work of Gordon and Mihm (1957, 1962), Williams, Davis and Cross (1968), and many others, it is apparent that there is a continuous spectrum of types between the genera *Mycobacterium*, *Nocardia* and *Streptomyces*. Thus the *Streptomyces* group as here defined may include certain nocardiae. Some strains, usually regarded as *Nocardia*, produce persistent, non-fragmenting mycelia, develop aerial hyphae, but do not produce conidiospores. Such organisms would be indistinguishable from asporogenous *Streptomyces*. In some cases, fragmentation of the aerial hyphae of *Nocardia* resembles spore production by *Streptomyces*. For further clarification of this complex group refer to the schematic illustrations of Williams *et al.* (1968).

Thermophilic species occur in several genera. *Micropolyspora faeni* and *Thermoactinomyces vulgaris* are of medical significance as two of the causative agents of the respiratory complaint known as farmers' lung disease. A description of the morphological and colonial characteristics of the thermophilic organisms is given by Cross (1968) who points out that while colonies of *T. vulgaris* form abundant aerial mycelia, *M. faeni* forms very small colonies and may easily be missed. The substrate mycelium of *T. vulgaris* is stable and non-fragmenting, whereas the mycelium of *M. faeni* may fragment. The health hazard arises when "self-heating" occurs in moist grain (or hay) during storage. Air sampling techniques may be used for isolation of the bacteria in such cases, as reported by Gregory *et al.* (1963).

2. *Mycobacterium* group

The genus *Mycobacterium* consists of non-motile, non-sporing, non-branching rods. Although the organisms are Gram-positive, staining by Gram's method is difficult, probably due to the high wax content of the mycobacteria. They are acid-fast, as determined by Ziehl-Neelsen's acid-fast staining technique. Some strains resist decolorisation with alcohol as well as acid, but as this characteristic is rather variable, Cowan and Steel 1965; Cowan, 1974) recommended the use of acid-alcohol as the decolorising agent, with no differentiation between organisms that are acid-fast and those that are acid- and alcohol-fast.

Mycobacteria can be divided into two main groups.

(*a*) *Parasites and pathogens of homothermic animals.*

This group is typified by *M. tuberculosis.* These mycobacteria grow very slowly *in vitro*, cultures requiring up to two weeks or more incubation to give visible growth, and require special media for their cultivation. They are incapable of growth on, or grow extremely poorly on, media such as nutrient agar. Media which are satisfactory for their growth include solid media in which coagulated serum or egg is the solidifying agent. Because of their pathogenicity these organisms require special precautions in handling.

(*b*) *Saprophytes, and parasites and pathogens of poikilothermic animals.*

These are less acid-fast than the members of group (*a*). They grow rapidly *in vitro*, with growth usually being visible after 2–5 days. They are capable of growing on media such as nutrient agar but usually grow better on media containing coagulated serum (e.g. Loeffler's serum medium) or egg. On solid media, particularly those containing glycerol, many saprophytic mycobacteria produce bright yellow, orange, pink or brick-red pigments, and the growth is dry and crumbly.

Thus acid-fast bacteria isolated in the laboratory during general microbiological examinations will exclude the members of group (*a*) except when techniques are employed specifically to detect them. There is not a distinct division between the mycobacteria of group (*b*) and *Nocardia.* Some strains regarded as *Nocardia* are slightly acid-fast and some organisms usually included in *Mycobacterium* are only weakly acid-fast. In addition, certain mycobacteria, such as *M. fortuitum*, produce branching filaments that fragment into shorter coccobacillary forms as the cultures age, thus corresponding to the morphological description of *Nocardia* given in the present work. For a detailed description of the saprophytic mycobacteria, see Gordon and Mihm (1959).

3. *Bacillus*

Species of the genus *Bacillus* are rods capable of producing endospores and of growing aerobically (although some are facultative) in most cases on nutrient agar, and are catalase positive. Difficulty may occasionally be experienced in inducing sporulation and in this case, additional manganese should be added to the growth medium as recommended by Charney, Fisher and Hegarty (1951). These authors found a marked stimulatory effect on the sporulation of *B. subtilis* at concentrations from 0·1–10·0 μg manganese per ml in all media, which is obtained by the addition of 0·1–10·0 ml of a 0·4 per cent solution of manganese sulphate ($MnSO_4 \cdot 4H_20$) per litre of medium. Most *Bacillus* spp. are motile by means of peritrichous flagella with the exception of some of the larger-celled species and *B. anthracis*. Although the genus *Bacillus* is regarded as typically Gram-positive, a number of species may be Gram-variable or even Gram-negative, particularly those species producing swollen sporangia and thus included in the morphological groups II and III of Smith, Gordon, and Clark (1952), and Gordon, Haynes and Pang (1973) described below.

Identification of *Bacillus* spp. can be carried out according to the methods of Smith *et al.* (1952) and Gordon *et al.* (1973). Preliminary subdivision of the genus into three morphological groups is made on the basis of the appearance of the sporangium: whether it is non-swollen, group I; swollen with an oval spore, group II; or swollen with a round spore, group III. Further identification is on the basis of cultural and biochemical tests. The validity of this system has been confirmed by other methods of investigation, for example by the work of Knight and Proom (1950) and Proom and Knight (1955), who found that nutritional requirements correlated well with individual species as delineated by the methods of Smith *et al.* (1946, 1952).

Bergey's Manual (Buchanan *et al.*, 1974) lists 48 species. The following table, adapted from the data of Gordon *et al.* (1973), allows differentiation of the species most frequently encountered in food microbiology.

Smith *et al.* (1952) recognised only two thermophilic *Bacillus* species, *B. coagulans* and *B. stearothermophilus*; cultures received as *B. calidolactis* from dairy sources were identified as one or other of these two species. *Bacillus coagulans* may be regarded as a facultative thermophile since it shows growth at both 37 and 55°C. It may be distinguished from *B. stearothermophilus* by its failure to grow at 65°C.

Identification of Commonly Encountered *Bacillus* Species

Morphological group I: ellipsoidal and cylindrical spores do not exceed diameter of the sporangium
II: ellipsoidal and cylindrical spores cause swelling of the sporangium
III: spherical spores cause swelling of the sporangium

Species	Group	Growth at 50°C	Growth at 60°C	Growth at 65°C	Voges-Proskauer	Starch hydrolysis	Growth in Hugh & Leifson's	Acid from mannitol	Nitrate reduction	Citrate utilisation
B. megatherium	I	—	—	—	—	+	A	+	—/+	+
cereus	I	—	—	—	+	+	F	—	+	+
licheniformis	I	+	—	—	+	+	F	+	+	+
subtilis	I	+	—	—	+	+	A	+	+	+
pumilis	I	+	—	—	+	—	A	+	—	+
firmus	I	—	—	—	—	+	A	+	+	—
coagulans	I, II	+	+	—	+	+	F	—/+	—/+	—/+
polymyxa	II	—	—	—	+	+	F	+	+	—
macerans	II	+	—	—	—	+	F	+	+	—/+
circulans	II	+	—	—	—	+	F/A	+	—/+	—/+
stearothermophilus	II	+	+	+	—	+	A	—/+	+/—	—
alvei	II	—	—	—	+	+	F	—	—	—
laterosporus	II	+	—	—	—	—	F	+	+	—
brevis	II	+	+	—	—	—	A	+/—	+/—	—/+
*pantothenticus**	II, III	V	—	—	—	+	F	—	+/—	—
sphaericus‡	III	—	—	—	—	—	A	—	—	—/+

* *B. pantothenticus* can grow in presence of 10 per cent NaCl, and may sometimes occur in canned cured meats.
‡ *B. psychrophilus* is similar to *B. sphaericus*, but has a maximum growth temperature of 25–30°C, and can sporulate at 0°C.

+: 85–100% strains positive. +/—: 50–84% strains positive. —/+: 15–49% strains positive. —: 0–14% strains positive.
A: 85–100% strains obligately aerobic. F: 85–100% strains facultatively anaerobic. F/A: 50–84% strains facultatively anaerobic.

(Adapted from Gordon, Haynes and Lang, 1973.)

4. *Clostridium*

Clostridium species are rods capable of producing endospores but may be differentiated from *Bacillus* spp. since they are catalase-negative and strict anaerobes apart from a few anaerobic, aerotolerant species capable of producing scant growth under aerobic conditions. The organisms are typically Gram-positive and may be either non-motile or motile by means of peritrichous flagella. The sporangium may be either non-swollen or swollen with oval or spherical spores in a terminal position, or in a central or subterminal position to give a spindle-shaped or clostridial form. Difficulty may be experienced in inducing sporulation of some species as for example, *Cl. perfringens* (*Cl. welchii*), which rarely sporulates *in vitro* unless special media (Ellner, 1956) are used.

A useful precaution with anaerobic culture, as pointed out by Willis (1962), is to set up a control plate incubated aerobically with each subculture of test organism. A comparison of aerobically and anaerobically incubated plates enables detection of contaminants, particularly *Bacillus* spp., capable of growing anaerobically.

A review of methods of classifying and identifying *Clostridium* species is given by Oakley (1956). In the system adopted in the 7th edition of *Bergey's Manual* (Breed *et al.*, 1957) the genus was first divided into strict anaerobic and anaerobic but aerotolerant species. The group of strict anaerobes was then divided into cellulose fermenters and cellulose non-fermenters and these subgroups in turn divided on the basis of pigment production.

The scheme in the 8th edition of *Bergey's Manual* (Buchanan *et al.*, 1974), divides the genus of 61 species into 4 groups by spore position and gelatin liquefaction. Species differentiation within these groups is then achieved by biochemical tests especially carbohydrate fermentation patterns, haemolysis, hydrogen sulphide production, nitrate reduction, Voges-Proskauer test, and proteolytic activity.

An alternative approach (Willis, 1962, 1965) for the identification of a limited number of species is to use Willis and Hobbs's lactose egg-yolk milk agar medium with neutral red as pH indicator. This medium has the advantage that it combines in one medium a method of detecting the proteolytic and saccharolytic activities of the isolate together with its effect on egg-yolk agar. The egg-yolk medium enables both lecithinase and lipase activities to be detected. Lecithinase activity results in a zone of opacity extending beyond the colony while lipase activity is indicated by a restricted zone of opacity and a pearly layer consisting of fatty acids, which give a surface sheen to the colony. The use of half-antitoxin plates as described by Willis (1962) enables lecithinase-producing organisms to be specifically identified.

Goudkov and Sharpe (1965) in an examination of milk and dairy products, found that clostridial spoilage was most likely to occur in low-acid sweet cheese of Swiss and Edam type and in processed cheese. Spoilage appeared to be mainly gas production by lactate fermenters such as *Cl. butyricum* or *Cl. tyrobutyricum* and proteolysis by proteolytic clostridia such as *Cl. sporogenes* and *Cl. perfringens* (*Cl. welchii*). Rosenberger (1956) and Gibson (1965) give accounts of the lactate-fermenting and other clostridia present in silage. Psychrotrophic clostridia (those capable of growth at 5°C) and their significance in food microbiology is discussed by Beerens, Sugama and Tahon-Castel (1965).

5. *Staphylococcus – Micrococcus* Group

Determination of the ability of an organism to bring about the anaerobic fermentation of glucose forms the basis of separation within this group. A standard procedure involving a modification of the Hugh and Leifson technique has been recommended for this purpose (Recommendations, 1965). The bromthymol blue indicator of the original medium is replaced by bromcresol purple and, since this detects only active acid producers, more clear-cut results are obtained. Catalase-positive Gram-positive cocci which produce acid throughout the medium, i.e. ferment glucose under anaerobic conditions, are regarded as staphylococci; those which produce no acid or produce acid only at the surface of the medium, i.e. oxidise glucose, are micrococci. Certain packet-forming, motile, Gram-positive cocci with an exclusively respiratory metabolism which have been isolated from sea water have been placed in the genus *Planococcus* (Buchanan *et al.*, 1974).

Staphylococcus. Separation of the three recognised species is on the basis of the coagulase test, acid production from mannitol, and sensitivity to novobiocin. For greatest accuracy the tube coagulase test method which tests for free coagulase is recommended, rather than the slide coagulase test.

	S. aureus	*S. epidermidis*	*S. saprophyticus*
Coagulase	+	–	–
Acid from mannitol aerobically	+	V	V
Acid from mannitol anaerobically	+	–	–
Novobiocin sensitivity*	S	S	R

* Resistance to novobiocin (R) = a minimum concentration of 2 μg per ml is required for inhibition. Sensitivity to novobiocin (S) = a concentration of less than 0·6 μg per ml is inhibitory. (Mitchell and Baird-Parker, 1967; Jeffries, 1969.)

The ability of the organism to ferment mannitol anaerobically is determined by substituting mannitol for glucose in the bromcresol purple version of Hugh and Leifson's medium. Most coagulase-positive staphylococci of human origin ferment mannitol, and conversely coagulase-negative staphylococci do not. The correlation appears to be less good for staphylococci of bovine origin. Bovine strains of *Staph. aureus* also differ from human strains in that on sheep- or calf-blood agar the β-haemolysin is characteristically produced. This is recognised by the production of dark zones around the colonies and for optimum development requires 24–48 h incubation at 37°C followed by overnight incubation at room temperature or in the refrigerator, i.e. "hot-cold lysis". Human strains of *Staph. aureus* characteristically produce the α-haemolysin on sheep- or calf-blood agar, recognisable by the clear zones which appear round the colonies within 24–48 h at 37°C. Further differentiation can be made on the basis of phage typing as described by Parker (1962).

Since there is no simple laboratory test for the detection of enterotoxin production by staphylococci, it is usual to regard all coagulase-positive staphylococci as potential enterotoxin producers, though enterotoxin production by coagulase-negative strains has been reported (Breckinridge and Bergdoll, 1971). The incidence of staphylococci and micrococci in milk and dairy products has been discussed by Sharpe, Neave and Reiter (1962).

Micrococcus. Preliminary observations should include the examination, under high-power and oil-immersion objectives, of a hanging drop preparation prepared from an 18-hour broth culture. This procedure was found by Pike (1962) to be more satisfactory for the detection of characteristic groupings of cocci and packet formation, i.e. cell division in 2 or 3 planes in space, than examination of stained preparations from agar cultures. It is also a useful method of ensuring that other coccal bacteria, e.g. streptococci, and soil-derived coryneforms of the genus *Arthrobacter* which show a coccal stage in mature culture, are not falsely identified as micrococci (Pike, 1962).

The three species of *Micrococcus* are differentiated by pigment production and acid production from glucose.

	M. luteus	*M. roseus*	*M. varians*
Yellow pigment	+	−	+
Red pigment	−	+	−
Acid from glucose in open tube of modified Hugh and Leifson's medium	−	V	+

6. *Nocardia* Group

The genus includes:

(*a*) Strains consisting of rods that rarely or never show branching; they produce soft, butyrous colonies. These organisms, particularly when slightly acid-fast (e.g. not decolorised by 1–5 per cent sulphuric acid), overlap with the saprophytic mycobacteria. On the other hand, non-acid-fast nocardias of this type are difficult to distinguish from some of the corynebacteria, which may even show rudimentary filament formation in slide cultures.

(*b*) Strains producing branching filaments or rudimentary mycelia that fragment into rods and coccoid forms. Colonies tend to be of a pasty or crumbly texture.

(*c*) Strains capable of producing persistent mycelia, sometimes with a slight aerial mycelium (see for example Gordon and Mihm, 1962). Colonies of these types tend to be leathery and resistant to being broken up with a wire loop. These strains, together with *Streptomyces*, present a continuous spectrum of morphological forms such that a dividing line cannot, with certainty, be drawn between the two genera.

Within this framework of morphological variation the nocardias are Gram-positive, obligately aerobic and largely saprophytic, although a few animal-pathogenic species of aerobic nocardias have been described. The saprophytic strains of *Nocardia* are frequently capable of utilising as an energy source such compounds as paraffin and phenol.

7. *Propionibacterium*

Propionibacteria are Gram-positive, non-acid-fast and non-motile rods which do not produce endospores. Although there is a tendency towards anaerobiosis and growth is improved in an atmosphere of 5 per cent carbon dioxide, both aerobically and anaerobically, cultures are catalase-positive, usually strongly so. Biochemically lactic acid (or lactate) and carbohydrates are fermented with the production of propionic acid, acetic acid and carbon dioxide. Production of carbon dioxide gas can be used under suitable conditions as an indicator of the presence of propionibacteria. It should be noted that, although some species of clostridia may also ferment lactate, they can be distinguished on the basis of morphological and cultural characters: lactate-fermenting clostridia are endospore-producing rods, usually motile and of a larger size than propionibacteria, catalase-negative and incapable of aerobic growth. The morphological appearance of propionibacteria depends very much on cultural conditions, in particular oxygen tension and pH of the medium, as shown by van Niel (1928). Under anaerobic conditions and in neutral media, the reaction of which remains so, the cells appear as very short rods with rounded ends sometimes resembling streptococci in appearance. Under aerobic conditions with adequate aeration and in a similar medium, long rod-shaped cells of irregular or even branched appearance may be formed.

A key for the identification of species devised by van Niel (1928) and used by Breed *et al.* (1957) makes use of the characteristic cell forms produced in acid media or in media which become acid with growth and fermentation (e.g. yeast glucose lemco broth). Under these circumstances, some propionibacteria appear as small spherical cells, $\sim 0 \cdot 5$ μm diameter, some as short rods $\sim 0 \cdot 7$ $\mu m \times 1 \cdot 0$–$1 \cdot 3$ μm and others as long rods $\sim 0 \cdot 7$ $\mu m \times$ 3–5 μm, frequently of irregular appearance. Pigment production is also a useful diagnostic criterion. The 8th edition of *Bergey's Manual* (Buchanan *et al.*, 1974) bases species differentiation primarily on biochemical reactions.

8. Saprophytic and Plant Coryneform Group

This group includes several different genera and groups of organisms at present poorly defined or defined by habitat rather than by definite characters and it is therefore difficult to construct a satisfactory scheme for separation of these organisms with the information at present available (see Buchanan *et al.*, 1974 and Jones, 1975).

The term "coryneform" is used here to imply non-acid-fast, non-spore-forming rods, typically Gram-positive and frequently showing features regarded as typically coryneform, such as beaded staining, club shapes and palisade- or Chinese letter-arrangements of the rods. In the exponential phase, irregular rods of variable length are the most common forms, with rudimentary branching often occurring, especially on richer media. In the stationary phase, cells are much shorter and more regular, even coccoid. Photomicrographs of typical coryneform organisms at the different morphological stages have been presented by Cure and Keddie (1973), and of *Arthrobacter* and *Brevibacterium linens* by Crombach (1974).

The genus *Corynebacterium* itself, as at present defined by Buchanan *et al.*, (1974), comprises three groups of organisms of diverse character, the animal coryneforms (*q.v.*), plant pathogens, and non-pathogenic corynebacteria. A number of the plant pathogens are motile and thus differ from other coryneform bacteria which, with a few exceptions, are non-motile. Da Silva and Holt (1965) presented evidence to suggest that the phytopathogenic corynebacteria are sufficiently different from the type species *C. diphtheriae* (and other animal commensals and pathogens) to be excluded from the genus *Corynebacterium* (see also Jones, 1975).

The genus *Microbacterium* consists of small Gram-positive rods commonly isolated from milk, dairy products and dairy equipment, and is characterised by a capacity to survive laboratory pasteurisation in milk (63°C for 30 min) or a heat treatment in milk of 72°C for 15 min, although such heat resistance appears *not* to be an invariable feature of these organisms. The genus *Microbacterium* includes two species found in such environments, *M. lacticum* and *M. flavum*. *Microbacterium lacticum* hydrolyses starch, whereas *M. flavum* fails to do so but produces a yellow pigment on normal media. A third species, *M. thermosphactum*, is found as a spoilage organism on refrigerated poultry and meat.

Microbacterium thermosphactum resembles organisms of the genus *Kurthia* except that the former is non-motile and facultatively anaerobic, whereas *Kurthia* is motile and obligately aerobic. Organisms of *Kurthia* (the single recognised species is *K. zopfii*) are rod-shaped in young cultures and coccoid in older cultures. They are peritrichously flagellate, catalase positive and strictly aerobic, although they produce an alkaline reaction in Hugh and Leifson's medium because of an inability to produce acid from glucose. *Kurthia* also is found on poultry and meat. Descriptions and photomicrographs of strains of *M. thermosphactum* isolated from meat have been given by Davidson, Mobbs and Stubbs (1968), and descriptions and photomicrographs of strains of *K. zopfii* isolated from meat have been given by Gardner (1969).

The genus *Cellulomonas* consists of soil coryneforms with an ability to digest cellulose—as seen by disintegration of a strip of paper in peptone water or nutrient broth. Some species of the genus are motile.

Arthrobacter species are soil coryneforms which do not decompose cellulose. As with other organisms in this group, in young cultures up to 18 h old, the cells appear as irregular rods which often show rudimentary branching, and which may be Gram-variable. In older cultures Gram-positive coccoid cells usually predominate. It is important that cultures of *Arthrobacter* in this condition are not confused with micrococci and, as pointed out by Pike (1962), examination of young broth cultures by the hanging drop method is helpful in this respect. An alternative procedure, and to follow morphogenesis, is to use the slide culture technique in which a sterile coverslip is placed over a surface-inoculated agar medium, either in a petri-dish as used by da Silva and Holt (1965) or in a slide culture. The development may then be followed microscopically, if possible by phase contrast. Older cultures of *Arthrobacter* may show the presence of enlarged coccoid forms sometimes known as cystites, but Stevenson (1963) concluded that these are of no particular significance in the life cycle of the organism and that the enlarged forms appear as a result of depleted cultural conditions. On transfer to fresh medium, rod forms are again produced.

The genus *Brevibacterium* is very broadly defined and includes an assemblage of typically short, unbranching, Gram-positive, non-acid-fast and non-sporing rods, isolated from dairy products and diverse other sources, including soil, salt water and fresh water. One species, *Bbm. linens*, that is found on the surface of bacterial-ripened soft cheeses such as Limburger (see page 207), has been considered by da Silva and Holt (1965) to be closely related to *Arthrobacter globiformis*. Jones (1975) demonstrated a 79 per cent similarity between *Bbm. linens* ATCC 9174 and *A. globiformis* NCIB 8602.

9. Animal Coryneform Group

The genus *Corynebacterium* includes animal parasites and pathogens, plant pathogens, and soil-dwelling (saprophytic) organisms. The plant pathogens and soil-dwelling organisms (see the preceding section) are so unlike the animal parasites and pathogens that they should be in separate genera. The animal parasites and pathogens grow poorly, if at all, on nutrient agar, plate count agar and similar media, but are capable of growing well on serum agar or egg yolk agar. Saprophytes and plant pathogens usually grow abundantly on nutrient agar, etc.

The appearance of *Corynebacterium* under the microscope is very characteristic of this genus and a few others (see also Group 8). The straight or slightly curved rods are frequently swollen at one or both ends. Stained by Gram's method or with methylene blue, the individual organisms usually stain irregularly, sometimes with the appearance of darkly staining granules. The arrangement of organisms is also very characteristic, with a snapping type of division causing V- or L-shaped formations of adjacent organisms. Groups or clusters of organisms thus tend to have the appearance of "Chinese characters" or cuneiform writing.

Two species of *Corynebacterium*, *C. pyogenes* and *C. bovis*, may be isolated from raw bovine milk plated on to blood agar, serum agar or egg yolk agar. Whereas *C. pyogenes* is associated with a particular and relatively uncommon type of suppurative mastitis, *C. bovis* can be isolated from most freshly drawn milk samples, whether from udders which are normal or from those presenting a subclinical or clinical mastitis.

Corynebacterium pyogenes is catalase-negative, haemolytic, and proteolytic—slowly liquefying coagulated serum (Loeffler's serum slopes) and gelatin (nutrient gelatin with 5 per cent serum should be used). It is urease negative. Acid is produced from a wide range of carbohydrates in Hiss's serum water sugars (see page 327). There is sparse or no growth on Tween agar with no zones of precipitation.

Corynebacterium bovis is catalase-positive and non-haemolytic. It liquefies neither coagulated serum nor gelatin. It is capable of hydrolysing urea when grown in Christensen's urea agar with added serum. Acid is produced from a number of substrates when *C. bovis* is grown in Hiss's serum water sugars, including glucose, fructose and glycerol. Maltose and/or galactose are attacked by some strains, but lactose, sucrose and salicin are not attacked. *C. bovis* grows luxuriantly on Tween 20 agar and Tween 80 agar with broad zones of precipitation. The description found in the 7th and earlier editions of *Bergey's Manual* differed markedly from this, emphasising the apparent metabolic inactivity of the organisms: this was due to the use of unsuitable media by most previous workers, since *C. bovis* is incapable of growing in carbohydrate peptone waters, and consequently does not decompose the substrates. The effect of the addition of serum, egg yolk or Tween to media has been examined and discussed by Cobb (1963, 1966) and Harrigan (1966).

Corynebacterium ulcerans has been reported (Jayne-Williams and Skerman, 1966) as being frequently isolated from milk freshly drawn from normal bovine udders. It has also been isolated from cases of diphtheria-like diseases in human beings. Unlike *C. bovis*, this organism is haemolytic and proteolytic, but it differs from *C. pyogenes* in hydrolysing urea.

10. *Streptococcus-Leuconostoc-Pediococcus* Group

These genera are Gram-positive, coccal or ovoid forms, typically catalase-negative and are included in the family Streptococcaceae (Buchanan *et al.*, 1974). Members of this family are characterised by their ability to ferment carbohydrates with the production of lactic acid. This may be either the sole product of the fermentation in which case it is described as homofermentative (as in the genera *Streptococcus* and *Pediococcus*) or it may be heterofermentative in which case a variety of fermentation products is produced including acetic acid, ethanol and carbon dioxide in addition to lactic acid (as in the genus *Leuconostoc*). These genera may be distinguished from *Staphylococcus* and *Micrococcus* by their negative catalase reaction, absence of pigmentation, and failure to produce abundant growth on nutrient agar. Optimal growth is only obtained in media containing a suitable fermentable substrate, e.g. glucose in yeast glucose lemco broth. Cell division of *Streptococcus* and *Leuconostoc* takes place in one plane only, which gives rise to the arrangement of cells in pairs or in short or long chains; cell division in the case of *Pediococcus* occurs in two planes to give tetrads.

The three genera may be differentiated on the type of lactic acid produced—dextrorotatory (*Streptococcus*), laevorotatory (*Leuconostoc*), or optically inactive (*Pediococcus*)—but often other methods of differentiation may be more practicable. In particular, the detection of heterofermentative activity by the ability to form carbon dioxide gas from glucose in Gibson's semi-solid medium enables *Leuconostoc* and *Streptococcus* to be differentiated.

Streptococcus

The *Streptococcus* genus is a complex one containing a number of species which may be differentiated into more limited groups by means of physiological tests as described by Sherman (1937), or into Lancefield serological groups by the precipitin test (see page 104). Determination of the ability of the organism to grow at 10° and 45°C, to grow in the presence of 6·5 per cent sodium chloride or at pH 9·6, or to survive heating at 60°C for 30 min should be made using a medium such as yeast glucose lemco broth.

The production of ammonia from arginine in the arginine broth of Abd-el-Malek and Gibson (1948a) (see page 69) serves both to differentiate individual *Streptococcus* species and at generic level to separate *Leuconostoc* from ammonia-producing *Streptococcus* species since no *Leuconostoc* has been shown to produce ammonia from arginine.

Differentiation of genus *Streptococcus*
(After Sherman, 1937; Breed *et al.*, 1957)

General characters of the genus (to be confirmed before applying these differential tests): Gram-positive cocci, in liquid media usually in pairs and/or short or long chains, forming small non-pigmented colonies on yeast glucose lemco agar, catalase negative. Carbon dioxide gas is not produced from glucose in Gibson's semi-solid medium (cf. *Leuconostoc*).

Test	Physiological group			
	Pyogenes	Viridans	Lactic	Enterococcus
Growth at 10°C	—	—	+	+
Growth at 45°C	—	+	—	+
Heat resistance (at 60°C for 30 min)	—	V	—	+
Growth in presence of 6·5 per cent NaCl	—	—	—	+
Growth at pH 9·6	—	—	—	+
Reduction of litmus before clotting, in litmus milk	—	—	+	+
Ammonia from arginine	+	—	V	+
Haemolysis on horse blood agar	β, α or—	α or—	—	α, β or—
Representative species and Lancefield group	*Str. pyogenes* (A) *Str. agalactiae* (B) *Str. dysgalactiae* (C)	*Str. bovis* (D) *Str. thermophilus**	*Str. lactis* (N) *Str. cremoris* (N)	*Str. faecalis* (D) *Str. durans* (D)

* *Strep. thermophilus* can be readily differentiated from *Str. bovis* by having a high optimum temperature of 40–45°C, also growing at 50°C, and surviving heating at 65°C for 30 min.

Detection of the type of haemolysis when streptococci are grown in, or anaerobically on, horse blood agar is best determined by the pour plate method, with the inoculum adjusted to give comparatively few colonies per plate, but for the detection of β-haemolysis a streak plate technique is sufficient.

A table for the differentiation of streptococci on the basis of Sherman's tests is given above, but the complete identification of *Streptococus* isolates can only be made by the application of further differential tests as indicated by the results obtained from primary grouping.

Suitable schemes for the identification of the pyogenes and enterococcus groups are presented by Sharpe and Fryer (1966). Schemes for the differentiation of all 21 species of *Streptococcus* are given in *Bergey's Manual* (Buchanan *et al.*, 1974), but further details of the lactic streptococci are given below.

Differentiation of group N lactic streptococci.

Only two species, *Streptococcus lactis* and *Strep. cremoris*, are described by Buchanan *et al.*, (1974), but a subspecies of *Strep. lactis*, namely *Strep. lactis* subsp. *diacetylactis*, is of great importance in the dairy industry. The subspecies differs from the type in the production of citratase (detected using semi-solid citrate milk agar in which citrate fermentation is indicated by gas production), and the production of a positive Voges-Proskauer reaction in milk or yeast glucose lemco broth.

Table for differentiating Group N lactic streptococci
(After Reiter and Møller-Madsen, 1963)

	Action in litmus milk	NH_3 from arginine	Growth at 40°C*	Growth in the presence of 4 per cent NaCl*	Growth at pH 9·2*	Gas from citrate in semi-solid citrate milk agar	Acetoin produced (Voges-Proskauer reaction)†
Strep. lactis	ACR	+	+	+	+	−	−
Strep. lactis subsp. *diacetylactis*	ACR	+	+	+	+	+	+
Strep. cremoris	ACR	−	−	−	−	−	−

* Tested using yeast glucose lemco broth or glucose lemco broth as a basal medium, *q.v.* with modification as required by the test.

† Tested in milk culture.

Leuconostoc

These are catalase-negative Gram-positive cocci showing cell division in one plane only and in this they resemble streptococci, but they differ in being heterofermentative and are thus capable of producing carbon dioxide from glucose. Other characters which are useful in the differentiation of *Leuconostoc* from certain other groups, in particular the lactic streptococci with which they may be associated in starter cultures, are their failure to produce ammonia from arginine and comparative inactivity in litmus milk. Garvie (1960), in an extensive study of the genus, found no strains giving reduction of litmus, a few capable of producing acid, but very few of clotting the milk, thus contrasting with the lactic streptococci of starter cultures which give a vigorous clot and reduction.

Leuconostoc is a comparatively restricted genus with 4 of the 6 species being closely related; *L. oenos* is found in wine, and unlike the other species it can initiate growth in media of pH 4·2–4·8. The following data are derived from Buchanan *et al.* (1974) and Garvie (1960).

	L. mesenteroides and *L. dextranicum*	*L. paramesenteroides*	*L. lactis*	*L. cremoris*	*L. oenos*
Growth at 37°C	V(+)	V	+	−	V
Acid from sucrose	+	+	+	−	−
Acid from trehalose	+	+	−	−	+
Dextran produced on sucrose agar	+	−	−	−	−
Growth initiated in media at pH 4·2–4·8	−	−	−	−	+

Pediococcus

These are homofermentative Gram-positive cocci with complex nutritional requirements and thus they resemble the streptococci, but differ from them in that cell division takes place in 2 planes (Gunther, 1959), so that tetrads and coccal arrangements resembling those of micrococci are produced. Gunther and White (1961) differentiated pediococci from micrococci on the basis of the failure of pediococci to grow in simple media, or to utilise ammonium salts as sole nitrogen source and failure to reduce nitrate or liquefy gelatin. In addition pediococci are microaerophilic.

Pediococci are particularly associated with fermenting plant materials, and as contaminants of brewers' yeast. Differentiation of *Pediococcus* from *Leuconostoc*, which may also be found in fermenting plant materials, can be made on the basis of morphology and on the failure of *Pediococcus* to produce carbon dioxide from glucose in Gibson's semi-solid medium. Like *Leuconostoc*, pediococci are comparatively inactive in litmus milk and rarely produce sufficient acid to cause clotting; in this they contrast sharply with the lactic streptococci.

For information on the differentiation of species within the genus refer to Gunther and White (1961), Sharpe and Fryer (1966), and Buchanan *et al.* (1974).

11. *Lactobacillus*

These are non-acid-fast, non-spore-forming Gram-positive rods. With very few exceptions they are non-motile, and the rods frequently appear in pairs or chains, particularly in liquid media. The lactobacilli are catalase-negative and microaerophilic, generally requiring layer plates for aerobic cultivation on solid media. Nutritional requirements are complex and optimal growth is only obtained in media containing a fermentable substrate, and adequate growth factors, as, for example, the MRS broth of de Man, Rogosa and Sharpe (1960). The genus includes both homofermentative and heterofermentative species, the optical activity of the lactic acid produced depending on the species. Lactobacilli are aciduric organisms and this property is utilised in the devising of selective media for their isolation as in acetate agar (Keddie, 1951) and Rogosa agar (Rogosa, Mitchell and Wiseman, 1951) both having a final pH of 5·4, and containing acetate as selective agent.

Orla-Jensen (1943) subdivided the lactobacilli on the basis of homofermentation or heterofermentation and on optimum temperature into three groups as follows: *Streptobacterium* (homofermentative, low optimum temperature), *Thermobacterium* (homofermentative, high optimum temperature) and *Betabacterium* (heterofermentative). Though no longer accorded taxonomic status, they form convenient descriptive terms, the value of which has not been diminished by more recent extensive work using a variety of physiological and cultural tests (see Sharpe, 1962; Buchanan *et al.*, 1974).

To identify *Lactobacillus* isolates, first determine to which physiological subgroup it belongs. The ability to produce carbon dioxide gas from glucose in Gibson's semi-solid medium indicates whether fermentation is homofermentative or heterofermentative. The ability to grow at 15°C and at 45°C (in MRS broth) should also be determined. Ability to produce ammonia from arginine in a medium containing 2 per cent glucose may also be a useful supplementary test since, as shown by Briggs (1953), most heterofermentative lactobacilli give a positive reaction. A suitable medium is

Differentiation of lactobacilli (After Naylor and Sharpe, 1958b; Sharpe, 1962; Buchanan *et al.*, 1974)

Group	Gas from glucose	Ammonia from arginine	Growth at 15°	Growth at 45°	Fermentation[3] of Raffinose	Melibiose	Arabinose	Cellobiose	Gluconate	Lactose	Salicin	Sucrose	
Streptobacterium (Homofermentative, low optimum temperature)	—	—	+	V	—	—	—	+	+	V	+	V	*L. casei*
					+	+	V	+	+	+	+	+	*L. plantarum*
Thermobacterium (Homofermentative, high optimum temperature)	—	—	—	+	V	V	—	+	—	+	+	+	*L. acidophilus*
					—	—	—	—	—	+	—	—	*L. bulgaricus*[1]
					—	—	—	—	—	+	+	+	*L. lactis*
					—	—	—	—	—	—	—	+	*L. delbrueckii*
					—	—	—	+	—	+	+	+	*L. leichmanni*
					—	—	—	—	—	+	—	—	*L. helveticus*[1]
Betabacterium[2] (Heterofermentative)	+	+	+	—	—/(+)	+	+	—	+	—/(+)	—	V	*L. brevis*
			—	+	+	+	V	—	+	+	—	+	*L. fermentum*
			—	+	—	—	—	—	+	—	—	V	*L. viridescens*
			—	—	V	—	—	—	+	V	—	V	*L. hilgardii*

[1] *L. helveticus* ferments maltose, *L. bulgaricus* does not.

[2] Includes *L. trichodes*, a wine spoilage organism which will grow in media containing 15–20 per cent ethanol.

[3] Determined using MRS fermentation medium (see page 328).

MRS broth containing 0·3 per cent arginine—ammonia production is detected by Nessler's reagent. The arginine test is also useful for differentiating heterofermentative lactobacilli from *Leuconostoc* since coccobacillary forms of *Lactobacillus* may be difficult to distinguish morphologically from *Leuconostoc*.

A table for differentiating the main groups of *Lactobacillus*, and identifying the species particularly likely to be isolated from food and associated samples is given on page 275. Fermentation tests should be carried out using MRS fermentation medium. The source of the isolate may be helpful in determining its possible identity, as indicated in the table below.

Lactobacillus species commonly associated with named habitats

Habitat	Species likely to be found
Cheese and dairy environments	*L. brevis* *L. bulgaricus* *L. casei* *L. lactis* *L. plantarum*
Emmental and Gruyere cheese	*L. helveticus*
Fermented and acid plant products, bakers' and brewers' yeast	*L. brevis* *L. delbrueckii* *L. fermentum* *L. leichmannii* *L. plantarum*
Cured meat products	*L. viridescens*
Wines and fruit juices	*L. brevis* *L. fermentum* *L. hilgardii* *L. trichodes*
Intestine of poultry, and also the intestine of human infants	*L. acidophilus* *L. brevis*
Human intestine	*L. casei* *L. plantarum*

IDENTIFICATION OF YEASTS AND MOULDS

In the following section descriptions are given to enable identification of the yeasts and moulds most commonly isolated from foods and dairy products. For more comprehensive descriptions of yeasts and moulds, refer to the works given in the bibliography.

The methods of examination described in Part I (page 106) are not always entirely satisfactory for purposes of identification. The preparation of wet-mounted stained slides from Petri-dish cultures almost invariably causes serious distortion or disintegration of sporing structures and disintegration of pseudomycelia or mycelia which are fragmenting into arthrospores. It is therefore more satisfactory to examine yeasts and moulds by slide culture, when less distortion and fragmentation can take place. The slide culture methods employed are somewhat different for yeasts than for moulds which form true mycelia. Certain fungi, such as *Endomyces* and *Candida*, which are intermediate in their structure between yeasts and mycelial fungi, may on first isolation have the colonial characterisation of either yeasts or moulds, depending on the cultural conditions such as medium, temperature and period of incubation. When subsequent examination suggests that the isolate may form a true mycelium *and* be capable of yeast-like growth, both methods of slide culture should be employed.

In addition to slide cultures, culture the fungi on a number of media in Petri-dish preparations. Since the nutritional requirements of fungi vary greatly, a variety of media should be used in an attempt to obtain sporulation, etc., on at least one medium. The media recommended are malt extract agar, Czapek-Dox agar, potato dextrose agar, and Davis' yeast salt agar. At least two media, one with a high and one with a low concentration of carbohydrate should be used. Cultures should be incubated at 25°C and examined every other day for up to two weeks, using a stereoscopic microscope with incident light.

Methods Used in the Examination of Yeasts

Yeasts are identified on the basis of morphological and cultural characteristics and ascospore and ballistospore production, but biochemical tests are also of great importance. The details that follow are intended to assist in the identification to generic level of the yeasts likely to be isolated from foods and dairy products, based primarily on morphological characters. For further characterisation and identification of yeast isolates, refer to the monographs by Lodder (1970) and Lodder and Kreger-van Rij (1952). A useful identification key, based on physiological tests only, has been produced by Barnett and Pankhurst (1974). Suitable media for the carbon-source and nitrogen-source assimilation tests are available in dehydrated form from Difco Laboratories Inc.

On the recommended agar media, e.g. on malt extract agar, young yeast colonies in 2–5 days at 25°C characteristically produce domed spherical colonies with entire edges, of a moist and and butyrous consistency, with a shiny, semi-matt or matt smooth surface. The colour of the colony is normally white, cream, pink or salmon. On further incubation the colonies of some species tend to become wrinkled and of a dry and crumbly consistency. The colonies therefore resemble those of some bacteria, but most bacteria tend to grow sparsely on the low-pH media or selective media used for counting or isolating yeasts and moulds. Subsurface colonies of yeasts in pour-plates often have a characteristic stellate appearance.

Microscopic examination of either wet-mounted preparations stained with Gram's iodine or Loeffler's methylene blue or Gram-stained heat-fixed smears normally reveal unicellular spherical, ellipsoidal, ovoid or cylindrical organisms. These may reproduce vegetatively by budding or by binary fission. The production of a pseudomycelium or mycelium is often revealed only by slide cultures. Slide cultures for the growth of yeasts should be carried out with malt extract agar or one of thè potato extract agars.

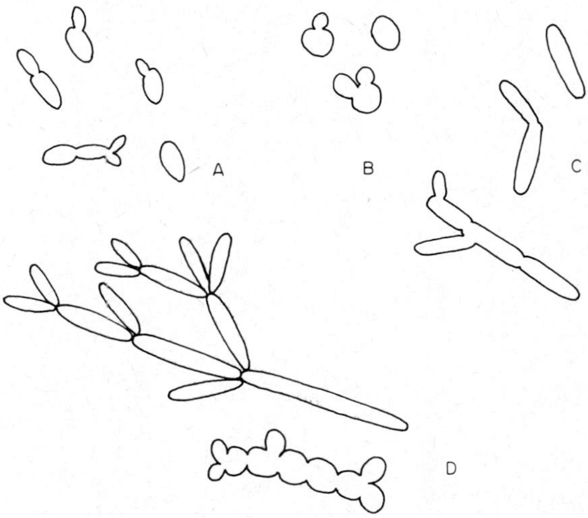

FIG. 1. A, Ovoid budding cells, e.g. *Saccharomyces lactis*, *S. cerevisiae* var. *ellipsoideus*. B, Spherical budding cells, e.g. *Saccharomyces*, *Debaryomyces*. C, Cylindrical budding cells, e.g. *Pichia*. D, Examples of pseudomycelia.

Preparation of Slide Cultures for the Examination of Yeasts

Sterilise a Petri-dish containing a filter paper disc in the base, on which is a U-shaped glass rod carrying two clean microscope slides. After sterilisation, aseptically add a few ml of a sterile 20 per cent solution of glycerol to the filter paper. This will help to keep the slide culture from drying out during incubation. With a Pasteur pipette, add molten malt extract agar or potato dextrose agar to the surface of the slides to form a very thin layer of agar medium. Inoculate the slides with the yeast to be examined in a series of streaks with a wire loop. Place a sterile cover slip over part of the inoculated medium. Incubate at 25°C for up to 7 days with the lid of the Petri-dish in place.

The slide cultures may be examined in the living state with the aid of a phase contrast microscope, and the types of growth under the coverslip and exposed to air can be compared. Alternatively, the coverslip can be gently removed from the slide culture and mounted in a drop of lactophenol, lactophenol-picric acid, or lactophenol-cotton blue on a slide. The edges of the cover slip can be sealed to the slide with paraffin wax, nail varnish, or a proprietary brand of ringing cement.

Morphological and Physiological Characters of Yeasts

Ascospore production. One method of inducing ascospore production has already been described (see page 108). Other methods include subculturing on to gypsum blocks, Gorodkowa agar or V-8 agar.

Ballistospore production. Lodder and Kreger-van Rij (1952) recommended the use of a modification of the slide culture technique to detect and examine ballistospore production by members of the Sporobolomycetaceae. In this method another very thin U-shaped glass rod is sterilised in a Petri-dish together with the filter paper, U-shaped glass rod and microscope slides. Cover only one slide with agar medium. Inoculate this slide with the yeast and invert the inoculated slide above the second slide with the thin U-shaped rod between the two. Replace the lid of the Petri-dish and incubate at 25°C. After incubation, determine the presence or absence of ballistospores by microscopic examination of the lower slide. In addition, examine the upper slide (the slide culture) to determine the morphological and cultural characteristics of the yeasts.

Fermentation of carbohydrates. Inoculate the yeast into tubes (with inverted Durham tubes) containing carbohydrate fermentation broths. The carbohydrate fermentation broths should consist of 0·5 per cent yeast extract with the addition of 2 per cent substrate (see page 327). A control containing no added carbohydrate should be used since the trehalose content of some batches of yeast extract (Jones, 1975) may be high enough for detectable fermentation to occur with trehalose-fermenting yeasts. Incubate at 25°C and examine every day for 10 days.

Utilisation of ethanol as a sole carbon source. Inoculate very lightly a tube of ethanol broth with the yeast under test, and at the same time inoculate a tube of the medium *without* the ethanol. Incubate for up to 3 weeks at 25°C, examining at frequent intervals for growth. Since yeast extract solution is used in the medium as a source of vitamins and growth factors, the growth of the yeast being examined *must* be compared in the tubes with and without ethanol to ensure that false positives are not recorded.

Resistance to Acti-dione (cycloheximide). Inoculate the yeast into filter-sterilised Yeast Nitrogen Base 0392 (Difco) containing 0·5 per cent glucose and 100 μg Acti-dione per ml. Incubate at 25°C for 4 weeks, examining weekly for growth. Good growth within 1 week is taken as indicating high resistance; scant or no growth in 3 or 4 weeks is taken as indicating high sensitivity.

Key to the Identification of Yeasts

Ascospores produced.......................... Ascosporogenous Yeasts
Ballistospores produced.................... Family Sporobolomycetaceae
Neither ascospores nor ballistospores produced...... Asporogenous Yeasts

Ascosporogenous yeasts

May form a pseudomycelium or a true mycelium. Vegetative reproduction is by budding, binary fission, or arthrospores.

1. Vegetative reproduction by fission, with a true mycelium and arthrospores formed. No budding occurs.............................. **1.** *Endomyces*, **2.** *Schizosaccharomyces*
Vegetative reproduction by multilateral budding and sometimes fission.. *2*
Vegetative reproduction by bipolar budding. Includes **3.** *Hanseniaspora*

2. True mycelium formed as well as budding cells...... **4.** *Endomycopsis*
No true mycelium but a pseudomycelium and/or loose collections of budding cells may be formed................. *3* and **5.** *Hansenula*

3. Dry dull pellicle develops rapidly in malt extract broth............ *4*
Dry pellicle develops very slowly if at all in malt extract broth....... *5*

4. Ellipsoidal to long cylindrical cells....................... **6.** *Pichia*
Round or short ovoid cells....................... **7.** *Debaryomyces*

5. Glucose always strongly fermented................ **8.** *Saccharomyces*
Glucose not fermented, or only weakly so.......... **7.** *Debaryomyces*

1. *Endomyces.* Forms a true mycelium, and is more fungal than yeast-like. Ascospores are spherical, ovoid or hat-shaped. The mycelium breaks up into arthrospores which are either cylindrical with rounded ends or ovoid. Vegetative reproduction is by fission.

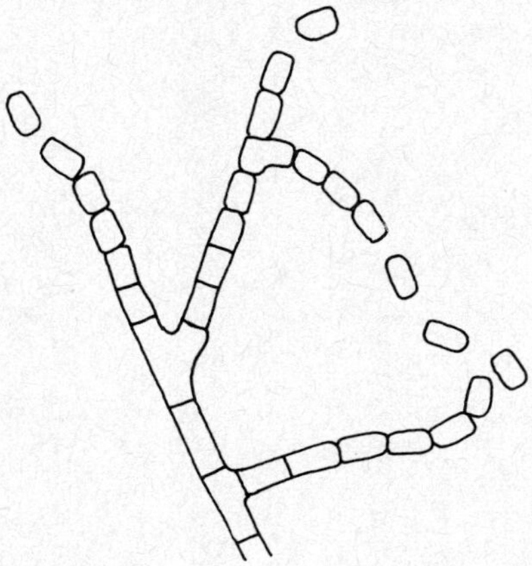

FIG. 2. *Endomyces.*

The equivalent genus in the Fungi Imperfecti is *Geotrichum.* The species *Endomyces lactis* (also known as *Geotrichum candidum* or *Oospora lactis*) is commonly found on dairy products and is therefore called the "dairy mould". The colonies are white in colour and are yeast-like and butyrous, particularly when older.

2. *Schizosaccharomyces.* Members of this genus are found in food products such as molasses, honey, dried fruits, and fruit juices.

The cells are cylindrical, ovoid, or spherical, but a mycelium that breaks up into arthrospores may be formed. Vegetative reproduction is by fission. Four or eight ascospores are formed per ascus. *Schizosaccharomyces* may have a similar appearance to *Endomyces* but in the former asci are formed after the fusion of two arthrospores, whereas in *Endomyces* asci develop in the mycelium at branches of the hyphae.

Schizosaccharomyces ferments carbohydrates. It cannot use ethanol as a sole carbon source.

3. *Hanseniaspora.* The cells are ovoid or lemon-shaped. A pseudomycelium is rarely formed. Vegetative reproduction is by budding at both poles. Ascospores (1–4 per ascus) are spherical, becoming hat-shaped or Saturn-shaped. *Hanseniaspora* can ferment carbohydrates. It cannot use ethanol as the sole carbon source.

4. *Endomycopsis.* This has typical yeast-like, budding cells and in addition may form either a true mycelium that gives rise to blastospores, or a pseudomycelium, in these respects resembling *Candida*.

Ascospores may be one of a variety of shapes, depending on species, including: hat-shaped, Saturn-shaped, sickle-shaped, spherical or ovoid.

A pellicle is usually formed on liquid media. *Endomycopsis* attacks sugars oxidatively and is only very weakly fermentative. It grows only slightly with ethanol as the sole carbon source. Members of the genus have been isolated from bread, fruit and vegetables. Von Arx (1972) has proposed the rejection of this genus.

5. *Hansenula*. This genus produces on solid media white or cream, dull or shiny colonies which are usually wrinkled.

The cells may be spherical, ovoid, ellipsoidal or cylindrical, and budding occurs. Pseudomycelia are frequently formed, particularly by species with cylindrically shaped cells. In liquid media a dry pellicle with a dull appearance is usually formed. Ascospores (1–4 per ascus) are spherical, hat-shaped or Saturn-shaped.

Glucose is vigorously fermented, other carbohydrates being fermented according to species. Some species produce aromatic esters amongst their fermentation products. Ethanol can be utilised as a sole carbon source. This genus is of importance as a contaminant in the brewing industry and may cause spoilage of pickling brines.

6. *Pichia*. Cells are ovoid, ellipsoidal or cylindrical, reproducing vegetatively by budding. A pseudomycelium may be formed. On solid media the colonies are white, or cream, dull, and usually wrinkled. In liquid media a wrinkled, dry pellicle is formed.

Ascospores (1–4 per ascus) are variously shaped, being spherical, hemispherical, hat-shaped, or Saturn-shaped.

Carbohydrates are not usually fermented, or only weakly so. Many species can use ethanol as a sole carbon source. Like *Hansenula*, *Pichia* is an important contaminant of beers and pickled products.

7. *Debaryomyces*. The cells are spherical or ovoid and show multilateral budding; sometimes a pseudomycelium is formed. On solid media the colonies are off-white, yellow or brown and dull to shiny with a wrinkled appearance. A dry pellicle may be produced in liquid media.

Ascospores (usually one per ascus) are spherical or ovoid.

Carbohydrates are not fermented, or only weakly so. Some species can use ethanol as a sole carbon source. *Debaryomyces* has a high salt tolerance and can be frequently isolated from pickled and salted foods.

8. *Saccharomyces*. The members of this genus usually grow as unicellular spherical, ovoid, or longer cells with rounded ends, but a pseudomycelium may sometimes be formed. Vegetative reproduction is by budding. On solid media, colonies are usually white or cream, domed, smooth, semi-matt to shiny, up to 5 mm in diameter, and of a butyrous consistency. In liquid media, sediment is formed, sometimes with ring growth at the air–liquid–glass junction; a pellicle is not formed.

Ascospores (1–4 per ascus) are usually spherical or ovoid.

Carbohydrates are fermented vigorously. Species are defined mainly on the basis of carbohydrate fermentation and carbon-source utilisation (see Lodder, 1970). Although morphological differences between species can be seen, there is usually too great a morphological variation within a species for characteristics such as cell shape to be used with any degree of reliability.

Differentiation of *Saccharomyces*

	Fermentation of							
	Glucose	Maltose	Galactose	Sucrose	Lactose	Fructose	Utilisation of ethanol as sole carbon source	Acti-dione resistance
Saccharomyces cerevisiae	+	+/(+)	+/(+)	+	—	+	V	—
S. cerevisiae var. *ellipsoideus*	+	+/(+)	+/(+)	+	—	+	V	—
*S. fragilis**	+	—	+	+	+	+	—	+
*S. lactis**	+	—	+	+	+	+	+	+
S. rouxii	+	+/(+)	—	—/(+)	—	+	V	—
S. bisporus (includes *S. bisporus* var. *mellis*	+	—	—	—/(±)	—	—	(+)	—
S. uvarum (*S. carlsbergensis*)	+	+/(+)	+/(+)	+	—	+	—	—

* These species are now included in the genus *Kluyveromyces* (Lodder, 1970). A significant difference between the genera is that the asci of *Kluyveromyces* rupture when mature to liberate the ascospores, whereas the asci of *Saccharomyces* do not so rupture.

S. cerevisiae is the "top yeast" used in the brewing industry. *S. cerevisiae* var. *ellipsoideus* is the yeast responsible for vinous fermentations, and is supposedly distinguished from *S. cerevisiae* on the basis of its more ellipsoidal shape, compared with the spherical or short ovoid cell characteristic of *S. cerevisiae*, although there appears to be little differentiation between the two. *S. cerevisiae* is also the typical bakers' yeast.

Several other species responsible for the production of fermented beverages have been described. *S. carlsbergensis*, for example, (which is included now in the species *S. uvarum*) is the "bottom yeast" used in the manufacture of lager.

Saccharomyces (*Kluyveromyces*) *fragilis* and *S.* (*Kluyveromyces*) *lactis* are found in certain fermented milks.

S. rouxii and *S. bisporus* var. *mellis* are osmophilic or osmotolerant and can cause spoilage of jams, honeys and similar products.

Sporobolomycetaceae

Ballistospores are formed and discharged forcibly into the air when ripe, and these may give rise to "mirror images" of the colonies on the lid of the Petri-dish or the lower slide when these yeasts are grown in plate cultures or 2-slide slide cultures. The ballistospores may be spherical, ovoid, kidney-shaped, bean-shaped or sickle-shaped.

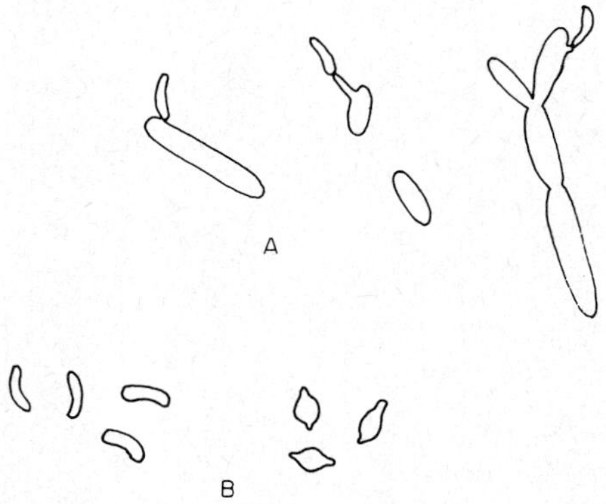

FIG. 3. A, Yeast cells bearing ballistopores. B, Two typical forms of ballistospore.

In liquid media growth is predominantly on the surface, although slight sediment may be formed. Neither *Sporobolomyces* nor *Bullera* can ferment carbohydrates.

Colonies red, pink or salmon-pink in colour, although sometimes very pale. A pseudomycelium or true mycelium may be formed. Ballistospores kidney-shaped or sickle-shaped.................... **9.** *Sporobolomyces*

Colonies colourless, white, cream or yellow. A pseudomycelium or true mycelium is never formed. Ballistospores spherical or ovoid. . **10.** *Bullera*

Asporogenous yeasts

The identification of an organism to generic level may be difficult in this family due to the relatively slight differences between genera and the extreme morphological variability which can be observed in a culture. It is recommended therefore that for confirmation of the identity of a culture and for differentiation at specific level, the monographs of Lodder (1970) and Barnett and Pankhurst (1974) are consulted.

1. Forms budding cells and pseudomycelia, and also a true mycelium that fragments into arthrospores . . . . **11.** *Trichosporon*
No arthrospores produced, although a true mycelium may be formed in addition to budding cells and pseudomycelia *2*
2. Pseudomycelium and/or a true mycelium . . . . **12.** *Candida*
Predominantly unicellular, although a very rudimentary pseudomycelium may be formed *3*
3. Brightly coloured red, orange or yellow . . . . **13.** *Rhodotorula*
Not brightly coloured (may be white, grey, cream, yellowish or brown) . *4*
4. Spherical or ovoid cells pointed at one end ("ogive"-shaped) . . . . **14.** *Brettanomyces*
Spherical or ovoid cells . . . . **15.** *Torulopsis*

11. *Trichosporon*. Morphologically very variable with budding cells, a pseudomycelium, or a true mycelium that fragments into arthrospores. Some species ferment carbohydrates.

12. *Candida*. Morphologically very variable ranging from budding unicellular organisms to pseudomycelia or sometimes a true mycelium. Blastospores are frequently formed on the mycelia.

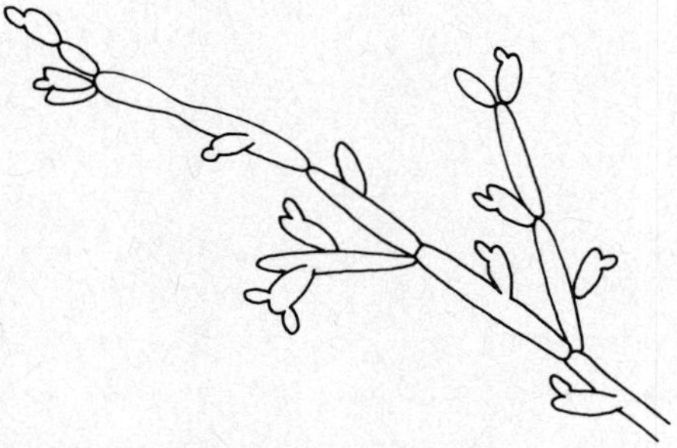

FIG. 4. *Candida* showing blastospore production.

Reproduction is by budding or by fission. On solid media colonies are off-white or cream. Species are differentiated partly by morphological characteristics and partly by carbohydrate fermentation patterns.

C. albicans is an animal parasite or pathogen and is capsulate. Other species may also be pathogenic for animals. A number of species may be isolated from foods and can cause spoilage of fermented beverages and pickled products. *Candida kefyr* (synonyms include *Torulopsis kefyr* and *Candida pseudotropicalis* var. *lactosa*) is involved in the production of kefir and may be isolated from kefir grains, and sometimes from buttermilk.

13. *Rhodotorula.* Usually consists of budding cells, but occasionally a very rudimentary pseudomycelium may be formed. The colonies on solid media are brightly coloured red, pink, orange or yellow. Members of this genus do not ferment carbohydrates. They are frequently involved in the spoilage of a wide range of food products.

14. *Brettanomyces.* Usually consists of budding cells that are "ogive"-shaped (i.e. spherical or ovoid, but pointed at one end). A pseudomycelium may be formed. On solid media, colonies are white to yellowish, and may be glistening, moist and smooth, or dull and wrinkled.

Carbohydrates are fermented. Acetic acid can be produced aerobically from glucose.

15. *Torulopsis.* Consists of spherical or ovoid, budding cells. A mycelium is not formed. A rudimentary pseudomycelium is rarely formed. It is rarely capsulate. On solid media, colonies are off-white, cream, yellowish or brownish, smooth, and may be shiny or dull.

Carbohydrates may be fermented. Some species utilise ethanol as a sole carbon source. Some are osmophilic or halophilic and can cause spoilage in pickled products and foods with a high sugar concentration.

Methods Used in the Examination of Moulds

Moulds are differentiated on the basis of morphological and cultural characteristics so that it is necessary to employ methods of culture and examination that, as far as possible, avoid distortion of sporing structures. In addition, the formulation of the medium on which a mould is grown has a profound influence on the colonial appearance and on the development of aerial sporing structures, chlamydospores, sclerotia, etc. It is advisable, therefore, during the identification of a mould, to grow it on a wide variety of media. As mentioned earlier, these may include malt extract agar, Czapek-Dox agar, potato dextrose agar, potato carrot dextrose agar and Davis's yeast salt agar.

Colonial and cultural characteristics are determined on plate cultures by centrally inoculating a poured plate with spores (obtained with a wire loop from the surface of a colony) or with a small piece of agar containing substrate mycelium from a colony of the mould.

Morphological characteristics sometimes can be determined by wet preparations from plate cultures as described on page 107). Very often, however, serious distortion of sporing structures results from such preparations, with the majority of spores being detached. If this prevents identification, slide cultures should be prepared. There are two main methods of slide culture which can be used. In addition, moulds which have fragmenting mycelia can be examined using the method of slide culture already recommended for yeasts. The methods described below do not include a method of fixing the fungal growth to the slides or coverslips as such fixing usually leads to distortion. If fixation is found necessary, brief heat fixation may be tried. Alternatively, fixation may be accomplished using Bouin's fixative for one hour before removing the agar from the slide or coverslip.

1. *Agar block slide culture.* Into a Petri-dish pour 10 ml of medium and allow it to set. Sterilise a Petri-dish containing a filter paper disc in the base, on which is a U-shaped glass rod carrying two clean microscope slides. After sterilisation, aseptically add a few ml of a sterile 20 per cent solution of glycerol to the filter paper, to keep the slide culture from drying up during incubation. From the solidified medium, aseptically cut blocks of 1 cm square and mount them on the microscope slides. Inoculate the four cut surfaces of each block with the mould to be examined and place a sterile coverslip over the block. Incubate the slide cultures in the Petri-dishes with the lids in place. The slide cultures may be removed from time to time and examined microscopically under the low-power or high-power dry objectives, the coverslips helping to prevent contamination at these times. When a satisfactory development of the mould is observed, wet preparations can be made since the mould tends to adhere to coverslip and slide. Carefully remove the coverslips from the agar blocks and very carefully mount the coverslips in lactophenol, lactophenol-cotton blue or lactophenol-picric acid on fresh slides. In addition, the agar blocks can be removed from the original slides and these slides also examined with the staining mountant and fresh coverslips.

2. *Johnson's slide culture method* (Johnson, 1946). Sterilise a Petri-dish containing filter paper, glass rod and slides, and add sterile 20 per cent glycerol as already described. Place aseptically two narrow parallel strips of sterile vaspar about 2 cm apart across the width of a slide. Centrally between the two strips, place a fungal inoculum on the slide and then cover with a coverslip placed across the vaspar strips (Fig. 5*a*). Using a sterile Pasteur pipette, run sterile melted agar medium (cooled to 55°C) between the slide and the coverslip, until the advancing front of medium just reaches the inoculum, (*b*). This thus provides the mould with a growing edge. Lastly, seal the edge of the coverslip over the agar medium, to prevent the medium from drying out, (*c*). The fourth side of the coverslip is left open.

Incubate the slide culture in the Petri-dish with the dish cover replaced.

The culture can be examined microscopically from time to time and when required, a wet, stained preparation of the coverslip can be examined. To remove the coverslip, cut off the Vaspar seal at the edge of the slide and carefully push the coverslip off the slide using another slide, (*d*). Then cut off

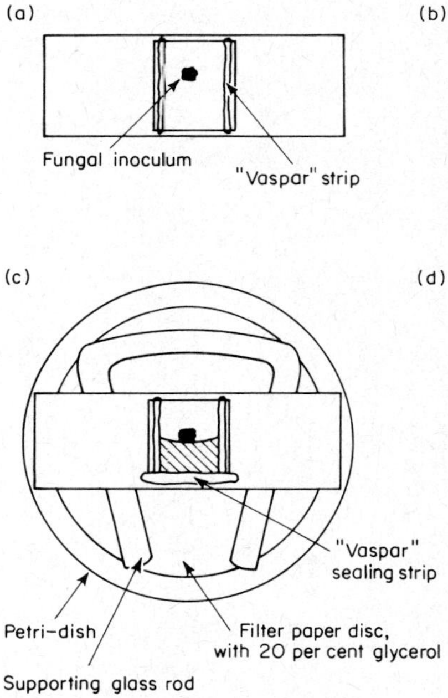

FIG. 5.

the Vaspar strips at each side. Carefully cut away the agar, leaving the fungal growth on the coverslip. Mount the coverslip in one of the lactophenol mounting media.

Key to the Identification of Moulds

The moulds described are those most commonly isolated from foods, dairy products, etc. and identification is primarily on the basis of the morphology of sporing structures.

1. Coenocytic mycelium (septa absent or only present in, for example, chlamydospore formation) *2*
Mycelium possessing many septa *4*
2. Sporangiophores each bear many cylindrical sporangia, each containing a chain of spores . . . . **1.** *Syncephalastrum*
Sporangiophores each bear terminally a single large globose sporangium containing many spores *3*

3. Stoloniferous type of spread, with sporangiophores developing from the nodes at which thick tufts of rhizoids develop...... **2.** *Rhizopus*

Long sporangiophores carry dichotomously-branched short sporangiophores bearing small sporangioles (containing few spores) usually in addition to a large terminal sporangium containing numerous spores.. **3.** *Thamnidium*

Sporangiophores each bear terminally a single large globose sporangium containing many spores. Stoloniferous growth does not occur **4.** *Mucor*

4. Vegetative hyphae clear and transparent, colourless or brightly coloured *5*

Vegetative hyphae darkly coloured, not clear and transparent......*12*

5. Asexually produced spores are unicellular..........................*6*

Asexually produced spores of 2 or more cells..................*11*

6. Spores borne singly on extremely short projections from near the ends of hyphae................................. **5.** *Sporotrichum*

Spores borne in clusters or chains................................*7*

7. Conidia formed in grape-like clusters. (Sclerotia, but not conidia, are formed on media of high C/N ratio, colonies being white and fluffy with black sclerotia at or below the surface of the medium. On media with a low C/N ratio, colonies are greenish-grey with conidia) **6.** *Botrytis*

Spores formed in branched chains...................... **7.** *Monilia*

Spores formed in unbranched chains...............................*8*

8. Conidiophores non-septate at first, but then become septate and produce arthrospores......................... **8.** *Sporendonema*

Non-septate conidiophores arising from thick-walled foot cells. Conidia borne by phialides arising from a terminal swelling on the conidiophore.................................... **9.** *Aspergillus*

Condiophores septate, no specialised foot cells.....................*9*

Conidiophores septate, with foot cells............. **10.** *Stachybotrys*

9. Spores truncated, each with a thickened basal ring................ **11.** *Scopulariopsis*

Spores produced on phialides, and not possessing thickened basal rings. Phialides in brush-like clusters.......................*10*

Spores produced on 3–7 swollen phialides at apex of conidiophore. Dark spores, usually becoming enveloped in slime............... **10.** *Stachybotrys*

10. Clusters of eight-spored asci are formed, without an outer retaining wall (peridium)............................ **12.** *Byssochlamys*

Ascospores, if formed, contained within perithecia............... **13.** *Penicillium*

11. Conidia 2-celled, pear-shaped .... **14.** *Trichothecium*
Conidia many-celled, spindle- or sickle-shaped .... **15.** *Fusarium*

12. Blastospores formed on any part of the mycelium; the colonies are at first slimy, and then become dark greenish-black and leathery...... **16.** *Aureobasidium*
Spores of 1 or 2 cells .... **17.** *Cladosporium*
Spores of more than 2 cells *13*

13. Spores with cross walls; spores usually bent or curved, with one or more of the middle cells thickened .... **18.** *Curvularia*
Spores with cross walls and longitudinal septa; spores usually pear-shaped .... **19.** *Alternaria*

1. *Syncephalastrum.* Possesses a non-septate mycelium of hyphae of large diameter. The sporangiophores, which are usually branched but with the branches not cut off by septa, have a terminal swelling. Each terminal swelling

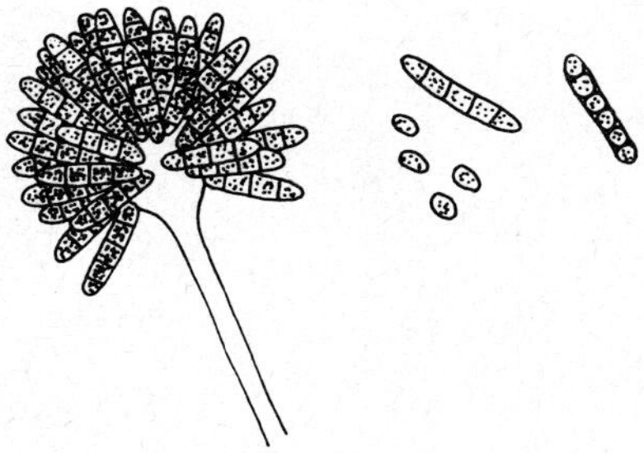

FIG. 6. *Syncephalastrum.*

bears a number of cylindrical sporangia, each of which contains a chain of spores.

The colonies are at first white, but later become grey to black as a result of the production of sporangia, which are black in colour.

2. *Rhizopus.* Mycelium of non-septate hyphae of large diameter. Growth on plate cultures is extremely rapid, by the development of stolons. Where a stolon touches the medium, a tuft of short, rather thick rhizoids grows into the medium, and from this "rooted" node sporangiophores develop. Each sporangiophore bears a terminal black spherical sporangium.

The mycelium is thick and similar to cotton wool, and when sporangia are present these are visible to the naked eye as black pin-heads. A Petri-dish frequently may be rapidly filled with the aerial mycelium. The commonest species is *R. nigricans.* Chlamydospores may be formed by some species, but not by *R. nigricans. Rhizopus* is a fairly common contaminant of foodstuffs, particularly bakery products.

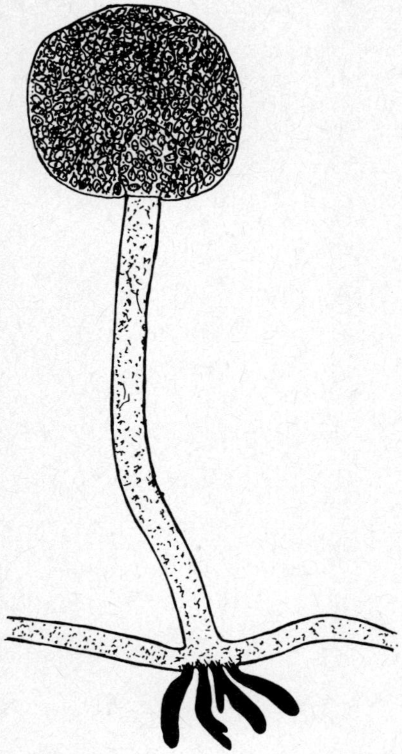

FIG. 7. *Rhizopus.*

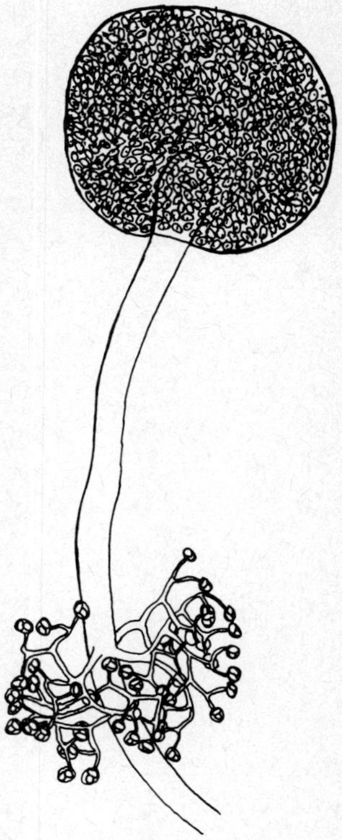

FIG. 8. *Thamnidium.*

3. *Thamnidium*. The hyphae are non-septate and of large diameter. The long main sporangiophores carry whorls of short, dichotomously-branched sporangiophores. Each long sporangiophore terminates in a large globose sporangium containing numerous spores, whereas the short sporangiophores terminate in small sporangioles, each containing 1–4 spores.

On solid culture media under favourable conditions the mycelium formed is loose and cotton-wool-like up to 1 cm or more in height, light grey at first, darkening as a result of the production of sporangia and sporangioles. This mould is occasionally found growing on meat carcasses kept in cold storage.

4. *Mucor*. The hyphae are non-septate and of large diameter, the younger the hyphae the smaller the diameter. The sporangiophores are erect and may be unbranched or branched, each sporangiophore or branch bearing a single globose sporangium, containing a large number of spherical or ellipsoidal spores. The sporangiophores never arise from nodes on stolons, and rhizoids are absent from the mycelium. Chlamydospores may be formed by some species.

Usually growth on a solid medium gives a loose cotton-wool-like aerial mycelium, at first white or grey in colour, later becoming darker as sporangia are produced. The sporangia are often just visible to the naked eye as darkly

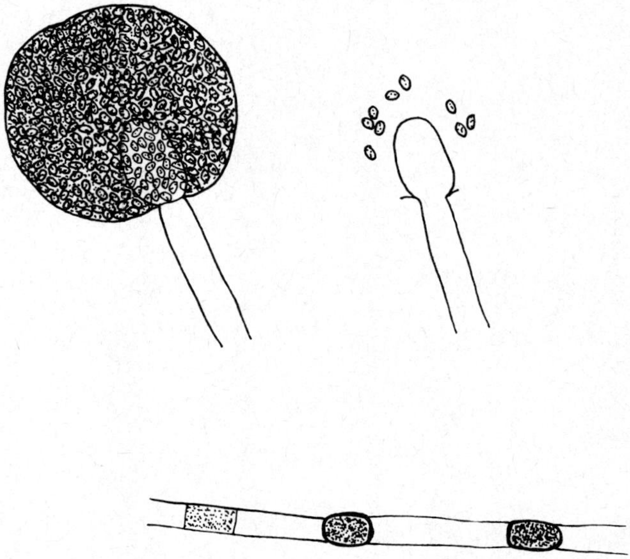

FIG. 9. *Mucor:* entire and dehisced sporangia, and chlamydospores.

coloured or black pin-heads scattered over the aerial mycelium. It should be noted that a few species do not produce a white or grey mycelium at first, being yellow, orange or bluish in colour. In addition, a few species produce yellow or orange sporangia.

Mucor is a very common soil-dwelling organism and therefore may be found contaminating a variety of food products.

5. *Sporotrichum* (*Aleurisma*). Proper conidiophores are not formed: conidia are borne singly on very short projections which arise toward the ends of the hyphae. The conidia may be spherical or ovoid, colourless or brightly coloured and transparent.

Colonies may be white, grey, yellow, green, pink or red, depending on species. As the conidia mature, the colonies change from being velvety to being powdery. The mycelium is fairly closely adpressed to the surface of the medium.

The genus includes soil-dwelling saprophytes and also dermatophytes.

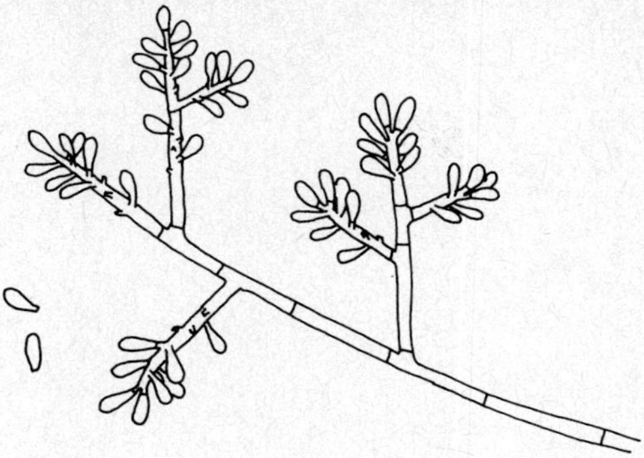

FIG. 10. *Sporotrichum.*

Sporotrichum has been found as a contaminant on the surface of meat carcasses being kept in cold stores.

6. *Botrytis.* Conidiophores are irregularly branched, and the apical cells are frequently enlarged. Conidia are borne singly on short sterigmata, but the positions and numbers of the sterigmata cause the conidia to be arranged in grape-like clusters.

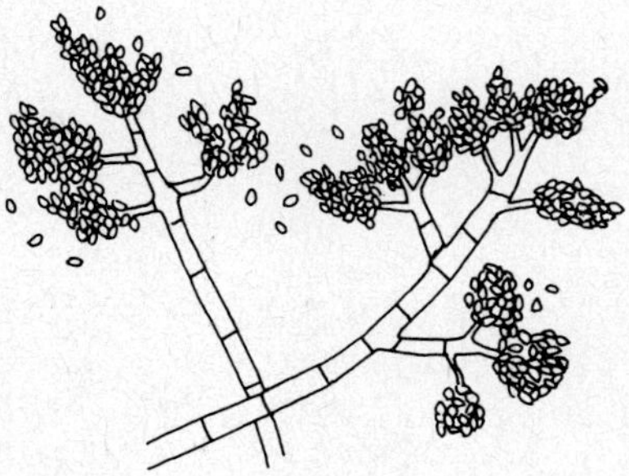

FIG. 11. *Botrytis.*

Botrytis cinerea is the species most commonly encountered. Colonies are frequently white and fluffy at first. On media with a low C/N ratio, conidia are formed. The conidia are usually grey or greenish-grey in mass, hyaline and 1-celled, and the colonies consequently change from being white and fluffy to being greenish-grey and dusty. Media with a high C/N ratio, on the other hand, suppress the production of conidia, but encourage the development of sclerotia. Colonies on such media tend to remain white and fluffy with numerous flat greenish-black sclerotia up to about 3 mm long being formed at or below the surface of the medium. Media with high and low C/N ratios can be made by using malt extract agar with modifications to the concentrations of malt extract and peptone. Reduction in the amount of peptone provides a medium with a high C/N ratio, whereas reducing the amount of malt extract and increasing the amount of peptone produces a medium with a low C/N ratio.

Botrytis cinerea is saprophytic or plant-parasitic. It causes "grey mould" of many plants, and a similar spoilage of fruit and vegetable products. As a saprophyte it is common in soils. Therefore, it is frequently the cause of mould spoilage of strawberries.

7. *Monilia.* Simple or branched chains of conidia occur, branching being due to budding of the conidia. The mycelium as it ages tends to form arthrospores.

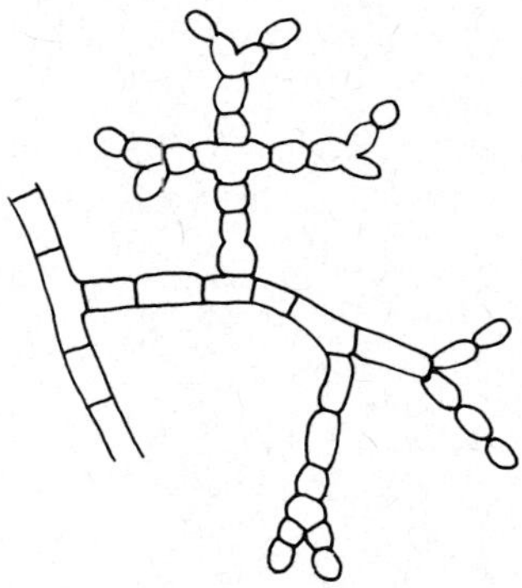

FIG. 12. *Monilia.*

Monilia sitophila is the commonest and most well-known species. It grows best on complex organic media. Colonies are at first white and very loose and fluffy, but they rapidly become pale pink to red in colour due to the production of masses of coloured conidia. The conidia can be produced in great profusion—covering the lid of the Petri-dish and even being found on the outside between the dish and its cover. *M. sitophila* can therefore easily become a contaminant in microbiological laboratories and incubators and its nuisance value is increased by its ability to grow well on nutrient agar and similar bacteriological media and at temperatures up to and including 37°C. It is most frequently associated with spoilage of bakery products—particularly wrapped, sliced bread—causing a pink fluffy growth which is extremely characteristic and easy to recognise.

It should be noted that the conidia are very easily detached and the formation and structure of the chains of conidia—in particular the characteristic branching due to conidial budding—can best be seen in slide culture.

8. *Sporendonema.* Conidiophores are usually non-pigmented, and produced erect from the substrate mycelium. The conidiophore terminates in a hypha which is non-septate at first, but develops septa, and produces endospores (or arthrospores) which are usually pigmented.

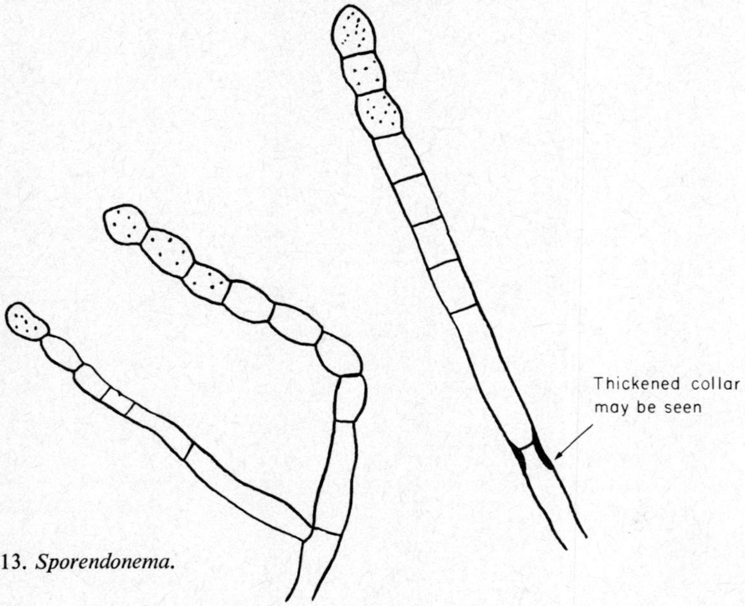

FIG. 13. *Sporendonema.*

Two species are associated with foods: *S. sebi* (synonym *Wallemia ichthyophaga*) which may be found in foods with high sugar or salt contents (it may be found in bakery goods, and in sweetened condensed milk); and *S. casei* which is found associated with dairy products, and may cause orange-coloured spots on cheese rinds.

The two species can be distinguished in plate cultures, by the colour of the arthrospores in the mass, *S. sebi* having brown colonies, and *S. casei* having orange colonies. *Sporendonema purpurascens* may be found on commercially grown mushrooms (Barron, 1968), the arthrospores being bright red in the mass.

9. *Aspergillus.* There are a large number of species in this genus and they are amongst the commonest isolates from soils, spoiled foods, etc. Identification of an isolate to the specific level is made difficult by the variation that can be found in a single strain according to the cultural conditions.

If identification of an isolate is required to specific level, the monograph of Raper and Fennell (1965) should be consulted.

Conidiophores are non-septate and arise from specialised, thick-walled hyphal cells (known as foot-cells). Each conidiophore ends in a terminal enlarged ellipsoidal, hemispherical or spherical swelling which bears phialides either at the apex or over the entire surface.

Conidia, which are unicellular, vary in their colour, shape, and wall marking, according to species. The conidia form unbranched chains arising from the tips of the phialides, the chains being arranged into radiate or mop-like heads or into columnar heads (the structure of the head once again being used as a species characteristic).

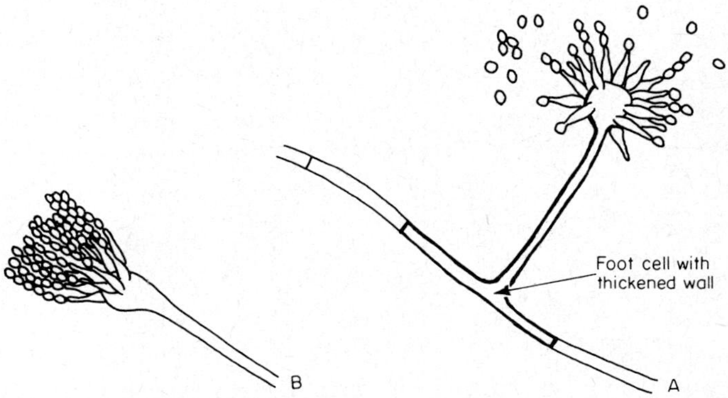

FIG. 14. *Aspergillus*. A, Radiate head. B, Columnar head.

Colonies may be of a wide variety of colours, depending, of course, on the colour of the conidia. The colours include green, brown, black, grey, orange, yellow and off-white.

Amongst the more important aspergilli are those of the *A. glaucus* group, which are frequently osmophilic or at least sugar-tolerant. They are capable therefore of causing spoilage of food products containing high sugar concentrations, for example jams, and such usually indestructible products as "plum puddings". Species in the *A. glaucus* group may be recognised by their greyish-green or green colonies in which bright yellow or orange cleistothecia develop, when grown on a medium such as Czapek-Dox agar with 20 per cent sucrose or glucose.

A. fumigatus is a species that causes a respiratory disease, aspergillosis, in poultry, horses and other farm and domestic animals and in man. The disease in man is usually occupational, being known as farmer's lung, tea-taster's cough, etc. The colonies of this species are dark green and usually velvety. The phialides are produced at the apex of the vesicle, thus giving rise to a columnar head of conidia. The conidia are spherical with rough surfaces.

The *A. niger* group is very common in a variety of habitats. As the name suggests, the heads are black or very darkly coloured resulting in colonies that are black or dark brown. The spherical vesicle bears phialides over its entire surface and thus gives rise to a "mop-like" head of conidia.

10. *Stachybotrys.* Although the vegetative mycelium has transparent hyphae, the conidiophores become dark as they mature and the spores are usually dark in colour. The phialides which bear the conidia are characteristically swollen, and borne as an apical cluster of only 3–7 on each conidiophore. The spores may be spherical, ovoid or cylindrical, and frequently become enveloped in slime.

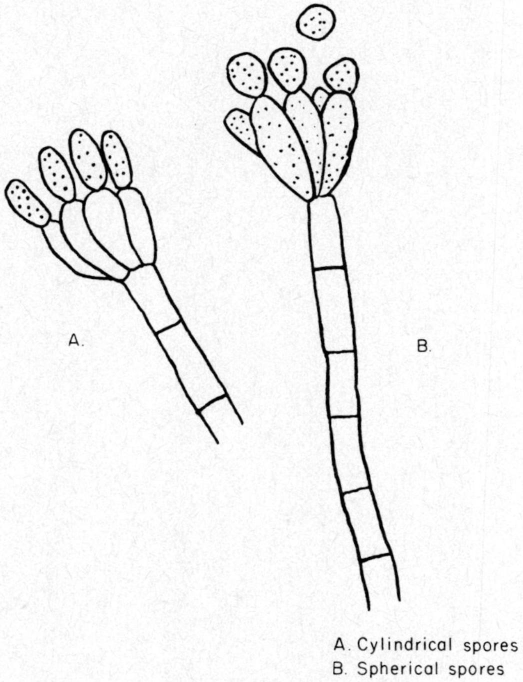

FIG. 15. *Stachybotrys.*

Some species produce spores which are not dark in colour, but cause the colonies to appear pink or salmon in colour. The common species of *Stachybotrys* are cellulolytic.

11. *Scopulariopsis.* The conidia are most characteristic, being truncated spheres, with a thickened basal ring around the truncation. The conidial wall is frequently roughened. The conidia tend to remain together in chains. Conidiophores bear a slight resemblance to those of *Penicillium* in the arrangement of the conidium-bearing structures (not true phialides), but single conidium-bearing "phialides" may occur scattered over the aerial hyphae.

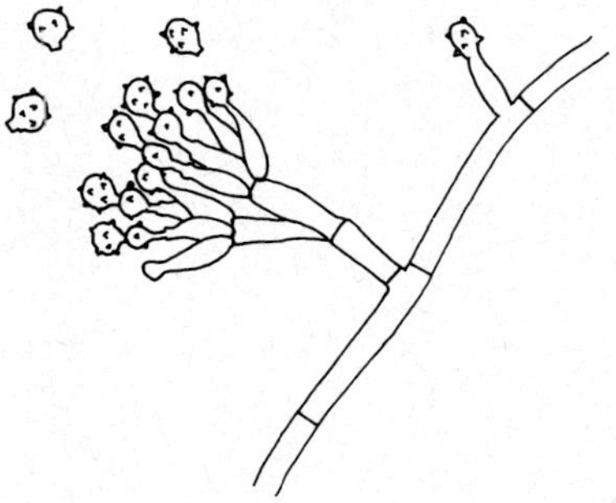

FIG. 16. *Scopulariopsis*.

The colonies may be coloured cream, yellow, brown, or chocolate-brown, but they are never green. *Scopulariopsis* grows well on media (and substrates) containing high concentrations of protein, and may thus cause spoilage of foods such as meat.

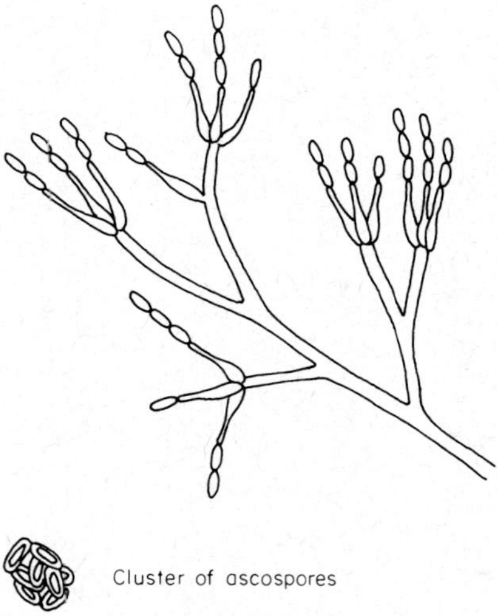

FIG. 17. *Byssochlamys*.

12. *Byssochlamys.* The conidiophore is branching and septate, with the ovoid conidia being borne on small brush-like clusters of phialides. Eight-spored asci are produced.

Byssochlamys fulva is the only species of interest. The colonies on solid media are loose and cotton-wool-like, becoming light brown or tawny when the conidia are produced.

The ascospores are more heat-resistant than any other fungal spores and this characteristic, together with the ability of *B. fulva* to grow in low oxygen concentrations, has led to the mould being the cause of spoilage of canned fruit products.

13. *Pencillium.* As in the case of the genus *Aspergillus* there is a very large

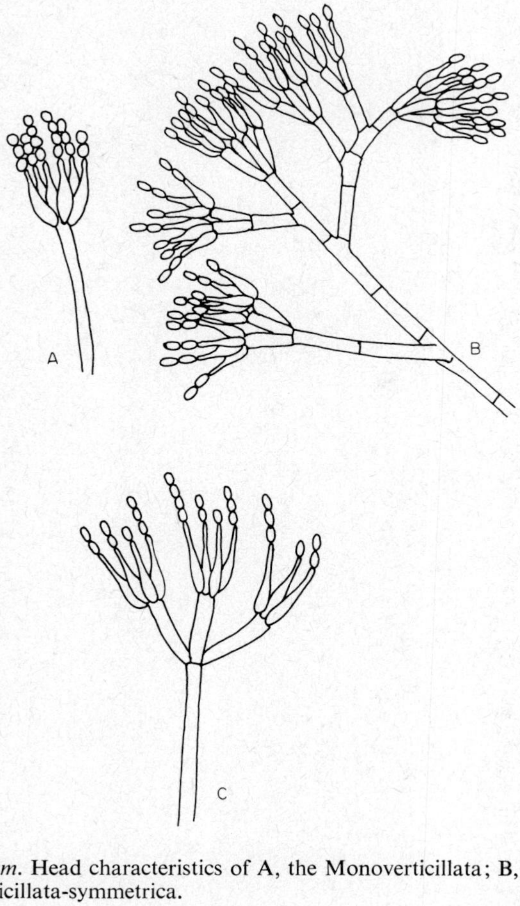

FIG. 18. *Penicillium.* Head characteristics of A, the Monoverticillata; B, the Asymmetrica, and C, the Biverticillata-symmetrica.

number of species, and for species differentiation the monograph of Raper and Thom (1949) should be consulted.

The conidiophore is branched to form a brush-like conidial head. According to the type of branching, the genus is divided into the Monoverticillata, in which each head consists of a single whorl of phialides; the Asymmetrica in which there is more than one branch in the conidiophore, branching being more or less asymmetrical; and the Biverticillata-Symmetrica in which the head consists of a compact whorl of metulae each bearing a number of phialides, the whole head being symmetrical about the conidiophore.

With the production of conidia colonies usually become green, grey-green, blue-green or yellow-green, although *P. caseicolum* remains white. Some species produce sclerotia or perithecia and this may cause colonies to develop areas of a different colour, for example pink, yellow or orange. Some species produce columnar chains of conidia that are readily visible under low-power magnification.

P. camemberti and *P. caseicolum* are of importance in the production of Camembert, Brie and similar cheeses. Colonies of *P. caseicolum* remain white, whereas those of *P. camemberti* gradually become pale grey or grey-green. *P. roqueforti* is used in the production of Roquefort, Gorgonzola and similar blue-veined cheeses.

14. *Trichothecium*. The long, slender septate conidiophores produce 2-celled ovoid or pear-shaped conidia at the tip, the conidia tending to remain together to form a small cluster or short chain. When chains are formed the conidia are not arranged end to end.

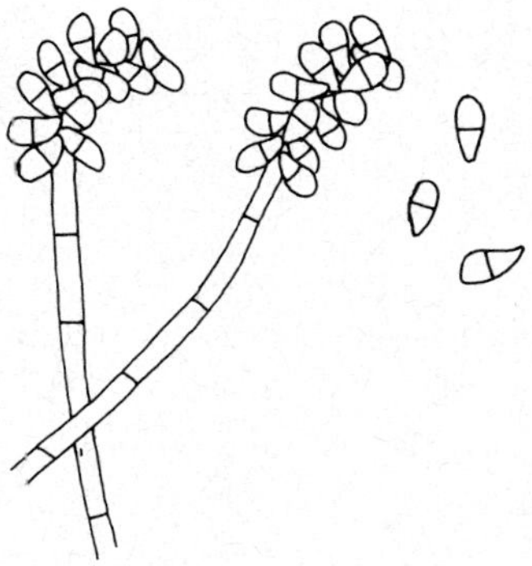

FIG. 19. *Trichothecium*.

Colonies may be white or pale pink. *T. roseum*, the single species, may cause spoilage of certain fruit and vegetable products.

15. *Fusarium*. The macroconidia produced by members of this genus are very characteristic, being multicellular and spindle-shaped or sickle-shaped. However, in addition to macroconidia, 1-celled ovoid microconidia may be formed. In the absence of macroconidia identification becomes rather difficult. The formation of conidia, and the colonial characteristics depend largely on the medium used, but no general recommendations can be made concerning the choice of medium since the correct cultural conditions for the formation of macroconidia vary according to the species. However, potato dextrose agar and potato carrot dextrose agar may be tried with and without the glucose.

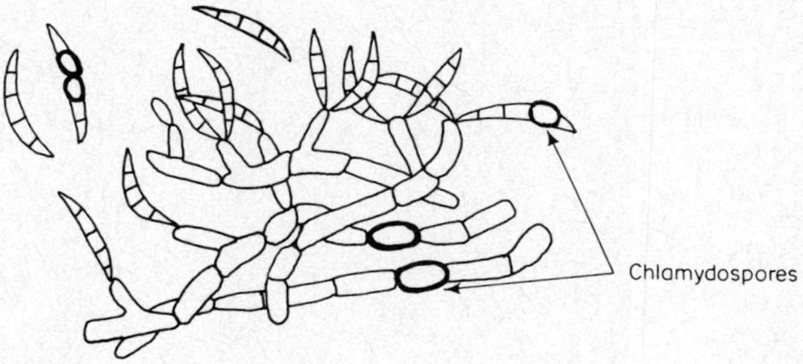

FIG. 20. *Fusarium.*

Colonies on suitable media may be fluffy and spreading, and often coloured rose-pink, salmon-pink, brown, purple or yellow. A typical appearance is for the central part of the colony to be fairly flat and pink in colour with fluffier, white peripheral parts. On media containing high concentrations of sugar, cultures tend to become slimy.

Chlamydospores are frequently formed in the mycelium and even in the conidia. For species differentiation see Gilman (1957).

16. *Aureobasidium*. The mature mycelium is dark-coloured – although

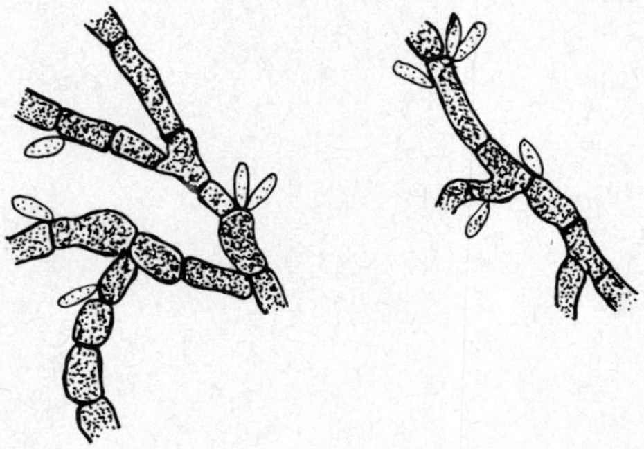

FIG. 21. *Aureobasidium.*

hyaline when younger—and very noticeably septate. Single blastospores are borne laterally on the hyphae. The blastospores may produce buds. The mature hyphae may show variation with some sections being thick-walled and dark, other sections being thin-walled and hyaline.

The colonies are most characteristic, being slimy and lightly coloured or dirty white when young, but rapidly becoming darker. The mature colony is a very dark greenish-black, leathery and shiny, adpressed to the medium. The single species, *A. pullulans*, is a fairly common isolate from soils and may be a contaminant on certain types of food product.

17. *Cladosporium*. The conidia are 1- or 2-celled and produced in branched chains on the conidiophores. The branching of the conidial chains is due to the ability of the conidia to reproduce by budding. Mycelium and conidia are dark in colour.

Colonies are darkly coloured green, olive-green, brown, or black and usually of a fairly thick, velvety nature.

This genus includes the forms producing 1-celled conidia, previously described as *Hormodendrum*.

Members of *Cladosporium* are common in soil, and as contaminants causing spoilage of a wide variety of foods, including meat kept in cold storage.

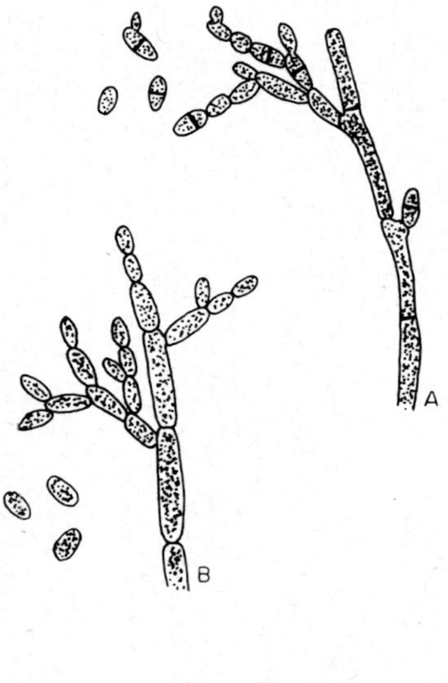

FIG. 22. A, "Cladosporium" type of *Cladosporium*. B, "Hormodendrum" type of *Cladosporium*.

18. *Curvularia.* The bent or curved conidia are 3- to 5-celled, with cross septa only. One or two of the central cells in each conidium are larger and darker than the other cells. These characteristically shaped conidia are produced in clusters at the tips of septate conidiophores which may be branched. Both mycelium and conidia are darkly coloured.

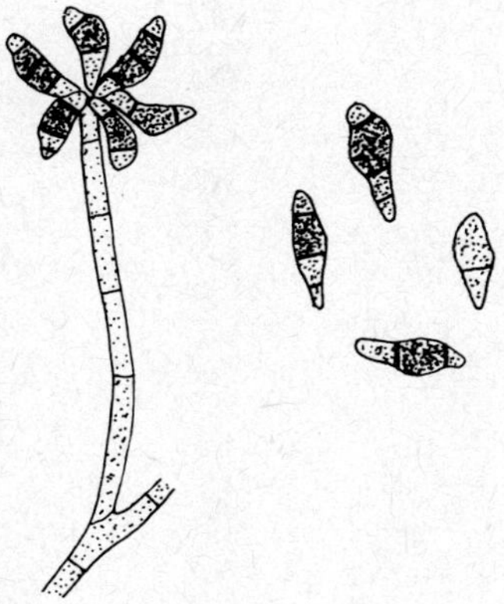

FIG. 23. *Curvularia.*

Colonies may be olive-green, brown or brownish-black; of either a velvety or a fluffy texture.

19. *Alternaria.* The conidia are pear- or club-shaped, multicellular, with both cross-walls and longitudinal septa and produced in chains. The mycelium and mature conidia are darkly coloured.

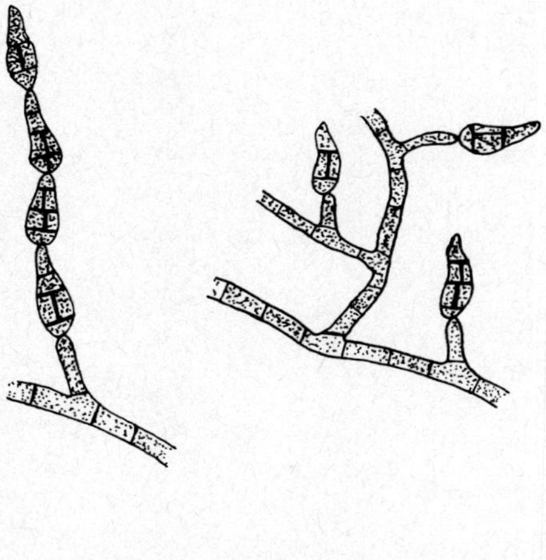

FIG. 24. *Alternaria.*

Colonies may be dark green, greenish-black, brown or yellowish-brown.

APPENDIX I

RECIPES FOR STAINS, REAGENTS AND MEDIA

Chemical Hazards

In the preparation of stains, reagents and media it is inevitable that many dangerous chemicals will be used. Correct handling procedures will minimise the hazards, whilst providing more pleasant working conditions. For example, it should be a strictly observed rule that balances and work benches are kept scrupulously clean. Remember that a worker will know the identity of the chemicals which he is using—and should know the hazards presented by them—but if the apparatus and work benches become contaminated, the next person to use them may not be as fortunate.

The following list should not be regarded as complete, but it indicates the more hazardous chemicals which are used; these are shown in the text by ®. It is as well to distinguish four types of hazardous chemical:

1. A number of these substances are primarily a fire hazard (e.g. ethanol, potassium nitrate).

2. Some are corrosive, cause burns and are dangerous if ingested because of the severe local damage which may be caused to the mouth, oesophagus and stomach (e.g. strong acids and alkalis).

3. Many are toxic, and note should be made of the possibilities and the insidious consequences of chronic toxicity effects. For example the worker frequently using Newman's stain gets accustomed to the odour of low concentrations of tetrachloroethane in the laboratory atmosphere, and may fail to use a fume cupboard, overlooking the long term harm which may ensue.

Great care should be taken when handling chemicals which may be absorbed through the skin (e.g. aniline, xylene, mercury salts).

4. A large number of chemical compounds are now suspect as possible carcinogens. It is a wise precaution to regard, for example, all dyes (e.g. carbol fuchsin) and indicators with suspicion, and to use forceps for handling microscope slides in order to avoid staining the hands. The incorporation of these compounds into media can usually be achieved using standard stock solutions so that the weighing of dry ingredients is minimised (this practice also facilitates the incorporation of the small quantities of such compounds usually required in media). Asbestos mats are largely unnecessary, and alternative heat-protecting materials should be used whenever possible.

Liquids should *not* be pipetted by mouth, and safety pipette fillers (e.g. the "Volac" manufactured by John Poulten, or the Fisher Safety Pipette Filler manufactured by Fisher Scientific Co.) should always be used. The accidental ingestion of poisons is unlikely to occur if this precaution is always observed.

In the unlikely event of such ingestion, vomiting should generally be induced in conscious victims *unless* the poison is corrosive or is a hydrocarbon. (Phenol is an exception in that it is both corrosive and toxic, and vomiting should be induced in this case.) Vomiting can be induced by drinking a tumbler (200 ml) of tepid water containing two tablespoons (30 g) of sodium chloride.

When corrosive poisons have been ingested, the victim should drink a demulcent such as milk, and large quantities of water.

The best treatment for chemical burns (or thermal burns) to the skin is irrigation by large quantities of water.

In all cases seek medical assistance immediately.

Compound	*Principal nature of hazard*
Acetic acid, glacial	Corrosive; causes burns; irritant vapour; flammable
Acetone	Toxic; harmful vapour; flammable
Acti-dione	Toxic; skin irritant
Ammonia (0·880)	Corrosive; causes burns; harmful vapour. Store in a cool place
Amyl alcohol	Irritant vapour; toxic; flammable
Aniline	Chronic toxicity; can be absorbed through the skin; toxic vapour
Chloral hydrate	Harmful vapour; dangerous by ingestion
Chromic acid	Corrosive, causes skin ulcers; powerful oxidising agent
Copper salts	Avoid inhalation of dust, or ingestion; acute toxicity
Ethanol	Highly flammable; intoxicant!
Formaldehyde	Irritant vapour; flammable; chronic toxicity. Avoid contact with chlorine-containing compounds as a volatile carcinogen may be formed. (Dewhurst and Cassels, 1976)
Hydrochloric acid, concentrated	Corrosive; causes burns; irritant gas
Hydrogen peroxide	Corrosive; causes burns; irritant

Compound	*Principal nature of hazard*
Iodine	Corrosive; causes burns; harmful vapour
Lactic acid	Irritant; causes burns
Lead acetate	Acute and chronic toxicity; avoid breathing dust
Mercury and mercury salts	Chronic toxicity; can be absorbed through the skin; harmful vapour (from elemental mercury)
Methanol	Chronic toxicity; harmful vapour; flammable
o-nitrophenyl-β-D-galactopyranoside	Chronic toxicity; may be absorbed through the skin
Orthophosphoric acid	Corrosive; causes burns
Oxalates	Toxic by ingestion
Phenol	Corrosive; causes burns; acute and chronic toxicity; can be absorbed through the skin; harmful vapour
Picric acid	Chronic toxicity; toxic on skin contact or by ingestion; explosive when dry
Potassium dichromate	Chronic toxicity; causes skin ulcers; harmful dust; oxidising agent—fire hazard
Potassium hydroxide	Corrosive; causes burns
Potassium nitrate	Oxidising agent—fire hazard
Potassium tellurite	Chronic toxicity; harmful dust
Silver nitrate	Corrosive; causes burns; toxic
Sodium azide	Toxic by skin absorption or ingestion; explosive
Sodium hydroxide	Corrosive; causes burns
Sodium hypochlorite	Corrosive; causes burns
Sodium hydrogen selenite	Chronic toxicity; possibly teratogenetic (Robertson, 1970); can be absorbed through the skin
Sulphuric acid concentrated	Corrosive; causes burns

Compound	*Principal nature of hazard*
Tetrachloroethane	Extremely harmful vapour; chronic toxicity
Thallium acetate	Chronic toxicity; can be absorbed through the skin
Xylene	Chronic toxicity; harmful vapour; can be absorbed through the skin; flammable

Note that many dehydrated media exist as very fine powders. Care should be taken in dispensing these that the dust is not inhaled, particularly if the medium is known to contain a toxic chemical.

The personnel trained in first aid should attempt to acquaint themselves with the recommended first aid procedures for all toxic substances in common use in the laboratory. A copy of "Hazards in the Chemical Laboratory" (ed. by G. D. Muir) should be displayed in a prominent position in every laboratory.

Ⓗ indicates a hazardous chemical from the above list.

STAINS

Acid-alcohol

(Decolorising agent in the acid-fast staining technique)

Concentrated hydrochloric acid®	3 ml
Ethanol (95 per cent)®	97 ml

(Cowan, 1974)

Bartholomew and Mittwer's Spore Stain

Stain

Saturated aqueous solution of malachite green

Counterstain

Safranin	0·25 g
Distilled water	100 ml

(Bartholomew and Mittwer, 1950)

Carbol Fuchsin, Dilute

(For Gram's staining method and for simple staining)

Ziehl-Neelsen's carbol fuchsin®	10 ml
Distilled water	150 ml

Carbol Fuchsin, Ziehl-Neelsen

(For acid-fast staining)

Basic fuchsin®	1·0 g
Ethanol (95 per cent)®	10 ml
Phenol®, 5 per cent aqueous solution	100 ml

Crystal Violet Stain

(For Gram's staining method and/or simple staining)

Crystal violet (gentian violet)	0·5 g
Distilled water	100 ml

Fontana's Flagella Stain (Modified)

Mordant

Tannic acid (10 per cent aqueous solution)	10 ml
Aluminium potassium sulphate (potash alum), saturated aqueous solution	5 ml

AnilineⒽ water (saturated aqueous solution of aniline)	1 ml
Ferric chloride (5 per cent aqueous solution)	1 ml

Mix the tannic acid and potash alum solutions, then add the aniline water, taking care not to add any drops of undissolved aniline. At this stage a precipitate forms. Shake the tube until the precipitate redissolves, and then add 1 ml of the ferric chloride solution. Allow the mordant to stand for at least 10 min before use. Use the same day.

Ammoniacal silver nitrate solution

Silver nitrateⒽ	5 g
Distilled water	100 ml

To 90 ml of the silver nitrate solution slowly add strong ammonia solutionⒽ (sp. gr. 0·880) until the precipitate which forms *just* redissolves. Add some of the remaining silver nitrate solution drop by drop, shaking between the addition of each drop, until the solution remains slightly turbid after shaking. Store the solution in dark bottles, and in a cool place protected from light. Do not store for too long (up to 3 weeks, if refrigerated), do not expose to sunlight during storage, and do not store at eye level—*this solution can become spontaneously explosive.*

(Rhodes, 1958)

Gram's Iodine Solution

IodineⒽ	1·0 g
Potassium iodide	2·0 g
Distilled water	300 ml

Lactophenol

(For wet microscopic preparations of moulds)

Lactic acidⒽ	100 ml
PhenolⒽ	100 g
Glycerol	200 ml
Water	100 ml

Dissolve the phenol in the water without heat, then add the lactic acid and the glycerol.

Lactophenol-Cotton Blue

(For wet-mounting and staining of moulds)

Cotton blue solution

Saturated solution of cotton blue® (soluble aniline blue)	10 ml
Glycerol	10 ml
Water	80 ml

Mix equal parts of lactophenol and cotton blue solution.

Lactophenol-Picric Acid

(For wet-mounting and staining of moulds)

Lactic acid®	100 ml
Phenol®	100 g
Glycerol	200 ml
Picric acid®, saturated aqueous solution	100 ml

Dissolve the phenol in the picric acid solution without heat, then add the lactic acid and the glycerol.

Leifson's Flagella Stain

Solution A

Basic fuchsin®	1·2 g
Ethanol (95 per cent)®	100 ml

Dissolve with frequent shaking. Store in a tightly stoppered bottle to prevent evaporation.

Solution B

Tannic acid	3·0 g
Distilled water	100 ml

If it is intended to keep this for some time as a stock solution, add phenol® to a concentration of 0·2 per cent to prevent microbial growth.

Solution C

Sodium chloride	1·5 g
Distilled water	100 ml

These solutions are stable at room temperature. To prepare the working solution, mix equal quantities of solutions A, B and C and store in a tightly stoppered bottle in the refrigerator (stable for several weeks) or deep freezer (stable for months). If stored deep frozen, the stain should be well shaken after thawing since the ethanol tends to separate from the water.

Counterstain

Methylene blue	1·0 g
Distilled water	100 ml

(Leifson, 1951, 1958)

Loeffler's Methylene Blue

(For simple staining)

Potassium hydroxide® solution, 1 per cent	1·0 ml
Methylene blue, saturated solution in 95 per cent ethanol®	30 ml
Distilled water	100 ml

Malachite Green

(For staining membrane filters)

Malachite green	0·1 g
Distilled water	1 litre

(Fifield and Hoff, 1957)

Newman's Stain, Modified

(For staining bacteria and bovine cells in milk)

Solution A

Methylene blue	1·0 g
Ethanol, 95 per cent®	54 ml
Tetrachloroethane®	40 ml
Glacial acetic acid®	6 ml

Add the ethanol to the tetrachloroethane and heat in a water bath in a fume cupboard, to a temperature not exceeding 70°C. Add the mixture to the methylene blue, and shake until the dye has completely dissolved. Cool, add the acetic acid very slowly and filter.

Solution B

Basic fuchsin®	0·25 g
Ethanol, 70 per cent	70 ml

Dissolve the dye in the ethanol and filter.

Mix the two solutions thoroughly and store in an air-tight glass stoppered bottle. This mixture will keep indefinitely if not allowed to evaporate. Because of the extremely serious chronic toxic effects of tetrachloroethane, this staining solution should be prepared, stored and used in a fume cupboard.

(Charlett, 1954)

REAGENTS

Andrade's Indicator

(Indicator for acid production from carbohydrates)

Acid fuchsin®	0·5 g
Sodium hydroxide®, N solution	16 ml
Distilled water	100 ml

Dissolve the acid fuchsin in the distilled water and add the sodium hydroxide solution. Allow to stand overnight at room temperature. If the fuchsin has not decolorised to a straw colour, add a further 1–2 ml of sodium hydroxide solution.

(Silverton and Anderson, 1961)

Benedict's Reagent

(For detecting reducing sugars, and used in the test for 3-ketolactose)

Copper sulphate, hydrated®	17·3 g
Sodium carbohydrate, anhydrous	100 g
Sodium citrate, hydrated	173 g
Distilled water	to 1 litre

Dissolve the sodium carbonate and sodium citrate in 600 ml of distilled water, filter and make up to 850 ml. Dissolve the copper sulphate in 100 ml of distilled water and make up to 150 ml. Add the copper sulphate solution to the carbonate and citrate solution very slowly and with constant stirring.

(Benedict, 1908–1909)

Bouin's Fixative

(For fixing slide cultures)

Picric acid®, a saturated aqueous solution	75 ml
Formaldehyde®, 40 per cent solution	25 ml
Glacial acetic acid®	5 ml

Chromic Acid Cleaning Solution

	Recipe 1	*Recipe 2*
Potassium dichromate®	63·0 g	100 g
Distilled water	35 ml	750 ml
Concentrated sulphuric acid®	960 ml	250 ml

Add the potassium dichromate to the water in a 2-litre flask. Slowly and carefully add the acid. Chromic acid cleaning solution can be used repeatedly until it begins to turn green, when it should be discarded. Great care should be taken when preparing and handling this cleaning solution as it is extremely corrosive. *Use protective clothing and eye protection.*

Griess-Ilosvay's Reagent (Modified)

(To test for the presence of nitrite)

Reagent 1

Sulphanilic acid	1 g
Acetic acid®, 5N	100 ml

Reagent 2

α-Naphthol	1 g
Ethanol, 95 per cent®	100 ml

(McLean and Henderson, 1966)

Indole Test Papers

(For the indole test with Kohn's Medium No. 2 and similar media)

p-Dimethylaminobenzaldehyde	5·0 g
Methanol®	50·0 ml
Orthophosphoric acid®	10·0 ml

Saturate 5 cm × 0·5 cm strips of filter paper in the prepared reagent, and heat at 50–70°C for the minimum time required for drying.

(Report, 1958*a*)

Kovacs's Indole Reagent

(To test for indole production in tryptone water)

Amyl or isoamyl alcohol®	150 ml
p-Dimethylaminobenzaldehyde	10 g
Concentrated hydrocholoric acid®	50 ml

Dissolve the aldehyde in the alcohol, then slowly add the acid. Store in the refrigerator.

(Report, 1958*a*)

Kovacs's Oxidase Test Reagent

(To test for the presence of cytochrome oxidase in a bacterial culture)

Tetramethyl-*p*-phenylenediamine hydrochloride®	0·1 g
Ascorbic acid	0·01 g
Distilled water	10·0 ml

The reagent may be kept in a *dark* bottle in a refrigerator, but must be discarded when auto-oxidation has caused it to become purple.

(Kovacs, 1956; Steel 1962)

Lactose Discs

(Inducer discs for ONPG test)

Prepare 1 cm diameter discs from filter paper (Whatman 3 MM). Dissolve 1·5 g lactose in 10 ml of distilled water. Dip each disc once, using forceps, and drain by touching the disc to the side of the bottle. Dry at 37°C for 3 h. Dip a second time and again dry at 37°C.

(Clarke and Steel, 1966)

Lead Acetate Papers

(To test for hydrogen sulphide production)

Soak 5 cm × 0·5 cm strips of filter paper in saturated lead acetate® solution and heat at 50–70°C until dry.

Mercuric Chloride Solution

(Used as a protein precipitant in detecting proteolysis)

Mercuric chloride®	15 g
Concentrated hydrochloric acid®	20 ml
Distilled water	100 ml

(Smith *et al.*, 1952)

Methyl Red Solution

(For the methyl red test)

Methyl red	0·1 g
Ethanol, 95 per cent®	300 ml
Distilled water	to 500 ml

Dissolve the methyl red in the ethanol and make up to 500 ml with distilled water.

Nessler's Reagent

(To test for the production of ammonia)

Potassium iodide	7 g
Mercuric iodide®	10 g
Potassium hydroxide®	10 g
Distilled water	to 100 ml

Dissolve the potassium iodide and mercuric iodide in 40 ml of distilled water. Dissolve the potassium hydroxide in 50 ml of distilled water, and allow to cool. Mix the two solutions and add distilled water to 100 ml. Allow the precipitate to settle, decant the clear supernatant liquid into a reagent bottle, and discard the precipitate.

ONPG Discs

(To test for β-galactosidase activity)

Prepare 1 cm diameter discs from filter paper (Whatman 3 MM). Dissolve 20 mg of o-nitrophenyl-β-D-galactoside® in 10 ml of distilled water. Using plastic forceps dip each disc once, and drain by touching the disc to the side of the bottle. Dry at 37°C for 3 h.

(Clarke and Steel, 1966)

Phenylalanine Discs

(To test for phenylalanine deaminase activity)

Prepare 1 cm diameter discs from filter paper (Whatman 3 MM). Dissolve 0·1 g of DL-phenylalanine in 10 ml of distilled water. Dip each disc once, drain by touching the disc to the side of the bottle. Dry at 37°C for 3 h.

(Clarke and Steel, 1966)

Voges-Proskauer Test Reagents

1. *O'Meara's modification* (O'Meara, 1931)
   (*a*) Sodium hydroxide®, 40 per cent solution
   (*b*) Creatine

2. *Barritt's modification* (Barritt, 1936)
   (*a*) Potassium hydroxide®, 16 per cent solution
   (*b*) α-Naphthol, 6 per cent solution in 95 per cent ethanol®

MEDIA

The methods for the preparation of the media used in this book are given below. In the case of salts with more than one hydrated form the hydrate to be used is specified in the recipe. With some media it is much more convenient to use the dehydrated version (e.g. as in the case of violet red bile agar) or the ready-prepared version (e.g. as in the case of Dorset's egg medium) which may be commercially available, unless large quantities are regularly required.

The media that are available in either dehydrated or ready-prepared form from Oxoid Ltd. are indicated after the name by (*). Similarly media available from Difco Laboratories are indicated by (†) and media available from Baltimore Biological Laboratories by (‡). Media available in a basal form that requires the addition of substrate or supplement are indicated by B. Dehydrated media frequently are provided in powder rather than granular form. If preparing such a dehydrated medium of a formulation that contains a toxic constituent (particularly if there is a hazard of chronic toxic effects) care should be taken to avoid inhaling or ingesting dust which may contaminate the air.

When preparing media, care should be taken that each ingredient or the dehydrated medium is adequately mixed into the water. If mixing is inadequate stratification of dense, concentrated solutions will occur and on heating there will be deleterious changes caused by caramelisation, hydrolysis, etc. When using dehydrated media it is good practice to allow the powder to soak in the water for 15 min with frequent mixing by agitation, before heating. Agar media being dissolved by the application of heat to the base of the container should be agitated to prevent the agar granules from settling out and being burned. The manuals provided by the manufacturers of dehydrated media give detailed recommendations on the procedures for reconstitution and sterilisation and these should be followed closely.

The recipes for some media recommend heat treatments insufficient to ensure sterility, often because the medium is susceptible to heat degradation. In such cases it is necessary to incubate the media for 2 days at 37°C and 2 days at 30°C to detect any residual contaminants; if the media are to be used for the growth of psychotrophs or thermophiles, pre-incubation at the appropriate temperature should also be carried out. Alternatively uninoculated controls should be incubated concurrently with the tests. Such procedures are an advisable precaution with *all* media.

In most recipes the type of closure to be used with test-tubes has not been specified. Non-absorbent cotton-wool, autoclavable polypropylene, alloy or stainless steel caps may be used according to personal preference. In some situations cotton-wool may be contra-indicated; for example there may be inhibitory effects on fastidious organisms because of the adventitious introduction of fatty acids (Meynell and Meynell, 1970), or turbidity measurements made to assess growth may be disturbed by the introduction of cotton-wool fibres into the medium.

Rubber bung closures (especially of the vented Astell seal design) are suitable for tests on many organisms, but the obstruction to gas exchange with the external atmosphere may affect the growth and biochemical test reactions, especially of obligate aerobes. Such closures do have the advantage of extending the shelf-life of media by preventing evaporation (as does the use of screw-capped bottles), but this in itself may disguise the unsuitability of media subjected to prolonged storage because of chemical changes.

Acetate Agar

(A selective medium for the isolation or enumeration of lactobacilli)

1. Basal medium

Peptone	10·0 g
Lab-Lemco meat extract (Oxoid)	10·0 g
Yeast extract (Difco)	5·0 g
Agar (Oxoid No. 3)	15·0 g
Tri-ammonium citrate	2·0 g
Salts solution§	5·0 ml
D-Glucose	10·0 g
Tween 80 (Atlas Chemicals)	0·5 ml
Distilled water	1 litre

pH 5·4

§*Salts solution*

Magnesium sulphate, hydrated ($MgSO_4.7H_2O$)	8·0 g
Manganese sulphate, hydrated ($MnSO_4.4H_2O$)	2·0 g
Distilled water	100 ml

Dissolve the peptone, meat extract, yeast extract and agar in 1 litre of distilled water by autoclaving at 121°C for 15 min. Then add the citrate and salts solution and mix well. Adjust the pH to 5·4 and filter if necessary. Add the glucose and Tween 80 ("Mycobacterium culture grade", available from Matheson Scientific, see p. 393 under Atlas Chemical Industries, Inc.) and mix thoroughly. Dispense in 90 ml amounts in screw-capped bottles. Sterilise by autoclaving at 121°C for 15 min.

2. Preparation of 2 M acetic acid-sodium acetate buffer at pH 5·4

Dissolve 23·3 g of sodium acetate (hydrated) and 1·7 g of glacial acetic acid® in distilled water and make up to 100 ml. Check the pH by diluting a portion 1 : 10 in distilled water before using the pH meter (dilution is necessary because the high concentration of sodium ions may affect the glass electrode). Distribute in screw-capped containers in 10 ml amounts, and sterilise by autoclaving at 115°C for 20 min.

3. Preparation of the medium for use

The medium may be used with *or* without the acetic acid-sodium acetate buffer (see the appropriate section in Part II). If the medium is to be used with buffer added, add aseptically 10 ml of the buffer solution to 90 ml of the medium which has been melted and cooled to 50°C. Mix gently by careful inversion to avoid frothing, and pour double layer plates in the usual way. The final pH of the complete medium should be 5·4 ± 0·05.

(Keddie, 1951 and personal communication)

Arginine Broth

(A medium to test for the production of ammonia from arginine by streptococci)

Tryptone	5·0 g
Yeast extract	2·5 g
D-Glucose	0·5 g
Dipotassium hydrogen phosphate	2·0 g
L-Arginine monohydrochloride	3·0 g
Distilled water	1 litre

pH 7·0

Dissolve all the ingredients in the water and sterilise at 121°C for 15 min.

(Abd-el-Malek and Gibson, 1948)

Arginine MRS Broth

(A medium to test for the production of ammonia from arginine by lactobacilli)

MRS broth + 0·3 per cent (w/v) L-arginine monohydrochloride. Add the arginine during the preparation of the MRS broth before sterilisation.

(Sharpe, 1962)

Arginine Tetrazolium Agar

(A medium to differentiate *Streptococcus cremoris* and *Strep. lactis*)

Tryptone	5·0 g
Yeast extract	5·0 g
L-arginine monohydrochloride	2·0 g
K_2HPO_4	2·0 g
Glucose	0·5 g
Agar	15·0 g
Distilled water	to 1 litre

pH 6·0

Dissolve the ingredients in the water by steaming. Mix, adjust the pH to 6·0, distribute in 100 ml amounts in screw-capped bottles, and sterilise by autoclaving at 121°C for 20 min.

For use melt 100 ml of medium, and cool to 50°C. Add 1 ml of a 0·5 per cent filter-sterilised solution of triphenyl tetrazolium chloride, mix well and pour plates. Dry plates before use, and spread 0·1 ml inocula.

(Turner, Sandine, Elliker and Day, 1963)

Baird-Parker's Medium (Egg-yolk Tellurite Glycine Agar)*†‡(B)

(A selective and diagnostic medium for isolation and enumeration of *Staphylococcus aureus*)

Tryptone (Difco)	10·0 g
Lab-Lemco meat extract (Oxoid)	5·0 g
Yeast extract (Difco)	1·0 g
Lithium chloride, hydrated	5·0 g
Agar (Difco)	20·0 g
Sulphadimidine sodium salt, 0·2 per cent solution	25·0 ml
Distilled water	975 ml

pH 6·8–7·0

Dissolve the ingredients by steaming. Adjust to pH 6·8. Dispense without filtration in 90 ml amounts in screw-capped bottles and sterilise by autoclaving at 121°C for 15 min.

Before pouring the plates, to each 90 ml of basal medium, melted and cooled to 50°C, add aseptically the following solutions (all previously sterilised by filtration) in the amounts given.

(*a*) 20 per cent (w/v) solution of glycine	6·5 ml
(*b*) 1 per cent (w/v) solution of potassium tellurite® (BDH Chemicals Ltd.)	1·1 ml
(*c*) Egg-yolk emulsion	5·4 ml

Mix well and pour into Petri-dishes in 12 ml amounts. These poured plates may be stored in plastic bags at 4°C for up to 1 month. Before use 0·5 ml of a 20 per cent (w/v) solution of sodium pyruvate (Koch-Light Laboratories) should be spread aseptically over the surface of each plate. The plates are then dried (with the medium *surface* uppermost) at 50°C for 1 h, prior to inoculation.

Note.

1. The 0·2 per cent solution of sulphadimidine sodium salt is prepared by dissolving 0·5 g of pure sulphadimidine (sulphamezathine, I.C.I. Ltd. Pharmaceuticals Division) in 25 ml of 0·1 N sodium hydroxide solution and making up to 250 ml with distilled water.

2. The method of preparation of the egg-yolk emulsion is given on page 334. It must be stored in a refrigerator.

3. The sodium pyruvate solution must be stored in a refrigerator and used within one month.

4. As indicated, this medium is available in dehydrated form, as a basal medium, to which must be added egg yolk, tellurite and the sulphadimidine salt. A ready-prepared egg-yolk tellurite emulsion is available from Oxoid Ltd. and from Difco Laboratories, which can be used in conjunction with these dehydrated media.

(Baird-Parker, 1962, 1969; Smith and Baird-Parker, 1964; Holbrook, Anderson and Baird-Parker, 1969)

Barnes's Thallium Acetate Tetrazolium Glucose Agar

(For the detection and enumeration of faecal streptococci)

Basal medium

Peptone (Difco)	10·0 g
Yeast extract	10·0 g
Agar	15·0 g
Distilled water	1 litre

pH 6·0

Dissolve the ingredients in the water by steaming, cool to 50–60°C and adjust to pH 6·0. Distribute in 95 ml amounts in screw-capped bottles, and autoclave at 121°C for 20 min.

Preparation of complete medium

To 95 ml of molten medium, cooled to 45–50°C, add the following:

2 ml of a 5 per cent aqueous solution of thallium acetate® (previously sterilised by autoclaving at 115°C for 15 min)

1 ml of a 1 per cent aqueous solution of 2,3,5-triphenyl tetrazolium chloride (previously sterilised by filtration or by steaming for 30 min)

5 ml of a 20 per cent solution of glucose (previously sterilised by filtration or by steaming for 30 min on 3 successive days)

Mix the medium well after each addition and finally pour the plates as required.

(Barnes, 1956)

Blood Agar*†‡(B)

Sterile defibrinated blood	5 ml
Nutrient agar (modified)	100 ml

The nutrient agar base should preferably contain an extra 3·5 g of sodium chloride per litre of medium. Blood agar bases are also available commercially as dehydrated media.

Horse-blood is most suitable for streptococci, but for studying the growth and haemolytic reactions of staphylococci, sheep-, rabbit-, or ox-blood may be required.

Liquefy the basal medium, cool to 45–50°C and add aseptically the sterile blood. Mix well and pour the plates. Layer plates may be prepared by pouring into each plate 5 ml of blood agar on top of a thin (5–10 ml) layer of the basal medium previously poured and allowed to solidify.

Brilliant Green Agar*†‡

(For the detection and isolation of *Salmonella*, other than *S. typhi*)

Peptone	10·0 g
Yeast extract	3·0 g
Sodium chloride	5·0 g
Lactose	10·0 g
Sucrose	10·0 g
Phenol red, 0·2 per cent aqueous solution	40·0 ml
Brilliant green, 1·0 per cent aqueous solution	12·5 ml
Agar	15·0 g
Distilled water	948 ml

pH 6·9

Dissolve the solid ingredients in the water by steaming, then add the solutions. Dispense in 100 ml amounts and autoclave at 121°C for 15 min. *Avoid over-heating.* The final pH should be approximately 6·9, and the medium will be orange in colour.

To use pour thick plates of 15 ml and dry the surface of the plates before inoculation. It should be noted that if bottles of medium have not been required after melting, a subsequent re-melting is likely to lead to a decrease in selectivity.

(Thatcher and Clark, 1968)

Brilliant Green Lactose Bile Broth*†‡

(Selective medium for isolating and counting coliform organisms and *Escherichia coli*)

Peptone	10·0 g
Lactose	10·0 g
Ox-bile (Oxoid L50 or equivalent)	20·0 g
Brilliant green, 1 per cent aqueous solution	13·3 ml
Distilled water	to 1 litre

pH 7·4

Dissolve the peptone and lactose in 500 ml of distilled water. Dissolve the ox-bile in 200 ml of distilled water. Mix the two solutions, add distilled water to 950 ml, and adjust to pH 7·4. Add 13·3 ml of a 1 per cent aqueous solution of brilliant green and then make up the volume to 1 litre with distilled water. Dispense into test-tubes containing inverted Durham tubes, plug and sterilise by autoclaving at 121°C for 15 min.

(Mackenzie, Taylor and Gilbert, 1948)

Bromcresol Green Ethanol Yeast Extract Agar

(For the differentiation of *Gluconobacter* (*Acetomonas*) and *Acetobacter*)

Basal medium

Yeast extract (Difco)	30·0 g
Bromcresol green, 2·2 per cent aqueous solution	1·0 ml
Agar	20·0 g
Distilled water	1 litre

Dissolve the yeast extract and agar in the water by steaming, and add the bromcresol green solution. Mix well and dispense in 6·5 ml amounts in 30 ml (1 oz) screw-capped bottles. Autoclave at 121°C for 15 min.

Preparation of complete medium

Melt the bottles of medium, cool to 45–50°C and add aseptically to each bottle 0·3 ml of 50 per cent ethanol (previously sterilised by filtration). Mix and allow to set in a sloped position.

(Carr, 1968)

Bromcresol Purple Milk

(For a simple test of the hygienic quality of cream)

Skim milk powder (Oxoid L31 or equivalent)	100·0 g
Distilled water	to 1 litre
Bromcresol purple, 1 per cent solution	10 ml

Mix the powder to a smooth paste with a small amount of water, and gradually add more distilled water to a final volume of 1 litre. Add and mix 10 ml of bromcresol purple solution. Dispense in 10 ml amounts in test-tubes. Sterilise by autoclaving at 121°C for *5 min* only, followed on two successive days by steaming for 20 min. Check sterility by incubating for 48 h at 30°C.

(Crossley, 1948)

Butter-fat Agar

Butter-fat	5 ml
Yeast extract agar or nutrient agar	10 ml

Before sterilisation of the basal medium, adjust the reaction to pH 7·8. Prepare the butter-fat by warming fresh, unsalted butter in a beaker in a 50°C water-bath until liquid. Separate the butter-fat from the curd by careful pipetting, by filtration, or with a separating funnel. Dispense the butter into screw-capped bottles and sterilise by autoclaving at 121°C for 20 min.

To 10 ml of molten, sterile basal medium at 45°C add aseptically 5 ml of molten sterile butter-fat. Emulsify the medium by shaking vigorously. Pour plates containing 15 ml amounts.

(Berry, 1933)

Calcium Lactate Yeast Extract Agar

(For the differentiation of *Gluconobacter* (*Acetomonas*) from *Acetobacter*)

Yeast extract (Difco)	20·0 g
Calcium lactate	20·0 g
Agar	15·0 g
Distilled water	1 litre

Mix the ingredients and dissolve by steaming. Mix and distribute as required. Sterilise by autoclaving at 121°C for 15 min.

(Carr, 1968)

Calgon-Ringer's Solution*

(For the suspension of alginate swabs)

Prepare as Ringer's Solution, Quarter-Strength (*q.v.*) and add sodium hexametaphosphate ("Calgon") at 10 g per litre. Dispense in 10 ml amounts in screw-capped bottles, and sterilise by autoclaving at 121°C for 15 min.

(Higgins, 1950)

Carbohydrate Fermentation Broths (*†‡ to various recipes, and (B))

Basal medium 1 (general purpose medium):

Tryptone	10·0 g
Sodium chloride	5·0 g
Bromcresol purple, 1 per cent solution§	2·5 ml
Distilled water	1 litre

§ Andrade's reagent to a final concentration of 1 per cent, or phenol red to a final concentration of 0·01 per cent may be used in place of the bromcresol purple if desired.

Add the ingredients (except the indicator) to the water and dissolve by steaming. Adjust to pH 7·2 and then add the indicator. Distribute in 3 ml or 5 ml amounts in screw-capped bottles or test-tubes, each bottle or tube being provided with an inverted Durham tube. Sterilise at 121°C for 20 min.

Basal medium 2 (recommended in Report, 1958a, for tests on the Enterobacteriaceae):

Nutrient broth: pH indicator, distribution, etc., as described above.

Basal medium 3 (recommended as the basal medium for tests on yeasts) 0·5 per cent solution of yeast extract powder: pH indicator, distribution, etc., as described above.

Basal medium 4 (recommended for tests on fastidious animal parasites and pathogens, e.g. *Corynebacterium*):

Mix one part of sterile bovine serum with 3 parts of sterile distilled water, adjust to pH 7·6; to each 400 ml of medium add 20 ml of a 0·2 per cent phenol red solution. Dispense in 2·5 ml amounts in screw-capped 7 ml (bijou) bottles and sterilise by steaming for 20 min on 3 successive days.

Phenol red solution is prepared by dissolving 1·0 g of phenol red in 28·4 ml of 0·1 N sodium hydroxide with gentle heating. Distilled water should then be added to approximately 400 ml, 28·4 ml of 0·1 N hydrochloric acid added, and the solution made up to 500 ml with distilled water and filtered before use.

Basal medium 5 (recommended for fermentation tests on lactobacilli). See MRS Fermentation Medium.

Substrate solutions

Prepare the substrates as 10 per cent solutions in distilled water (salicin has low solubility and should be prepared as a 5 per cent solution), and sterilise by filtration.

Preparation of complete fermentation broth

To each 2·5 ml, 3 ml, or 5 ml amount of sterile basal medium, add aseptically 0·25 ml, 0·3 ml or 0·5 ml respectively of the desired substrate solution in the case of media *1*, *2* and *4*. In the case of fermentation broths for tests on yeasts using medium *3*, and for tests on lactobacilli using MRS fermentation medium the substrate should be added to a final concentration of 2 per cent. Salicin solution should be added in double the above quantities in order to achieve an adequate final concentration of substrate.

Cheese Agar

(For the isolation of *Brevibacterium linens*)

Ripened cheese	100·0 g
Potassium citrate	10·0 g
Peptone	10·0 g
Sodium chloride	50·0 g
Sodium oxalate®	2·0 g
Agar	15·0 g
Distilled water	1 litre

pH 7·4

Dissolve the potassium citrate in 300 ml of distilled water and add the cheese. Warm the suspension to 50°C, then transfer to a funnel with rubber tubing and spring clip and allow to stand for 30 min to allow separation of the fat. Run off the aqueous portion and add to the remainder of the ingredients dissolved in 700 ml of distilled water. Adjust the pH to 7·4 and distribute as required. Sterilise at 121°C for 25 min. When using the medium it must be thoroughly agitated to distribute suspended cheese solids.

(Albert, Long and Hammer, 1944)

Christensen's Urea Agar*†‡ (B)

(For differentiation within the Enterobacteriaceae)

Peptone	1·0 g
Sodium chloride	5·0 g
Potassium dihydrogen phosphate	2·0 g
D-Glucose	1·0 g
Phenol red, 0·2 per cent aqueous solution	6·0 ml
Agar	20·0 g
Distilled water	1 litre

pH 7·0

Dissolve the ingredients in the water by steaming. Distribute in bottles or test-tubes in sufficient quantities to enable slants to be prepared, and sterilise by autoclaving at 121°C for 15 min. Cool to 50°C and to each tube add aseptically sufficient sterile 20 per cent urea solution (previously sterilised by filtration) to give a final concentration of 2 per cent. Allow the medium to set in the sloped position.

(Christensen, 1946)

CPS Medium

(For the isolation and enumeration of aquatic bacteria)

Soluble casein (BDH)	0·5 g
Peptone (Difco)	0·5 g
Soluble starch	0·5 g
Dipotassium hydrogen phosphate	0·2 g
Magnesium sulphate, hydrated ($MgSO_4.7H_2O$)	0·05 g
Ferric chloride, 0·01 per cent solution	4 drops
Glycerol	1 ml
Agar (Difco)	15·0 g
Distilled water	1 litre

pH 6·9–7·0

Dissolve the casein, peptone, starch and agar in the water by steaming. Add the remaining ingredients to the hot medium, mix well and filter. Distribute as required and sterilise by autoclaving at 121°C for 20 min.

(Collins *et al.*, 1973)

Crossley's Milk Peptone Medium*

(For sterility testing of evaporated milk and other canned foods)

Skim milk powder (Oxoid L31 or equivalent)	100 g
Peptone	10 g
Bromcresol purple, 1 per cent aqueous solution	10 ml
Distilled water	1 litre

Mix the milk powder to a smooth paste with a little of the water and gradually stir in the rest of the water. Dissolve the peptone and mix the bromcresol purple into the medium. Distribute in 10 ml amounts in test-tubes or as required. Sterilise by autoclaving at 121°C for 5 min, followed by steaming for 30 min on each of two successive days. Incubate at 37°C for 2 days and at 30°C for two days prior to use to check sterility.

(Crossley, 1941)

Crystal Violet Agar

(A medium for the isolation of Gram-negative bacteria)

Yeast extract agar, plate count agar, or nutrient agar	100 ml
Crystal violet, 0·05 per cent aqueous solution	0·4 ml

Add aseptically the crystal violet solution (previously sterilised by filtration) to the sterile molten medium immediately before pouring the plates. This gives a final concentration of crystal violet of 2 ppm.

(Holding, 1960)

Cystine Broth

(For detecting the ability to produce hydrogen sulphide)

Add cystine to a final concentration of 0·01 per cent to a basal medium of peptone water or nutrient broth. Distribute in 10 ml amounts in test-tubes, plug and sterilise at 121°C for 15 min.

Czapek-Dox Agar*†‡

(For the culture of yeasts and moulds)

Sodium nitrate	2·0 g
Potassium chloride	0·5 g
Magnesium sulphate, hydrated ($MgSO_4.7H_2O$)	0·5 g
Dipotassium hydrogen phosphate	1·0 g

Ferrous sulphate, hydrated ($FeSO_4.7H_2O$)	0·01 g
Sucrose	30·0 g
Agar	15·0 g
Distilled water	to 1 litre

Dissolve ingredients in water by steaming. Dispense as required, and sterilise at 115°C for 20 min.

Davis's Yeast Salt Agar*

(For the isolation and enumeration of yeasts and moulds)

Ammonium nitrate	1·0 g
Ammonium sulphate	1·0 g
Sodium monohydrogen phosphate, anhydrous	4·0 g
Potassium dihydrogen phosphate	2·0 g
Sodium chloride	1·0 g
D-Glucose	10·0 g
Yeast extract powder	1·0 g
Agar	20·0 g
Distilled water	1 litre

pH 6·6

Dissolve all the ingredients by steaming. Check the reaction and adjust if necessary to pH 6·6. Distribute in 100 ml amounts in screw-capped bottles and sterilise by autoclaving at 115°C for 20 min.

The pH can be adjusted to 3·5 if required, by the addition aseptically of 5·7 ml of sterile 10 per cent citric acid solution (previously sterilised by filtration) to each 100 ml of molten sterile medium at 50°C immediately before pouring the plates.

Buffered Yeast Agar (Oxoid) is a manufactured dehydrated medium rather similar to Davis's yeast salt agar. This medium also can be acidified by the addition of sterile citric acid or lactic acid solution immediately before pouring the plates.

As an alternative to acidification for the inhibition of bacteria, antibiotics may be added as sterile solutions to the molten medium immediately before pouring the plates (see pages 106–107).
(Davis, 1958)

Davis's Yeast Salt Broth

This should be prepared in a similar way to the agar medium to the same recipe, or as a double-strength medium as required, except that the agar is omitted from the recipe. Distribute the single-strength medium in test-tubes, and the double-strength medium in suitable containers, insert inverted Durham tubes, and sterilise by autoclaving at 115°C for 20 min.

(Davis, 1970)

Differential Reinforced Clostridial Medium

(For the detection and enumeration of sulphite-reducing clostridia by a multiple tube technique)

Basal medium

Peptone	10·0 g
Lab-Lemco meat extract (Oxoid)	10·0 g
Sodium acetate, hydrated	5·0 g
Yeast extract (Difco)	1·5 g
Soluble starch	1·0 g
D-Glucose	1·0 g
L-Cysteine monohydrochloride	0·5 g
Distilled water	1 litre

pH 7·1–7·2

Dissolve the peptone, meat extract, sodium acetate and yeast extract in 800 ml of distilled water. Mix the soluble starch to a paste with a little of the remaining 200 ml of water, boil the rest of the 200 ml of water and stir it into the paste. Mix the two solutions and steam for 30 min to complete the solution of the ingredients. After the steaming add the glucose and cysteine, mix and adjust the pH to 7·1–7·2. Filter through hot paper pulp (*q.v.*), dispense in 25 ml amounts in 30 ml (1 oz) screw-capped (McCartney) bottles. Sterilise by autoclaving at 121°C for 15 min.

Sodium sulphite solution

Prepare a 4 per cent (w/v) solution of sodium sulphite (anhydrous) and sterilise by filtration. This solution may be stored in fully filled screw-capped bottles in the refrigerator for up to 14 days.

Ferric citrate solution

Prepare a 7 per cent (w/v) solution of ferric citrate (scales), heating briefly to dissolve. Sterilise by filtration. It may be stored in fully filled screw-capped bottles in the refrigerator for up to 14 days.

Preparation of complete medium

When the DRCM is required, steam and cool the basal medium and then add aseptically to each bottle of medium 0·5 ml of a freshly prepared mixture of equal volumes of sodium sulphite solution and ferric citrate solution.

The medium may be made more selective for sulphite-reducing *Clostridium* spp. by also adding aseptically to each bottle 0·9 ml of a solution of 10,000 units of polymyxin B sulphate (Wellcome Reagents Ltd.) in 5 ml of sterile distilled water.

(Gibbs and Freame, 1965; Freame and Fitzpatrick, 1971)

E C Broth†‡

(For the detection of coliform bacteria and/or *Escherichia coli*)

Tryptose (or trypticase)	20·0 g
Lactose	5·0 g
Bile salts (No. 3: Oxoid L56 or equivalent)§	1·5 g
Dipotassium hydrogen phosphate	4·0 g
Potassium dihydrogen phosphate	1·5 g
Sodium chloride	5·0 g
Distilled water	1 litre

pH 6·8–6·9

Dissolve the ingredients in the water with minimal steaming, distribute in 5 ml amounts in 150 × 16 mm test-tubes with inverted Durham tubes. Autoclave at 121°C for 15 min.

§ See footnote on page 56.

(Hajna and Perry, 1943)

Edwards's Aesculin Crystal Violet Blood Agar*

(For the isolation of animal-parasitic streptococci)

Nutrient agar (pH 7·4)	100 ml
Crystal violet, 0·05 per cent solution (sterile)	0·4 ml
Defibrinated ox blood (sterile)	5·0 ml
Aesculin	0·1 g

Add the aesculin to the molten nutrient agar and sterilise by steaming for 30 min on each of 3 successive days.

To 100 ml of molten sterile aesculin nutrient agar cooled to 50°C, add with aseptic precautions 0·4 ml of sterile 0·05 per cent crystal violet solution (previously sterilised by autoclaving or by filtration) and mix well. Next add aseptically 5 ml of sterile defibrinated ox blood, mix thoroughly and pour into Petri-dishes.

(Edwards, 1933)

Egg-yolk Agar

(For the detection of lecithinase production)

Egg-yolk emulsion	10 ml
Sodium chloride	1·0 g
Yeast extract agar or nutrient agar	100 ml

To the basal medium add the extra salt and sterilise by autoclaving at 121°C for 20 min. Cool to 45°C and add aseptically the sterile egg-yolk emulsion, mix well and pour plates.

The egg-yolk emulsion (*q.v.*) may be prepared after the method of Billing and Luckhurst (1957). Alternatively, egg-yolk emulsions are available commercially from most suppliers of dehydrated media and media supplements.

Egg-yolk Broth

(For the detection of lecithinase production)

Nutrient broth	100 ml
Sodium chloride	1·0 g
Egg-yolk emulsion, sterile (*q.v.*)	10 ml

Add the sodium chloride to the nutrient broth, distribute in 5 ml amounts in 150 × 16 mm test-tubes, and sterilise at 121°C for 15 min. When cold, add with aseptic precautions 0·5 ml of sterile egg-yolk emulsion to each tube of medium.

The egg-yolk emulsion may be prepared after the method of Billing and Luckhurst (see below), or obtained as a commercial preparation from most suppliers of media.

Egg-yolk Emulsion*†

Separate the yolks from the whites of the eggs by pipette and add 4 parts of distilled water to 1 part of egg yolk. Mix thoroughly and heat in a water-bath at 45°C for 2 h. Centrifuge to remove the precipitate (alternatively, stand the mixture overnight in the refrigerator). Decant the supernatant liquid. Sterilise by filtration through a Ford Sterimat Grade SB (or stacked graded membrane filters, see page 89).

(Billing and Luckhurst, 1957)

Ethanol Broth

(For the determination of the utilisation of ethanol by yeasts)

Ammonium sulphate	1·0 g
Potassium dihydrogen phosphate	1·0 g

Magnesium sulphate, hydrated ($MgSO_4.7H_2O$)	0·5 g
Distilled water	1 litre

Dissolve the salts in the water, distribute in 5 ml amounts in test-tubes and sterilise at 121°C for 15 min. When cool, add aseptically to each tube 0·3 ml of sterile 50 per cent ethanol and *one drop* of a sterile 5 per cent solution of yeast extract. The ethanol and yeast extract solution are sterilised beforehand by filtration.

Growth of the yeast being examined should be compared with growth in the same medium with the ethanol omitted.

(Lodder and Kreger-van Rij, 1952; Lodder, 1970)

Ferrous Chloride Gelatin

(For the simultaneous determination of hydrogen sulphide production and liquefaction of gelatin)

Ferrous chloride, 10 per cent solution	0·5 ml
Nutrient gelatin, sterile	100 ml

Heat the sterile nutrient gelatin to 100°C in a steamer, and add the *freshly prepared* ferrous chloride solution. Immediately dispense aseptically in sterile 130 × 12 mm test-tubes to give 8 cm depth of medium. Seal with sterile air-tight stoppers (e.g. rubber bungs or Astell roll-tube closures, sterilised by immersion in boiling water for 20 min), and cool the tubes rapidly by partial immersion in cold water.

(Report, 1958a)

Fortified Nutrient Agar

(For spore production by *Bacillus stearothermophilus*)

Nutrient agar powder (Difco or equivalent)	15 g
Plain agar powder (Difco or equivalent)	5 g
D-Glucose	0·5 g
Manganese sulphate, hydrated ($MnSO_4.4H_2O$)	30 mg
Distilled water	1 litre

Dissolve the ingredients in the water by steaming. Dispense in Roux bottles or in 500 ml screw-capped medical flat bottles in sufficient amount to provide an approx. 2 cm layer when the bottles are laid on their sides. Sterilise by autoclaving at 121°C for 25 min, and place the bottles on their sides so that the medium sets to provide the maximum surface area.

(Finley and Fields, 1962)

Frazier's Gelatin Agar

(For the determination of gelatinase production (proteolysis))

Gelatin	0·4 g
Yeast extract agar or nutrient agar	100 ml

Add the gelatin to the molten agar medium and steam with occasional mixing until completely dissolved. Sterilise by autoclaving at 115°C for 20 min.

(Smith, Gordon and Clark, 1952)

Gelatin Charcoal Discs in Peptone Water

(For the determination of gelatinase production (proteolysis))

Dispense peptone water in 1 ml amounts in 75 mm × 10 mm test-tubes. Sterilise by autoclaving at 121°C for 15 min. Immediately before use add aseptically one gelatin charcoal disc to each tube. The gelatin charcoal discs can be prepared as described by Kohn (1953) or can be obtained ready to use from Oxoid Ltd.

(Greene and Larks, 1955)

*Preparation of gelatin charcoal discs**

Add 3–5 g of finely powdered charcoal (e.g. as obtainable from Oxoid Ltd.) to 100 ml of melted nutrient gelatin. Mix thoroughly while cooling until just above gelling temperature. Pour into sterile Petri-dishes to form layers 3 mm thick. In order to facilitate the subsequent removal of the gelatin, first lightly smear the dishes with petroleum jelly. When set, remove the sheets of charcoal gelatin and place in 10 per cent formaldehyde® for 24 h. Remove from the formaldehyde and use a cork borer to punch discs of 1 cm diameter from the gelatin. Wash the discs in running tap water for 24 h. Place the washed gelatin charcoal discs in distilled water in screw-capped bottles. Sterilise by steaming for 20 min on each of 3 successive days.

(Kohn, 1953)

Gibson's Semi-solid Tomato Juice Medium

(For the detection of carbon dioxide production from glucose by lactic-acid bacteria)

Yeast extract	2·5 g
D-Glucose	50·0 g
Tomato juice, pH 6·5§	100 ml

Reconstituted skim milk	800 ml
Nutrient agar	200 ml
§, Or, Manganese sulphate ($MnSO_4.4H_2O$), 0·4 per cent solution	10 ml

Mix the tomato juice or manganese sulphate solution with the reconstituted skim milk, add the yeast extract and glucose and heat in the steamer. While still hot, add the molten nutrient agar and mix well. Check the pH and adjust if necessary to pH 6·5. Distribute in test-tubes to a depth of 5–6 cm, and sterilise by steaming for 30 min on each of 3 successive days. (Gibson and Abd-el-Malek, 1945; Stamer, Albury and Pederson, 1964)

Gluconate Broth

(For the determination of gluconate utilisation, especially by members of the Enterobacteriaceae)

Peptone	1·5 g
Yeast extract	1·0 g
Dipotassium hydrogen phosphate	1·0 g
Potassium gluconate	40·0 g
Distilled water	1 litre

pH 7·0

Dissolve the ingredients in the water, distribute in 5 ml amounts in test-tubes, and sterilise at 115°C for 10 min.

(Shaw and Clarke, 1955)

Glucose Azide Broth†‡

(For the enumeration of faecal streptococci by the multiple tube technique)

Peptone (Difco)	10·0 g
Sodium chloride	5·0 g
Dipotassium hydrogen phosphate	5·0 g
Potassium dihydrogen phosphate	2·0 g
D-Glucose	5·0 g
Yeast extract	3·0 g
Sodium azide®	0·25 g
Bromcresol purple, 1·0 per cent solution	3 ml
Distilled water	1 litre

pH 6·6–6·8

Dissolve the ingredients in the water. Distribute in 5 ml amounts in 150 × 16 mm test-tubes. Sterilise by autoclaving at 121°C for 15 min. For inocula of large amounts of sample or diluent (5 ml or more) a double strength medium should be prepared and distributed in amounts equal in volume to the inocula to be used. Double strength medium is prepared in a similar manner to that described above, the ingredients being dissolved in half the quantity of distilled water.

Note: the effectiveness of this medium should be checked from time to time using stock cultures at low inoculum levels; it has been found that variations in the nutrients composition (particularly the peptone) from batch to batch may affect selectivity and/or sensitivity.

(Hannay and Norton, 1947)

Glucose Lemco Broth, pH 9·2

(For the differentiation of lactic streptococci)

Prepare this by admixture of glucose lemco broth and buffer solution in the same way as glucose lemco broth, pH 9·6 (*q.v.*). In this case, adjust the pH of the complete medium to 9·4 prior to overnight storage in the cold and to filtration. Check that the final pH is 9·2, both prior to inoculation and after incubation.

(Garvie, 1953; Shattock and Hirsch, 1947)

Glucose Lemco Broth, pH 9·6

(For the differentiation of the primary groups of streptococci)

Glucose lemco broth

D-Glucose	10·0 g
Lab-Lemco meat extract (Oxoid)	10·0 g
Peptone	10·0 g
Sodium chloride	5·0 g
Distilled water	1 litre

Dissolve ingredients and sterilise by autoclaving at 121°C for 20 min.

Buffer solution

Glycine (glycocoll)	7·505 g
Sodium chloride	5·85 g
Distilled water, freshly boiled	1 litre

Dissolve ingredients and sterilise by autoclaving at 121°C for 20 min. Mix 6 parts with 4 parts of 0·1 N NaOH by volume.

Preparation of complete medium

To 900 ml of glucose lemco broth add 100 ml of buffer solution and adjust to pH 9·8 with N NaOH. Store overnight in a stoppered flask in the cold to complete precipitation. Sterilise by filtration and dispense immediately and aseptically into sterile screw-capped bottles to leave a minimum of air space. The medium should be used within 48 h of preparation, and the pH of uninoculated control tubes should be checked electrometrically (with a glass electrode) immediately before and after incubation. The initial pH should be 9·6 ± 0·02 and should not drop during incubation by more than 0·04 units.

(Shattock and Hirsch, 1947)

Glucose Phosphate Broth*†‡

(Used for the methyl red test and Voges-Proskauer test, particularly for the differentiation of the Enterobacteriaceae)

Peptone	5·0 g
D-Glucose	5·0 g
Dipotassium hydrogen phosphate	5·0 g
Distilled water	1 litre

pH 7·5

Dissolve the ingredients in the water. Adjust to pH 7·5 and distribute in 5 ml amounts in 150 × 16 mm test-tubes. Sterilise by autoclaving at 115°C for 20 min.

Glucose Tryptone Agar (Dextrose Tryptone Agar)*†‡

(For the detection and enumeration of "flat-sour" spoilage organisms)

Tryptone	10·0 g
D-Glucose	5·0 g
Bromcresol purple, 1 per cent solution	4·0 ml
Agar	15·0 g
Distilled water	1 litre

pH 7·0

Add the ingredients to the water and heat in a steamer until dissolved. Adjust to pH 7·0, distribute as required and sterilise at 121°C for 15 min.

(Hersom and Hulland, 1963)

Glucose Tryptone Broth (Dextrose Tryptone Broth)*

(For the detection and enumeration of "flat-sour" spoilage organisms)

Prepare in the same way as glucose tryptone agar, except that the agar is omitted. Dispense in 5 ml or 10 ml amounts in test-tubes or other containers, and sterilise at 121°C for 15 min. For the detection of low concentrations of organisms in foods, double strength medium may be prepared, and distributed in amounts equal in volume to the inocula to be used.

(Hersom and Hulland, 1963)

Glucose Yeast Chalk Agar

(Used in the 3-ketolactose test)

Yeast extract	10·0 g
D-Glucose	20·0 g
Calcium carbonate	20·0 g
Agar	15·0 g
Distilled water	1 litre

Add the ingredients to the water and dissolve by steaming. Distribute in 7 ml amounts in 150 × 16 mm test-tubes, with frequent mixing to retain the calcium carbonate in suspension. Sterilise at 115°C for 20 min, and set in the slanted position.

(Bernaerts and De Ley, 1963)

Gorodkowa Agar (Modified)

(For the sporulation of yeasts)

Peptone	10·0 g
D-Glucose	1·0 g
Sodium chloride	5·0 g
Agar	20·0 g
Distilled water	1 litre

Dissolve the ingredients in the water by steaming, distribute as required and sterilise at 121°C for 15 min, set as slants.

(Lodder and Kreger-van Rij, 1952; Lodder, 1970)

Gypsum Blocks

(For the promotion of sporulation by yeasts)

Prepare 3 cm × 3 cm × 1 cm plaster of Paris blocks from a mixture of 8 parts of calcium sulphate hemihydrate and 3 parts of water. After setting place the blocks in sterile Petri-dishes and sterilise in a hot air oven at 120°C for 2 h.

(Lodder and Kreger-van Rij, 1952; Lodder, 1970)

Hajna's GN Broth†‡

(For the selective enrichment of enterobacteria)

Tryptose	20·0 g
D-Glucose	1·0 g
Mannitol	2·0 g
Sodium citrate	5·0 g
Sodium desoxycholate	0·5 g
Dipotassium hydrogen phosphate	4·0 g
Potassium dihydrogen phosphate	1·5 g
Sodium chloride	5·0 g
Distilled water	1 litre

Dissolve the ingredients in the water with minimum heating. Adjust the pH of the medium to 7·0 if necessary. Distribute as required and sterilise at 115°C for 15 min. It is important to avoid excessive heating of this medium.

(Croft and Miller, 1956)

Hayes's Medium

(For the demonstration of gliding motility)

Lab-Lemco beef extract (Oxoid)	1·0 g
Peptone	2·5 g
Sodium chloride	5·0 g
Agar	20·0 g
Distilled water	1 litre

pH 7·2

Dissolve the ingredients in the water by steaming, distribute and sterilise at 121°C for 20 min.

For the examination of marine organisms the sodium chloride and distilled water should be replaced by aged sea water 750 ml + distilled water 250 ml.

(Hayes, 1963)

Holding's Inorganic-nitrogen Medium

(For the differentiation of certain Gram-negative bacteria)

D-Glucose	5·0 g
Sodium citrate, hydrated	1·0 g
Sodium acetate, hydrated	1·0 g
Sodium succinate, hydrated	1·0 g
Calcium gluconate, hydrated	1·0 g
Ammonium dihydrogen phosphate	1·0 g

Dipotassium hydrogen phosphate	0·08 g
Potassium dihydrogen phosphate	0·02 g
Potassium nitrate®	1·0 g
Distilled water	1 litre

Dissolve the ingredients in the water, heat to 107°C, and as soon as the desired pressure of 34·5 kilonewtons per m² (5 lb/in²) is reached stop heating. Filter, distribute in 5 ml amounts in 150 × 16 mm test-tubes, and sterilise at 115°C for 20 min.

(Holding, 1960)

Hugh and Leifson's Medium

(For differentiating oxidative and fermentative metabolism of carbohydrates, by Gram-negative bacteria)

Recipe 1 (Hugh and Leifson, 1953)†‡

Peptone	2·0 g
Sodium chloride	5·0 g
Dipotassium hydrogen phosphate	0·3 g
Bromthymol blue, 1 per cent aqueous solution	3·0 ml
Agar	3·0 g
Distilled water	1 litre

pH 7·1

Recipe 2 (Scholefield, 1964)

Tryptone	1·0 g
Yeast extract	1·0 g
Sodium chloride	5·0 g
Dipotassium hydrogen phosphate	0·3 g
Bromthymol blue, 1 per cent aqueous solution	3·0 ml
Acid fuchsin, 1 per cent solution	1·5 ml
Agar (Oxoid No. 3)	4·5 g
Distilled water	1 litre

pH 7·1

Add the ingredients, except indicators, to the water and dissolve by steaming. Adjust to pH 7·1, add the indicator/s and mix well. Dispense in 10 ml amounts in 150 × 16 mm test-tubes, and close with Astell seals (Astell-Hearson). Leave the closures loose, and sterilise by autoclaving at 121°C for 15 min. To each tube of molten medium add aseptically 1 ml of a sterile 10 per cent solution of the desired substrate solution, mix well (without aeration) and allow to set. The carbohydrate normally employed in this medium is glucose, and it may be incorporated into the medium at the time of preparation if differential studies of reactions with a range of substrates are not being undertaken. The second recipe has been found to allow the use of a single tube only, provided that the tubes of media are stood in a boiling water bath for 10 min and then cooled rapidly immediately prior to inoculation; in addition a very clear colour change from blue-green to orange-red results from the production of acid.

Hugh and Liefson's Medium, Modified

(For differentiating oxidative and fermentative metabolism of glucose and mannitol by *Staphylococcus* and *Micrococcus*)

Tryptone	10·0 g
Yeast extract (Difco)	1·0 g
D-Glucose or mannitol	10·0 g
Bromcresol purple, 1 per cent solution	4 ml
Agar	2·0 g
Distilled water	1 litre

pH 7·0

Dissolve the ingredients in the water by steaming. Adjust to pH 7·0, and dispense in 150 × 16 mm test-tubes, filling them two-thirds full. Autoclave at 115°C for 20 min.

Immediately before use, steam the medium for 10–15 min to expel dissolved oxygen, and solidify by placing the tubes in cold or iced water.

(Recommendations, 1965)

King, Ward and Raney's Medium "B"†

(For demonstration of fluorescin production by *Pseudomonas*)

Proteose peptone No. 3 (Difco)	20·0 g
Dipotassium hydrogen phosphate	1·5 g
Magnesium sulphate, hydrated ($MgSO_4.7H_2O$)	1·5 g
Glycerol	10·0 g
Agar	15·0 g
Distilled water	1 litre

pH 7·2

Dissolve the ingredients in the water by steaming. Dispense as required—the medium may be used as poured plates or as slants in test-tubes—and sterilise by autoclaving at 121°C for 15 min.

(King, Ward and Raney, 1954)

Knisely's Chloral Hydrate Agar

(For the identification of *Bacillus cereus*)

Distribute Heart Infusion Agar (Difco) in 100 ml amounts and sterilise as recommended by the manufacturer.

Chloral hydrate solution: prepare a 10 per cent solution of chloral hydrate®, and sterilise by filtration.

To prepare the complete medium add 2·5 ml of chloral hydrate solution (using a safety pipette) to each 100 ml of molten medium at 45–50°C immediately before pouring the plates.

(Knisely, 1965)

Kohn's Two-tube Media*

(For differentiating *Salmonella* and *Shigella* from other isolates obtained with selective media)

Medium 1 (Kohn, 1954)

Beef extract	2·0 g
Proteose peptone No. 3 (Difco)	15·0 g
Yeast extract	2·0 g
D-Glucose	1·0 g
Mannitol	10·0 g
Phenol red, 0·2 per cent solution	15 ml
Agar	15·0 g
Distilled water	985 ml

pH 7·2

Add the ingredients to the water and heat in a steamer until dissolved. Adjust to pH 7·2. Distribute in 8 ml amounts in 150 × 16 mm test-tubes and sterilise by autoclaving at 115°C for 15 min. Cool to 60°C and add 0·2 ml of a sterile 40 per cent solution of urea (previously sterilised by filtration). Slope the tubes to give a 3 cm butt.

Medium 2 (Gillies, 1956)

Peptone	10·0 g
Tryptone	10·0 g
Sodium chloride	5·0 g
Disodium hydrogen phosphate, hydrated	0·25 g
Sucrose	10·0 g
Salicin	10·0 g
Bromthymol blue 1 per cent solution	1·0 ml
Sodium thiosulphate, hydrated	0·025 g
Agar	3·0 g
Distilled water	1 litre

pH 7·4

Dissolve the ingredients in the water by steaming, and adjust the pH if necessary to 7·4. Distribute in 10 ml amounts in 150 × 16 mm test-tubes and autoclave at 121°C for 15 min. Allow to solidify in a vertical position.

Koser's Citrate Medium*†‡

(For determination of the ability to use citrate as sole carbon source)

Sodium ammonium hydrogen phosphate	1·5 g
Potassium dihydrogen phosphate	1·0 g
Magnesium sulphate, hydrated ($MgSO_4.7H_2O$)	0·2 g
Sodium citrate	2·0 g
Distilled water	1 litre

pH 7·0

Flasks used in preparing the medium and the test-tubes into which it is dispensed must be chemically clean.

Dissolve the ingredients in the water and adjust the pH to 7·0. Filter if necessary to obtain maximum clarity of the medium, and dispense in 5 ml amounts in 150 × 16 mm test-tubes. Sterilise by autoclaving at 121°C for 15 min.

Lactose Resuscitation Broth*†‡

(For non-selective resuscitation of enterobacteria)

Peptone	5·0 g
Beef extract	3·0 g
Lactose	5·0 g
Distilled water	1 litre

pH 6·9

Dissolve the ingredients in the water by steaming, distribute as required and sterilise at 121°C for 15 min.

(North, 1961)

Lactose Yeast Extract Agar

(Used in the 3-ketolactose test)

Lactose	10·0 g
Yeast extract	1·0 g
Agar	20·0 g
Distilled water	1 litre

Dissolve the ingredients in the water by steaming. Distribute as required and sterilise at 115°C for 20 min.

(Bernaerts and De Ley, 1963)

Lauryl Sulphate Tryptose Broth*†‡

(For detection and enumeration of coliform organisms)

Tryptose	20·0 g
Dipotassium hydrogen phosphate	2·75 g
Potassium dihydrogen phosphate	2·75 g
Sodium chloride	5·0 g
Lactose	5·0 g
Sodium lauryl sulphate	0·1 g
Distilled water	1 litre

Dissolve the ingredients in the water by heating in a steamer. Distribute into tubes or bottles each containing an inverted Durham tube and sterilise in a steamer at 100°C for 30 min on each of 3 successive days, or by autoclaving at 121°C for 15 min.

(Mallman and Darby, 1941)

Liquid Paraffin

(Sterile, for covering cultures)

Dispense liquid paraffin in Erlenmeyer flasks in shallow layers and sterilise in the hot air oven at 160°C for 2 h.

Litmus Milk*†‡

Add sufficient litmus solution to reconstituted skim milk to give a pale mauve colour (10 ml of 4 per cent litmus solution (BDH Chemicals Ltd.) per litre of milk). Dispense as required (in relatively small volumes) and sterilise at 121°C for 5 min, followed by steaming for 30 min on each of the two following days. Autoclaving on the first day has been found to reduce substantially the number of spoiled tubes or bottles resulting from residual spores. Always pre-incubate prior to use, as described on page 319.

Loeffler's Serum*†‡

(For the culture of fastidious animal parasites)

Nutrient broth + 1 per cent D-glucose	250 ml
Sterile serum	750 ml

Sterilise the nutrient broth with 1 per cent added glucose by steaming at 100°C for 30 min on each of three successive days, or by autoclaving at 115°C for 20 min. Mix aseptically with the sterile serum in a sterile flask and distribute aseptically into sterile 7 ml ($\frac{1}{4}$-oz or bijou) screw-capped bottles.

Place the bottles (with caps screwed on tightly) in a sloping position in an oven or inspissator and heat slowly to 85°C to coagulate the serum. The medium may be sterilised by heating at 85°C for 20 minutes on each of three successive days.

Note. An alternative method of inspissation is to heat the medium to 80°C and maintain this temperature for 2h.

MRS Agar*

(For the culture of Lactobacilli)

MRS broth + 1·5 per cent agar.

Dissolve ingredients for MRS broth (*q.v.*) with the exception of glucose by heating in the steamer, and adjust pH to 6·2–6·6. Add the agar and dissolve at 121°C for 5 min. Dissolve the glucose in the molten agar medium and distribute as required. Sterilise at 121°C for 15 min.

(de Man, Rogosa and Sharpe, 1960)

MRS Broth*

(For the culture of lactobacilli)

Peptone (Oxoid)	10·0 g
Lab-Lemco meat extract (Oxoid)	10·0 g
Yeast extract	5·0 g
D-Glucose	20·0 g
Tween 80 (Atlas)	1·0 g
Dipotassium hydrogen phosphate	2·0 g
Sodium acetate	5·0 g
Triammonium citrate	2·0 g
Magnesium sulphate, hydrated ($MgSO_4.7H_2O$)	0·2 g
Manganese sulphate, hydrated ($MnSO_4.4H_2O$)	0·05 g
Distilled water	1 litre

Dissolve the ingredients in the distilled water by steaming. Adjust the pH to 6·2–6·6. Distribute as required and sterilise at 121°C for 15 minutes. The final pH after sterilisation should be 6·0–6·5.

(de Man, Rogosa and Sharpe, 1960)

MRS Fermentation Medium

(For fermentation studies of lactobacilli)

Prepare MRS broth as usual, but omit the glucose and meat extract. Adjust the pH to 6·2–6·5. Add 0·004 per cent chlorphenol red as indicator. Distribute in test-tubes or as required.

Prepare 10 per cent solutions of the test substrates and sterilise by filtration.

Add aspetically the required substrate to give a final concentration of 2 per cent.

(de Man, Rogosa and Sharpe, 1960)

MacConkey's Agar*†‡

(For detection and isolation of enterobacteria, especially the coliform bacteria)

Peptone	20·0 g
Bile salts	5·0 g
Sodium chloride	5·0 g
Lactose	10·0 g
Neutral red, 1 per cent aqueous solution	7·0 ml
Agar	15·0 g
Distilled water	1 litre

Dissolve the peptone, bile salts and sodium chloride in the water by steaming. Cool and adjust to pH 7·4. Add the agar and dissolve by autoclaving. Filter through hot paper pulp (*q.v.*). Adjust the pH to 7·4. Add the lactose and neutral red and steam until dissolved. Mix well and distribute in bottles or test-tubes as required. Sterilise at 115°C for 15 min.

(Report, 1969)

Brilliant green MacConkey's agar: add 3·3 ml of 1 per cent brilliant green solution at the same time as the lactose and neutral red. (Harvey and Price, 1974)

MacConkey's Broth*†‡

(For the detection and enumeration of lactose-fermenting enterobacteria by the multiple tube technique)

Peptone	20·0 g
Bile salts	5·0 g
Sodium chloride	5·0 g
Lactose	10·0 g
Bromcresol purple, 1 per cent solution	1 ml
Distilled water	1 litre

pH 7·4

Add to the water the peptone, bile salts and sodium chloride and heat in a steamer for 1–2 h. Add the lactose and dissolve by heating for a further 15 min. Cool and filter. Adjust to pH 7·4. Add the indicator. Mix well. Distribute in 5 ml amounts in 150 × 16 mm test-tubes provided with inverted Durham tubes. Sterilise by autoclaving 115°C for 15 min.

Double strength medium can be prepared in a similar manner, by dissolving the ingredients in half the quantity of distilled water. Distribute in amounts equal in volume to the inocula to be added.

(Report, 1969)

MacConkey's Broth (for Membrane Filtration)*

(For the detection and enumeration of lactose-fermenting enterobacteria by membrane filtration)

Peptone	10·0 g
Bile salts	4·0 g
Sodium chloride	5·0 g
Lactose	30·0 g
Bromcresol purple, 1 per cent solution	12 ml
Distilled water	1 litre

pH 7·4

Prepare in a manner similar to standard MacConkey's broth, but adjust the pH to 7·4, and distribute in suitable storage containers (e.g. screw-capped bottles) before sterilising.

(Taylor, Burman and Oliver, 1955)

Malonate Broth†‡

(For the determination of malonate utilisation by the Enterobacteriaceae)

Yeast extract	1·0 g
Ammonium sulphate	2·0 g
Dipotassium hydrogen phosphate	0·6 g
Potassium dihydrogen phosphate	0·4 g
Sodium chloride	2·0 g
Sodium malonate	3·0 g
Bromthymol blue, 1 per cent solution	2·5 ml
Distilled water	1 litre

Dissolve the ingredients in the water, distribute in 5 ml amounts in 150 × 16 mm test-tubes and sterilise at 121°C for 15 min.

(Report, 1958a)

Malt Extract Agar*†‡

(For the culture of yeasts and moulds)

Malt extract	30·0 g
Mycological peptone	5·0 g
Agar	15·0 g
Distilled water	1 litre

pH 5·4

Dissolve the ingredients in the water by steaming. Distribute as required and sterilise by autoclaving at 121°C for 15 min.

In order to inhibit the growth of bacteria, antibiotics may be added as sterile solutions to the molten medium immediately before pouring the plates (see page 106); or the medium may be acidified to pH 3·5. Acidification may be achieved by adding aseptically sterile 10 per cent lactic acid (or citric acid) solution to the molten medium immediately before pouring the plates. The exact amount of acid to be added will depend upon the make or even the batch of the constituents used.

Sucrose (20 per cent) may be added to this medium to make it suitable for osmophilic counts.

Maltose Azide Broth

(For detection and enumeration of faecal streptococci in foods, etc., using the multiple tube technique)

Proteose peptone No. 3 (Difco)	10·0 g
Yeast extract (Difco)	10·0 g
Sodium chloride (A.R. grade)	5·0 g
Sodium glycerophosphate, hydrated	10·0 g
Maltose	20·0 g
Lactose	1·0 g
Sodium azide®	0·4 g
Sodium carbonate (A.R. grade)	0·636 g
Bromcresol purple, 1 per cent aqueous solution	1·5 ml
Distilled water	1 litre

pH 7·2

Dissolve the ingredients in the water by steaming. Adjust to pH 7·2. Distribute in 10 ml amounts in 150 × 16 mm test-tubes and sterilise at 121°C for 10 min.

For large volumes of inocula (5 ml and over) use the appropriate amounts of 1½-strength broth, e.g. for 10 ml inocula use 20 ml amounts of 1½-strength broth.

(Kenner, Clark and Kabler, 1961)

Maltose Azide Tetrazolium Agar

(For detection and enumeration of faecal streptococci)

This has the same basic recipe as maltose azide broth with the addition of 2 per cent agar. Distribute in 100 ml amounts and sterilise at 121°C for 15 min. Immediately before pouring, add aseptically to each 100 ml of molten agar medium 1 ml of a sterile 1 per cent (w/v) solution of triphenyltetrazolium chloride (previously sterilised by filtration). The TTC solution should be stored in a dark bottle in the refrigerator, and boiled for 5 min each time immediately before use.

(Kenner, Clark and Kabler, 1961)

Mannitol Egg-yolk Phenol Red Polymyxin Agar†(B)

(For the detection and differentiation of *Bacillus cereus*)

Basal medium

Peptone	10·0 g
Meat extract	1·0 g
D-Mannitol	10·0 g
Sodium chloride	10·0 g
Phenol red, 0·2 per cent solution	12·5 ml
Agar	15·0 g
Distilled water	887·5 ml

pH 7·1

Dissolve all the ingredients in the water by steaming. Distribute in 90 ml amounts in screw-capped bottles, and sterilise at 121°C for 15 min.

Polymyxin B sulphate solution. Dissolve 50 mg of polymyxin B sulphate (e.g. "Aerosporin" from Wellcome Reagents Ltd.) in 50 ml of distilled water. Sterilise by filtration.

Preparation of complete medium. To 90 ml of molten medium, cooled to 45–50°C, add with aseptic precautions 10 ml of egg-yolk emulsion and 1 ml of sterile polymyxin B sulphate solution. The final concentration of antibiotic in the medium is thus 10 μg per ml of medium. Mix well, and pour plates with about 15 ml medium in each plate. Dry for 1 h at 45°C before use in order that the medium will absorb the inoculum liquid during surface colony court procedures.

(Mossel, Koopman and Jongerius, 1967)

Milk Agar (10 per cent Milk)

(For detecting proteolytic activity)

Reconstituted skim milk	10 ml
Yeast extract agar or nutrient agar	100 ml

Add the milk to the molten agar medium, mix well, dispense as required and sterilise by autoclaving at 115°C for 20 min. Alternatively, add aseptically 1 ml of sterile skim milk (sterilised as for litmus milk—*q.v.*) to 10 ml of sterile molten medium, mix well, and pour into a Petri-dish.

Milk Agar (30 per cent Milk)

(For the detection of caseolytic (proteolytic) activity)

Mix aseptically 10 ml of a hot sterile 2·5 per cent solution of agar with 5 ml of hot sterile reconstituted skim-milk and pour into a Petri-dish. This medium may be overlaid as a thin layer on a layer of 10 ml of plain agar previously poured and allowed to solidify.

(Smith *et al.*, 1952)

Minimal Nutrient Recovery Medium

(For the resuscitation of injured bacteria from foods prior to selective isolation)

Disodium hydrogen phosphate	7·0 g
Potassium dihydrogen phosphate	3·0 g
Sodium chloride	0·5 g
Ammonium chloride	1·0 g
Magnesium sulphate, hydrated ($MgSO_4.7H_2O$)	0·25 g
D-Glucose	2·0 g
Distilled water	1 litre

Dissolve the ingredients in the water, dispense as required, and sterilise by autoclaving at 121°C for 15 min.

(Gomez *et al.*, 1973; Wilson and Davies, 1976)

Moller's Decarboxylase Medium*†‡

(For the determination of decarboxylase activity, especially by members of the Enterobacteriaceae)

Peptone	5·0 g
Beef extract	5·0 g
D-Glucose	0·5 g
Pyridoxal	5 mg
Bromthymol blue, 1 per cent aqueous solution	1·0 ml
Cresol red, 0·2 per cent aqueous solution	2·5 ml
Distilled water	1 litre

pH 6·0

Dissolve the solid ingredients in the water by steaming. Adjust to pH 6·0 and add the indicator solutions. Prepare the complete media as follows:

For the control medium (no added amino acid) distribute in 3 ml amounts in 100 × 12 mm test-tubes, add a 5 mm depth of liquid paraffin and sterilise by autoclaving at 115°C for 10 min.

To prepare the arginine, lysine or ornithine medium, add the appropriate amino acid as the hydrochloride to a final concentration of 1 per cent w/v (if the L-amino acid is used) or 2 per cent w/v (if the DL- amino acid is used). Readjust to pH 6·0, distribute in 3 ml amounts in 100 × 12 mm test-tubes. Add a 5 mm depth of liquid paraffin and sterilise at 115°C for 10 min.

(Møller, 1955)

Neutral Red Chalk Lactose Agar

(For the detection of lactic streptococci in milk and milk products)

Peptone	3·0 g
Lab-Lemco meat extract (Oxoid)	3·0 g
Yeast extract	3·0 g
Agar	15·0 g
Lactose	10·0 g
Calcium carbonate, precipitated	15·0 g
Neutral red, 1 per cent aqueous solution	5 ml§
Distilled water	1 litre

§ Or 2·5 ml of a 1 per cent aqueous solution of bromcresol purple for bromcresol purple chalk lactose agar.

Dissolve the peptone, meat extract and yeast extract and agar in the distilled water by steaming. Allow to cool and adjust if necessary to pH 6·8. Filter if necessary through hot paper pulp (*q.v.*). Add the lactose, calcium carbonate and indicator and mix well to dissolve the lactose. Dispense in 100 ml or 10 ml amounts in screw-capped bottles, keeping the chalk dispersed. Sterilise by autoclaving at 121°C for 20 min.

When using the medium, the chalk must be resuspended before pouring the plates.

(Chalmers, 1962)

Nitrate Peptone Water†

(For detecting nitrate-reducing ability by bacteria)

Potassium nitrate® (AR grade)	0·2 g
Peptone water	1 litre

Dissolve the potassium nitrate in the peptone water, distribute in 5 ml amounts in 150 × 16 mm test-tubes each provided with an inverted Durham tube. Sterilise by autoclaving at 121°C for 15 min.

For the determination of nitrate reduction by anaerobes (e.g. as in the identification of *Clostridium perfringens*), 0·3 per cent agar should also be incorporated into the medium.

Nutrient Agar*†‡

(A general purpose culture medium for bacteria)

Nutrient broth	1 litre
Agar	15·0 g

pH 7·2

Dissolve the agar in the nutrient broth by autoclaving at 121°C for 20 min. Adjust the pH to 7·2. Filter through paper pulp (*q.v.*) Distribute as required and sterilise at 121°C for 20 min.

Nutrient Broth*†‡

(A general purpose culture medium for bacteria)

Peptone	10·0 g
Lab-Lemco meat extract (Oxoid)	10·0 g
Sodium chloride	5·0 g
Distilled water	1 litre

pH 7·2

Dissolve the ingredients in the water by steaming. Cool, adjust to pH 7·6, and autoclave at 121°C for 15 min. Filter and adjust to pH 7·2. Distribute as required, and sterilise at 121°C for 20 min.

Nutrient Broth (for Membrane Filtration)

Peptone	40·0 g
Yeast extract	6·0 g
Distilled water	1 litre

pH 7·2

Dissolve the ingredients in the water in a steamer. Cool and adjust to pH 7·2. Distribute into screw-capped bottles and sterilise at 121°C for 15 min.

(Taylor, Burman and Oliver, 1955)

Nutrient Gelatin*†‡

(For detection of proteolytic activity)

Gelatin	150·0 g
Nutrient broth	1 litre

pH 7·2

Add the gelatin to the nutrient broth and steam until dissolved. Adjust to pH 7·2. Distribute in 100 × 12 mm test-tubes, and sterilise by autoclaving at 115°C for 20 min.

Olive-oil Agar

(Used for detection of lipolytic activity)

Olive-oil	5 ml
Yeast extract agar or nutrient agar	100 ml

Before sterilisation of the basal medium, adjust the reaction to pH 7·8.
Dispense the olive-oil in screw-capped bottles and sterilise by autoclaving at 115°C for 20 min.

To 100 ml of sterile basal medium, melted and cooled to 45°C, add aseptically 5 ml of sterile olive-oil. Emulsify the medium by shaking vigorously, and pour plates containing about 15 ml amounts.

(Berry, 1933; Jones and Richards, 1952)

ONPG Peptone Water

(For the detection of β-galactosidase activity)

ONPG Solution

o-nitrophenyl-β-D-galactopyranoside®	0·6 g
0·01 M sodium phosphate buffer pH 7·5	100 ml

Dissolve at room temperature and sterilise by filtration.

Preparation of medium

Add aseptically 1 part of ONPG solution to 3 parts of sterile peptone water (prepared to pH 7·5). Distribute aseptically in 2 ml amounts into sterile 100 × 12 mm test-tubes. (Alternatively the peptone water; pH 7·5, can be prepared and distributed and sterilised in 1·5 ml amounts in the test-tubes, and the ONPG solution added as 0·5 ml amounts.)

Check for sterility of medium by incubating at 37°C for 24 h.

(Lowe, 1962)

Orange Serum Agar†‡

(For isolation and enumeration of yeasts and moulds in foods)

This medium is based on orange serum, and since such an extract prepared in small amounts in the laboratory will show considerable batch-to-batch variation, the authors recommend that the commercially available dehydrated media are used, particularly if the medium is being used for enumeration of microfungi in foods.

(See Murdock, Folinozzo and Troy, 1951; Hayes and Reister, 1952)

Osmophilic Agar

(For the growth of osmophilic and osmotolerant yeasts and moulds)

This medium is prepared by dissolving a dehydrated Wort Agar (*, †, ‡) in a 45° Brix syrup containing 35 g of sucrose and 10 g of glucose in 100 ml of solution (for preparation of the medium from basic ingredients, see Wort Agar). Sterilise the medium by autoclaving at 108°C for 20 min, cool to 45–50°C and pour plates as required. It is advisable to avoid unnecessary heating of this medium, so it is preferable to make the medium as required.

(Scarr, 1959; Beech and Davenport, 1969)

Packer's Crystal Violet Azide Blood Agar*†‡(B)

(For the detection and enumeration of faecal streptococci in foods)

Basal medium

Tryptose	15·0 g
Meat extract	3·0 g
Sodium chloride	5·0 g
Agar	15·0 g
Distilled water	1 litre

pH 6·8

Dissolve the ingredients in the water by steaming. Distribute in 100 ml amounts in screw-capped bottles, and sterilise by autoclaving at 121°C for 15 min.

Preparation of the complete medium:

To 100 ml of molten basal medium cooled to 45–50°C add aseptically

(*a*) 0·4 ml of a 0·05 per cent aqueous solution of crystal violet (previously sterilised at 121°C for 20 min, or by filtration);

(*b*) 1·0 ml of a 5 per cent aqueous solution of sodium azide® (previously sterilised by filtration);

(*c*) 5·0 ml of fresh defibrinated sheep blood.

Mix well and pour plates.

Note. An Azide Blood Agar Base is available as a commercially dehydrated medium (*, †, ‡); this can be made into a similar medium to the above by the addition of crystal violet and blood as described above, but the sodium azide concentration is 0·02 per cent, compared with 0·05 per cent in Packer's medium.

(Mossel, van Diepen and de Bruin, 1957; Thatcher and Clark, 1968; Packer, 1943)

Paper Pulp for Filtration of Media

Soak two large (46 cm × 57 cm) sheets of Whatman No. 1 filter paper in water and mash to a pulp. Bring to the boil in a large beaker and pour into a Buchner-type filter funnel while applying suction. Replace the receiving flask by a clean flask and immediately filter the agar medium.

Peptone Water*†‡

(Suitable for the indole test)

Tryptone or tryptose	10·0 g
Sodium chloride	5·0 g
Distilled water	1 litre

pH 7·2

Dissolve the peptone and sodium chloride in the water by steaming. Adjust to pH 7·2, and dispense in 5 ml amounts in 150 × 16 mm test-tubes and sterilise by autoclaving at 121°C for 15 min.

Peptone Water Diluent

(For enumeration and isolation techniques)

Peptone	1·0 g
Distilled water	1 litre

pH 7·0

Dissolve the peptone in the water, adjust to pH 7·0, dispense as required and sterilise by autoclaving at 121°C for 20 min.
(Straka and Stokes, 1957b)

Phenolphthalein Phosphate Agar

(For the detection of phosphatase production, particularly by staphylococci)

Phenolphthalein phosphate, 1 per cent solution	1 ml
Plate count agar, yeast extract agar or nutrient agar	100 ml

To the sterile molten basal medium cooled to 45°C, add aseptically 1 ml of a sterile 1 per cent solution of phenolphthalein phosphate (previously sterilised by filtration), mix, and pour plates as required.

Phenolphthalein phosphate agar with polymyxin (Gilbert *et al.*, 1969)

The above medium can be made suitable for the detection of phosphatase-producing staphylococci in foods, by the addition, immediately before pouring the plates, of 12,500 units of polymyxin B sulphate (e.g. "Aerosporin" from Wellcome Reagents Ltd.) per 100 ml of medium.

Phenylalanine Malonate Medium†

(A combined medium for the determination of phenylalanine deamination and malonate utilisation)

Ammonium sulphate	2·0 g
Dipotassium hydrogen phosphate	0·6 g
Potassium dihydrogen phosphate	0·4 g
Sodium chloride	2·0 g
Sodium malonate	3·0 g
DL-Phenylalanine	2·0 g
Yeast extract	1·0 g
Bromthymol blue, 1 per cent solution	2·5 ml
Distilled water	1 litre

Dissolve the ingredients in the water by steaming. Distribute in 5 ml amounts in 150 × 16 mm test-tubes, and sterilise at 115°C for 10 min.

(Shaw and Clarke, 1955)

Phosphate Buffer, 0·1 M

Solution A Dissolve 13·6 g of potassium dihydrogen phosphate (KH_2PO_4) in distilled water, and make up to 1 litre of solution.

Solution B Dissolve 26·8 g of disodium hydrogen phosphate heptahydrate ($Na_2HPO_4.7H_2O$) in distilled water, and make up to 1 litre of solution.

	Volume (ml) of solution	
pH	A	B
6·70	52	48
6·81	48	52
7·00	34	66
7·10	28	72
7·30	20	80
7·42	16	84

Distribute and sterilise as required.

Physiological Saline

Sodium chloride	8·5 g
Distilled water	1 litre

Dissolve the sodium chloride in the water, distribute as required, and sterilise by autoclaving at 121°C for 15 min.

Plate Count Agar (Tryptone Glucose Yeast Extract Agar)*†‡

(A non-selective medium for general viable counts of bacteria in foods)

Tryptone	5·0 g
Yeast extract	2·5 g
D-Glucose	1·0 g
Agar	15·0 g
Distilled water	1 litre

pH 7·0

Dissolve the ingredients in the water by steaming. Adjust to pH 7·0 dispense in 10 ml or 100 ml amounts in screw-capped bottles, and sterilise by autoclaving at 121°C for 15 min.

Note. This medium is also known as Standard Methods Agar.

(American Public Health Association, 1960)

Polypectate Gel Medium

(For the detection of pectinolytic micro-organisms)

Peptone	5·0 g
Dipotassium hydrogen phosphate	5·0 g
Potassium dihydrogen phosphate	1·0 g
Calcium chloride, hydrated ($CaCl_2.2H_2O$)	0·6 g
Sodium polypectate (Polygalacturonic acid, sodium salt)	70·0 g
Distilled water	to 1 litre

Heat 500 ml of water using a magnetic-stirrer-hotplate, and dissolve the first four ingredients. Next gradually add the sodium polypectate and add water to make 1 litre of medium. Heat the medium in a steamer for 15 min, then distribute as required and sterilise by autoclaving at 121°C for 15 min.

The medium may be made partially selective for Gram-negative bacteria by incorporating crystal violet (see page 55), and yeasts and moulds may be inhibited by the addition of Acti-dione (see page 218).

Note. After pouring plates they should be stored overnight before use.

(American Public Health Association, 1966)

Potato Dextrose Agar*†‡

(For the growth of microfungi)

Potatoes, peeled and diced	200 g
D-Glucose	20 g
Agar	15 g
Distilled water	to 1 litre

Boil 200 g of peeled, diced potatoes for 1 h in 1 litre of distilled water. Filter, and make up the filtrate to one litre. Add the glucose and agar and dissolve by steaming. Distribute in 10 ml or 100 ml amounts in screw-capped bottles and sterilise by autoclaving at 121°C for 20 min.

Note. Potato carrot dextrose agar can be prepared by substituting 50 g carrot for 50 g potato in the above recipe.

Reinforced Clostridial Medium*‡

(For the growth of anaerobes, and for use as a diluent in the enumeration of anaerobes)

Yeast extract	3·0 g
Peptone	10·0 g
Lab-Lemco meat extract (Oxoid)	10·0 g
D-Glucose	5·0 g
Sodium acetate, hydrated	5·0 g
Cysteine	0·5 g
Soluble starch	1·0 g
Distilled water	1 litre

pH 7·1–7·2

If required as an agar medium, agar to 1·5 per cent should be included; if required as a semi-solid medium, include agar to 0·5 per cent. The medium without agar is suitable as a diluent.

Add all the ingredients to the water, and dissolve by steaming. Filter through hot paper pulp (*q.v.*). Adjust to pH 7·4, dispense in screw-capped bottles as required, and sterilise by autoclaving at 121°C for 15 min. The pH after autoclaving should be 7·1–7·2.

(Hirsch and Grinsted, 1954; Gibbs and Hirsch, 1956)

Ringer's Solution, Quarter-strength*

((Diluent and suspending liquid)

Sodium chloride	2·25 g
Potassium chloride	0·105 g

Calcium chloride, hydrated	0·12 g
Sodium hydrogen carbonate	0·05 g
Distilled water	1 litre

Dissolve the salts in the water, distribute as required, and sterilise by autoclaving at 121°C for 20 min.

(Wilson, 1935)

Robertson's Cooked Meat Medium*†‡

(For growth of anaerobes, and maintenance of stock cultures of bacteria, especially micro-aerophilic and anaerobic bacteria)

Mince 500 g of fresh fat-free bullock's heart and simmer for 20 min in 500 ml of boiling 0·05 N sodium hydroxide. After cooking, adjust to pH 7·4. Strain off the liquid and dry the meat by spreading on filter paper. Distribute the meat in 30 ml (1 oz McCartney) screw-capped bottles to a depth of 3 cm. Add 10 ml of peptone water or nutrient broth. A 2 cm depth of liquid paraffin may be added if required. Sterilise, with the screw-caps slightly loosened, by autoclaving at 121°C for 20 min. After autoclaving, tighten the caps.

(Lepper and Martin, 1929)

Note. The authors have found this medium to be more satisfactory than equivalent dehydrated formulations, particularly for growth and maintenance of *Clostridium*.

Rogosa Agar*†

(For the isolation and enumeration of lactobacilli)

Tryptone or trypticase	10·0 g
Yeast extract	5·0 g
D-Glucose	20·0 g
Tween 80 (Atlas)	1·0 g
Potassium dihydrogen phosphate	6·0 g
Ammonium citrate	2·0 g
Sodium acetate, hydrated	25·0 g
Glacial acetic acid®	1·32 ml
Magnesium sulphate, hydrated ($MgSO_4.7H_2O$)	0·575 g
Manganese sulphate, hydrated ($MnSO_4.4H_2O$)	0·14 g
Ferrous sulphate, hydrated ($FeSO_4.7H_2O$)	0·034 g

Agar	15·0 g
Distilled water	to 1 litre

Final pH 5·4

Suspend all the ingredients in the water and dissolve by heating in a steamer. Distribute into sterile containers as required and heat in a steamer for a further 50 min. Store the medium under refrigeration until required. Use the minimum amount of heating necessary to melt the medium and avoid overheating (e.g. avoid melting more medium than is required, and attempting to re-store the surplus).

(Rogosa *et al.*, 1951)

Rogosa Agar Modified (Mabbitt and Zielinska)

(For the isolation and enumeration of lactobacilli in fermented milk products)

Milk digest

Mix 1 litre of separated milk or reconstituted milk, adjusted to pH 8·5, 5 g of trypsin®, (BDH Chemicals) and 10 ml of chloroform®. Incubate at 37°C for 24 hours, then steam for 20 min and filter while hot. Adjust the pH to 6·65 ± 0·02 with glacial acetic acid® (about 0·5 ml per litre) using a pH meter.

Nutrient solution

Yeast extract (Difco)	12·0 g
Di-ammonium hydrogen citrate	4·8 g
Potassium dihydrogen phosphate	14·4 g
D-Glucose	48·0 g
Tween 80 (Atlas)	2·4 g
Salts solution§	12·0 ml
Distilled water	200 ml

§*Salts solution*

Magnesium sulphate, hydrated ($MgSO_4.7H_2O$)	11·5 g
Manganese sulphate, hydrated ($MnSO_4.4H_2O$)	2·8 g
Ferrous sulphate, hydrated ($FeSO_4.7H_2O$)	0·08 g
Distilled water	100 ml

Dissolve ingredients in water by gentle heating. Add 60 ml of 4 M sodium acetate/acetic acid buffer at pH 5·37 ± 0·02. Make up the volume to 200 ml with distilled water. The solution has a final pH of 5·0 and is not sterilised; it should be stored in the refrigerator until required.

Preparation of complete medium

Dissolve 19 g agar in 700 ml of milk digest at 121°C for 20 min, and while hot mix with 185 ml of nutrient solution previously warmed to 50°C. Make up the volume to 1 litre with hot digest.

Distribute the medium aseptically in 10 ml amounts into sterile screw-capped bottles, and store in the refrigerator (without sterilising) until required.

The medium should be prepared for use with as little heating as possible to avoid darkening and the formation of a precipitate.

(Mabbitt and Zielinska, 1956)

Rogosa Broth Modified (Mabbitt and Zielinska)

This is prepared in a similar manner to Rogosa agar (modified), described above, but the agar is omitted.

(Mabbitt and Zielinska, 1956)

Rose Bengal Agar

(For the isolation of moulds in the presence of large numbers of bacteria)

D-Glucose	10·0 g
Mycological peptone	5·0 g
Potassium dihydrogen phosphate	1·0 g
Magnesium sulphate, hydrated ($MgSO_4.7H_2O$)	0·5 g
Rose bengal	0·035 g
Streptomycin	0·03 g
Agar	15·0 g
Distilled water	1 litre

Dissolve all the ingredients *except the streptomycin* in the water by steaming. Distribute as required and sterilise by steaming for 30 min on each of three successive days.

To the molten, sterile medium cooled to 45°C aseptically add the streptomycin as a sterile solution immediately before pouring the plates. Chlortetracycline at a concentration of 10 mg per litre may be substituted for the streptomycin (Jarvis, 1973).

(Martin, 1950)

Salt Meat Broth*

(For the liquid enrichment of staphylococci from low-salt foods)

Peptone	10·0 g
Beef extract	10·0 g
Sodium chloride	100·0 g
Distilled water	1 litre

pH 7·3

Dissolve the ingredients in the water by steaming. Adjust to pH 7·3, distribute as required, and sterilise by autoclaving at 121°C for 20 min.

Sea Water Yeast Peptone Agar

(For the growth of marine organisms in general, and the detection of luminous marine bacteria in particular)

Yeast extract	3·0 g
Peptone	5·0 g
Agar	15·0 g
Aged sea water	750 ml
Distilled water	250 ml

pH 7·4

Dissolve the ingredients in the water by steaming. Adjust to pH 7·4. Distribute as required and sterilise by autoclaving at 121°C for 20 min.

Note. The sea water should be aged for at least 3 weeks in the dark, and then filtered.

(Hendrie, Hodgkiss and Shewan, 1970)

Selenite Broth†‡

(An enrichment medium for *Salmonella*)

Peptone	4·0 g
Lactose	4·0 g
Sodium hydrogen selenite®	5·0 g
Disodium hydrogen phosphate, anhydrous	5·0 g
Potassium dihydrogen phosphate	5·0 g
Cystine	10 mg
Distilled water	1 litre

pH 6·9

Dissolve the ingredients in the water. The reaction of the medium should be 6·9 without the need for adjustment. Dispense in suitable amounts (depending on the amount of inoculum to be used) and sterilise in a boiling water bath for 10 min only.

(North and Bartram, 1953)

Note. The dehydrated medium available from Oxoid Ltd. contains no cystine.

Semi-solid Citrate Milk Agar

(For the detection of citrate fermentation by lactic streptococci)

(1) Dispense 500 ml of reconstituted skim milk into 150 × 16 mm test-tubes in 10·5 ml quantities. Sterilise in the autoclave by heating at 100°C for 30 min and then at 115°C for 10 min.

(2) Prepare 100 ml of a 10 per cent solution of sodium citrate dihydrate in distilled water, distribute in 10 ml amounts and sterilise by autoclaving at 121°C for 20 min.

(3) Dissolve 2·0 g of agar in 100 ml of distilled water by steaming, distribute in 4 ml amounts in 150 × 16 mm test-tubes. Close loosely with Astell rubber stoppers and sterilise at 121°C for 20 min.

Preparation and use of complete medium. Add 0·5 ml of sterile 10 per cent sodium citrate solution to each tube of skim milk. Invert to mix and allow to stand for 30 min. When required for use, inoculate 0·1 ml of the test culture into the citrated milk and mix by rotation of the tube. Pour the citrated milk culture onto 4 ml of molten 2 per cent agar (cooled to 45–50°C), replace the Astell seal, invert to mix and incubate at the desired temperature.

(Crawford, 1962)

Serum Agar

(For the growth of fastidious animal parasites)

Sterile serum	7·0 ml
Nutrient agar, sterile	100 ml

Liquefy the basal medium and cool to 45–50°C. Add the serum with aseptic precautions, mix well and pour plates as required.

Simmon's Citrate Agar*†‡

(For the determination of citrate-utilisation, particularly for differentiation of enterobacteria)

Sodium chloride	5·0 g
Magnesium sulphate, hydrated ($MgSO_4.7H_2O$)	0·2 g
Ammonium dihydrogen phosphate	1·0 g

Dipotassium hydrogen phosphate	1·0 g
Sodium citrate	5·0 g
Bromthymol blue, 1 per cent aqueous solution	8 ml
Agar	15·0 g
Distilled water	990 ml

pH 7·0

Add all the ingredients except the indicator solution to the water and dissolve by steaming. Adjust to pH 7·0 and add the bromthymol blue. Mix and dispense in test-tubes or screw-capped bottles with sufficient medium in each tube or bottle to form a slope with a 3-cm butt. Sterilise by autoclaving at 121°C for 15 min, and allow to set in a sloped position.

(Report, 1958a)

Sodium Chloride (15 per cent) Diluent

(A diluent for detection or enumeration of halophiles)

Sodium chloride	150·0 g
Distilled water	to 1 litre

Dissolve the sodium chloride in distilled water, and make up to 1 litre. Distribute in 9 ml or 90 ml amounts in screw-capped bottles and sterilise at 121°C for 20 min.

Starch Agar†

(For the detection of starch-hydrolysing ability)

Soluble starch	0·2 g
Yeast extract agar or nutrient agar	100 ml

Dissolve the starch in the molten basal medium, and sterilise by autoclaving at 121°C for 15 min.

Alternatively add aseptically 0·2 ml of a sterile 10 per cent solution of soluble starch to 10 ml of sterile molten basal medium, mix well, and pour into a Petri-dish.

Starch Milk Agar

(For the detection of spores in heated milk and milk products)

Nutrient agar	100 ml
Reconstituted skim milk	1 ml
Soluble starch, 10 per cent solution	1 ml

Add the skim milk and the soluble starch solution to molten nutrient agar. Mix well and sterilise by autoclaving at 121°C for 20 min.

(Grinsted and Clegg, 1955)

Stuart's Transport Medium*†‡

(For transport of swabs, etc. particularly for examination for fastidious animal parasites, and for anaerobes)

Sodium thioglycollate	1·0 g
Agar	3·0 g
Sodium glycerophosphate, 20 per cent solution	50 ml
Calcium chloride, 1 per cent solution	10 ml
Methylene blue, 0·1 per cent solution	2 ml
Distilled water	950 ml

Add the agar and sodium thioglycollate to the water and dissolve by steaming. Adjust to pH 7·2, then add the sodium glycerophosphate and calcium chloride solutions. Mix and adjust to pH 7·4. Add the methylene blue solution, mix and place in the steamer for 10 min. Distribute in 30 ml (1 oz) bijou or screw-capped bottles, filling them completely. Screw on the caps tightly and sterilise by autoclaving at 115°C for 20 min.

This medium should keep well, but occasionally an insufficiently tightened cap will allow oxygen to be absorbed, indicated by the methylene blue regaining its blue colour. When in this condition the medium should not be used until the oxygen has been expelled by steaming with the cap slightly loosened, after which the cap is tightened fully once more until use.

(Moffett, Young and Stuart, 1948; Stuart, Toshach and Patsula, 1954)

Sucrose Agar (1)

(For detecting polysaccharide production from sucrose)

Sucrose	10 g
Yeast extract agar or nutrient agar	100 ml

Add the sucrose to the molten basal medium, mix well and sterilise at 115°C for 20 min.

(Evans *et al.*, 1956)

Sucrose Agar (2)

(For detecting dextran production by *Leuconostoc*)

Tryptone	10·0 g
Yeast extract	5·0 g
Dipotassium hydrogen phosphate	5·0 g
Triammonium citrate	5·0 g
Sucrose	50·0 g
Agar	15·0 g
Distilled water	1 litre

Dissolve the ingredients in distilled water by heating in a steamer, with frequent mixing. Distribute as required and sterilise at 121°C for 15 min.

(Garvie, 1960)

Sucrose Diluent (20 per cent)

(Diluent for studies involving osmophilic or osmotolerant organisms)

Sucrose	200·0 g
Distilled water	to 1 litre

Dissolve the sucrose in distilled water and make up to one litre. Distribute in 90 ml or 9 ml amounts in screw-capped bottles, and sterilise by autoclaving at 115°C for 20 min.

Sulphite Polymyxin Sulphadiazine Agar†‡

(For the detection and enumeration of sulphite-reducing clostridia in foods)

Basal medium

Tryptone	15·0 g
Yeast extract	10·0 g
Ferric citrate scales	0·5 g
Agar	15·0 g
Distilled water	1 litre

pH 7·0

Dissolve the ingredients by steaming, and adjust the pH to 7·0. Distribute in 100 ml amounts in screw-capped bottles. Sterilise by autoclaving at 121°C for 15 min.

Before pouring the plates, aseptically add the following solutions (all previously sterilised by filtration) to each 100 ml of medium, melted and cooled to 45°C:

(*a*) 10 per cent solution of sodium sulphite, freshly prepared	0·5 ml
(*b*) 0·1 per cent solution of polymyxin B sulphate (e.g. "Aerosporin" from Wellcome Reagents Ltd.)	1·0 ml
(*c*) 1·2 per cent solution of sulphadiazine, sodium salt	1·0 ml

Mix well and pour plates.

(Angelotti *et al.*, 1962)

Taylor's Xylose Lysine Desoxycholate (XLD) Agar†‡

(For the selective isolation of *Salmonella* and *Shigella*)

Basal medium

Yeast extract	3·0 g
Xylose	3·75 g
L-Lysine monohydrochloride	5·0 g
Lactose	7·5 g
Sucrose	7·5 g
Sodium chloride	5·0 g
Phenol red, 0·2 per cent solution (see page 327)	40 ml
Agar	15·0 g
Distilled water	960 ml

Dissolve the ingredients in the water by steaming, cool to 50°C, add the indicator solution, and adjust to pH 6·9. Distribute in 100 ml amounts in screw-capped bottles and sterilise by autoclaving at 121°C for 15 min.

Thiosulphate- citrate solution

Sodium thiosulphate, hydrated	34·0 g
Ferric ammonium citrate	4·0 g
Distilled water	to 100 ml

Dissolve the salts in the water, make up to 100 ml and sterilise by filtration.

Sodium desoxycholate solution

Dissolve 10 g of sodium desoxycholate in sufficient distilled water to make 100 ml of solution and sterilise at 121°C for 15 min.

Preparation of complete medium

To 100 ml of basal medium, melted and cooled to 50°C, add aseptically 2·0 ml of thiosulphate-citrate solution and 2·5 ml of sodium desoxycholate solution. If necessary the pH should be readjusted to pH 6·9 using sterile reagents, the medium mixed, and plates poured as required. Since it may be necessary to readjust pH just before pouring plates, the procedure is made easier if the basal medium and supplements are made in reasonably large batches and kept under refrigeration until required, so that any adjustment necessary may be determined with the first bottle of medium, and all other bottles in the batch marked accordingly.

(Taylor, 1965; Taylor and Schelhart, 1967)

Tetrathionate Broth*†‡(B)

(For the enrichment of *Salmonella*)

Solution A

Sodium thiosulphate, hydrated	24·8 g
Distilled water	to 100 ml

Sterilise by steaming for 30 min on each of 3 successive days.

Solution B

Iodine®	12·7 g
Potassium iodide	20·0 g
Distilled water	to 100 ml

Preparation of complete medium

To 78 ml of sterile nutrient broth add 2·5 g of calcium carbonate and sterilise by steaming for 30 min on each of 3 successive days. When cool, add aseptically 15 ml of solution A and 4 ml of solution B. Store refrigerated for not more than one week before use.

The components—solutions A and B and the sterile chalk nutrient broth may conveniently be stored separately until immediately before use.

(Knox, Gell and Pollock, 1942)

Thornley's Semi-solid Arginine Medium

(For the production of ammonia from arginine by pseudomonads)

Peptone	1·0 g
Dipotassium hydrogen phosphate	0·3 g
Sodium chloride	5·0 g
L-Arginine monohydrochloride	10·0 g
Phenol red, 0·2 per cent solution (see page 327)	5 ml
Agar	3·0 g
Distilled water	1 litre

pH 7·2

Dissolve the ingredients in the water by steaming. Mix well, and dispense in 7 ml ($\frac{1}{4}$-oz bijou) screw-capped bottles to a depth of 2 cm. Sterilise at 121°C for 15 min.

(Thornley, 1960)

Tomato Juice Lactate Agar

(For the detection and enumeration of citrate-fermenting lactic-acid bacteria)

Basal medium

Tomato juice agar, dehydrated (Oxoid, Difco or BBL)	15·0 g
Calcium lactate, hydrated	5·0 g
Agar	9·0 g
Distilled water	1 litre

Dissolve the ingredients in the distilled water by heating at 121°C for 5 min. Distribute the medium in 15 ml amounts in 30 ml screw-capped bottles (Universal containers). Sterilise at 121°C for 15 min.

Calcium citrate suspension

Add 10 g of calcium citrate to 100 ml of distilled water. Shake well and, keeping the particles in suspension, distribute in 10 ml amounts in 30 ml screw-capped bottles (Universal containers). Sterilise at 121°C for 15 min.

Preparation of complete medium

The complete medium is prepared by adding aseptically 1 ml of well-shaken calcium citrate suspension to 15 ml of molten tomato juice lactate agar, inverting to mix and pouring into Petri-dishes.

(Skean and Overcast, 1962)

Note. The authors have not found it necessary to suspend the calcium citrate in a 1·5 per cent solution of carboxymethylcellulose as used in the original version of this medium by Galesloot *et al.*, (1961).

Tributyrin Agar*

(For the detection of lipolytic activity)

Tributyrin	10·0 g
Yeast extract agar or nutrient agar, adjusted to pH 7·5	100 ml

Add the tributyrin to the molten basal medium in an electric blender and mix until completely emulsified. Dispense into screw-capped bottles in 10 ml amounts and sterilise by steaming for 30 min on each of three successive days.

Tryptone Glucose Yeast Extract Broth

(Non-selective medium for general viable counts by the multiple tube technique)

Prepare to the same recipe as Plate Count Agar, except that the agar should be omitted. Distribute in bottles or test-tubes as required (double strength medium can be prepared for large volumes of inocula) and sterilise by autoclaving at 121°C for 20 min.

Tryptone Soya Broth*†‡

(General purpose non-selective liquid medium)

Tryptone or trypticase	17·0 g
Soya peptone or phytone	3·0 g
Dipotassium hydrogen phosphate	2·5 g
Sodium chloride	5·0 g
D-Glucose	2·5 g
Distilled water	1 litre

pH 7·3

Dissolve ingredients in the water by steaming. Cool and adjust to pH 7·3. Distribute as required and sterilise at 121°C for 15 min.

The medium may also be prepared without the glucose, or as a double-strength medium for membrane filtration.

(American Public Health Association, 1960)

Tween Agar

(A lipolysis test medium)

Peptone	10·0 g
Calcium chloride, hydrated	0·1 g
Sodium chloride	5·0 g
Tween (Atlas)	10·0 g
Agar	15·0 g
Distilled water	1 litre

pH 7·0–7·4

Dissolve all the ingredients in the water by steaming. Check that the pH is within the indicated range, and adjust if necessary. Dispense as required and sterilise by autoclaving at 115°C for 20 min.

Tween 80 (an oleic acid ester) is the Tween most often used. Although liquid at ambient temperatures, it is fairly viscous, and it will prove easier to dispense by warming to 45–50°C.

(Sierra, 1957)

V-8 Agar

(A medium for the inducement of sporulation by yeasts)

Adjust the pH of canned V-8 vegetable juice (manufactured by Campbell's Soup Co.) to pH 6·8. To 100 ml of the juice add 40 g of baker's yeast, and heat for 10 min in a steamer. Readjust the pH to 6·8 and add an equal amount of a melted 4 per cent solution of agar in distilled water. Mix, and dispense in 7 ml amounts in 150 × 16 mm test-tubes and sterilise at 121°C for 15 min. Not more than 8 h before use, prepare as slopes in the test-tubes by melting the medium in a steamer, and allowing to set in a slanted position.

(Wickerham, Flickinger and Burton, 1946)

Vaspar

(For sealing media for anaerobic growth studies)

Melt and mix together equal amounts of petroleum jelly ("Vaseline") and paraffin wax. Distribute into wide-mouthed screw-capped bottles, and sterilise by autoclaving at 121°C for 15 min.

Victoria Blue Butter-fat (or Margarine) Agar

(For the detection of lipolytic activity by micro-organisms)

Preparation of Victoria blue base

Boil 2 g of powdered Victoria blue (BDH Chemicals Ltd.) in 200 ml of distilled water until thoroughly dispersed. Slowly add a 10 per cent solution of sodium hydroxide® with constant mixing until the colour disappears from the solution. Allow the water-insoluble precipititate (the basic dye) to settle out. Filter off the precipitated basic dye and wash with distilled water made slightly alkaline with ammonium hydroxide. Dry the dye at 30°C.

Preparation of fat

Obtain separated butter-fat or margarine fat as already described (see page 326).

Preparation of dye/fat mixture

Heat 100 g of fat in a conical flask with 100 ml of distilled water and some glass beads. When boiling, slowly add Victoria blue base with constant mixing until the fat is saturated with dye (the fat will be deep red, with some particles of undissolved dye at the bottom of the flask). Boil for 30 min. Separate the fat from the bulk of the water, and filter overnight at 37°C. Separate the filtered fat from any residual water, using a separating funnel, dispense in 30 ml screw-capped bottles in 10 ml amounts and sterilise by autoclaving at 121°C for 15 min.

Preparation of basal medium

The basal medium may consist of yeast extract agar, nutrient agar, or Tryptose Blood Agar Base (Oxoid). Whichever basal medium is used the agar content should be increased to 2 per cent (w/v), and the pH adjusted to 7·8. Distribute the medium in 20 ml amounts in 30 ml screw-capped bottles, and sterilise by autoclaving at 121°C for 20 min.

Preparation of the complete medium

Add aseptically 1 ml of the molten dye/fat mixture to 20 ml of sterile molten basal medium which has been cooled to 45°C. Emulsify by vigorously shaking for one minute and pour into a Petri-dish. Surface air bubbles can be eliminated by immediately flaming the surface rapidly with a Bunsen flame.

If the plates must be stored before use they should be refrigerated. The colour of the medium may change on storage, or during incubation (an uninoculated plate should always be incubated as a control). The medium should be pinkish-mauve; if it is bright blue it is unsuitable for use.

(Jones and Richards, 1952; Paton and Gibson, 1953; Harrigan, 1967)

Violet Red Bile Agar*†‡

(For the enumeration of lactose-fermenting enterobacteria in foods)

Peptone	7·0 g
Yeast extract	3·0 g
Bile salts§ (No. 3, Oxoid L56 or equivalent)	1·5 g
Sodium chloride	5·0 g
Lactose	10·0 g
Neutral red, 1 per cent aqueous solution	3 ml
Crystal violet, 0·05 per cent solution	4 ml
Agar	15·0 g
Distilled water	1 litre

pH 7·4

§ See footnote on page 56.

Dissolve the peptone, yeast extract, bile salts, agar and sodium chloride in the water by steaming. Cool to 50°C and adjust to pH 7·4. Add the lactose, neutral red and crystal violet and mix until dissolved. Distribute in 15 ml amounts in screw-capped bottles and sterilise by autoclaving at 115°C for 15 min. This medium should not be reheated to melt more than once.

Violet Red Bile Glucose Agar*

(For the enumeration of the Enterobacteriaceae in foods)

This medium should be made to the same recipe as Violet Red Bile Agar (see above), but 10 g of glucose per litre of medium incorporated at the same time as the lactose.

This medium should be melted once only for use.

(Mossel, Mengerink and Scholts, 1962)

Willis and Hobbs's Lactose Egg-yolk Milk Agar

(For the differentiation of clostridia)

Basal medium

Nutrient broth	800 ml
Agar	12·0 g
Lactose	9·6 g
Neutral red, 1 per cent solution	2·6 ml

Dissolve the agar in the nutrient broth by heating, then add the lactose and neutral red solution. Mix well and distribute in 80 ml amounts in screw-capped bottles. Sterilise by autoclaving at 121°C for 20 min.

Egg-yolk emulsion

Separate aseptically an egg-yolk from the egg-white and drain the yolk into a sterile measuring cylinder. Mix the yolk with an equal volume (*c.* 20 ml) of 0·85 per cent sterile saline solution, and transfer aseptically to a sterile screw-capped bottle. This emulsion has been found satisfactory for routine purposes; it can be tested for sterility by plating out 1 ml quantities.

Preparation of complete medium

Melt the basal medium and cool to 50–55°C. Add 3 ml of egg-yolk emulsion and 12 ml of sterile reconstituted skim milk to 80 ml of molten basal medium. Mix well and pour into Petri-dishes.

(Willis, 1962)

Wilson and Blair's Bismuth Sulphite Agar (for *Salmonella* and *Shigella*)*†‡

Nutrient agar (containing 3 per cent agar)	100·0 ml
Sulphite-bismuth-phosphate solution	20·0 ml
Ferric citrate-brilliant green solution	4·5 ml

The basal medium consists of a nutrient agar containing 3 per cent agar to ensure solidification after the addition of the stock solutions.

To the molten nutrient agar at 50°C add the two stock solutions previously warmed to 50°C, mix well and pour plates containing 20 ml of medium per plate. *Refrigerate the plates for 4–5 days at 4–7°C before use.* This storage before use is advisable as it reduces the inhibitory action shown by the freshly prepared medium towards certain strains of *Salmonella*.

Preparation of the sulphite-bismuth-phosphate solution

Bismuth ammonium citrate scales	6·0 g
Sodium sulphite, anhydrous	20·0 g
Disodium hydrogen phosphate, anhydrous	10·0 g
D-Glucose	10·0 g
Distilled water	200·0 ml

Dissolve the bismuth ammonium citrate in 50 ml of boiling distilled water, and dissolve the sodium sulphite in 100 ml of boiling distilled water. Mix and, while still boiling, add the disodium hydrogen phosphate, and stir until dissolved. Cool and then add the glucose previously dissolved by warming in the remaining 50 ml of distilled water.

Preparation of ferric citrate-brilliant green solution

Ferric citrate scales	2·0 g
Brilliant green, 1 per cent solution	25·0 ml
Distilled water	200·0 ml

These stock solutions may be stored refrigerated for months without significant deterioration.

(Wilson, 1938)

Wilson and Blair's Sulphite Medium (for *Clostridium*)

Nutrient agar (containing 3 per cent agar)	100 ml
Sodium sulphite-glucose solution	15 ml
Ferrous sulphate, hydrated, 8 per cent solution	1 ml

The nutrient agar base is of the usual composition except that the concentration of agar is increased to 3 per cent. To 100 ml of the molten nutrient agar cooled to 55°C, add, immediately before use, the solutions of sodium sulphite-glucose and ferrous sulphate, and mix. In use the predetermined amount of water sample or other inoculum to be examined should be mixed with an equal quantity of the molten medium (just prepared) in sterile Miller-Prickett tubes or other suitable containers.

Preparation of sodium sulphite-glucose solution

Sodium sulphite, anhydrous	20·0 g
D-Glucose	10·0 g
Distilled water	150·0 ml

Dissolve the sodium sulphite in 100 ml of boiling distilled water, and dissolve the glucose in the remaining 50 ml of boiling distilled water. When cool, mix the two solutions together.

(Wilson, 1938)

Wort Agar*†‡

(For growth of yeasts and moulds)

Malt extract	15·0 g
Peptone	0·78 g
Maltose	12·75 g
Dextrin	2·75 g
Glycerol	2·35 g
Dipotassium hydrogen phosphate	1·0 g
Ammonium chloride	1·0 g
Agar	20·0 g
Distilled water	1 litre

pH 4·8

Dissolve all the ingredients in the water by steaming. Adjust to pH 4·8, and distribute in 15 ml amounts in test-tubes or screw-capped containers. Sterilise by autoclaving at 115°C for 15 min.

This medium will normally prove too soft for inoculation by streaking, and the low pH makes it unsuitable for repeated reheating.

For Osmophilic Agar (*q.v.*) the ingredients of the medium should be dissolved in a 45° Brix syrup consisting of 350 g of sucrose and 100 g of glucose in 1 litre of solution: sterilise at 108°C for 20 min.

Yeast Glucose Chalk Litmus Milk

(For maintenance of cultures of lactic-acid bacteria)

Prepare 1 litre of yeast glucose litmus milk. Distribute calcium carbonate (precipitated) in 0·5 g quantities in 30 ml (Universal) screw-capped containers. Sterilise in the autoclave at 121°C for 20 min. Allow to cool and add 10 ml of yeast glucose litmus milk to each container. Sterilise in the steamer at 100°C for 30 min on each of three successive days. Incubate at 37°C for 2 days and at 30°C for 2 days (or at 55°C if to be used for thermophilic organisms) prior to use to check sterility of the medium.

Yeast Glucose Lemco Agar

Peptone	10·0 g
Lab-Lemco meat extract (Oxoid)	10·0 g
Sodium chloride	5·0 g
D-Glucose	5·0 g
Yeast extract	3·0 g
Agar	15·0 g
Distilled water	1 litre

pH 7·0

Dissolve the ingredients, except the glucose and agar, by steaming for ½–1 h. Cool to room temperature and adjust to pH 7·0. Add the agar and dissolve at 121°C for 15 min. Filter through hot paper pulp (*q.v.*) and dissolve the glucose in the filtrate. Distribute as required and sterilise by autoclaving at 121°C for 15 min.

(Naylor and Sharpe, 1958a)

Yeast Glucose Lemco Broth

Prepare in a similar manner to yeast glucose lemco agar, except that the agar is omitted from the recipe.

(Naylor and Sharpe, 1958a)

This medium is also used as the basal medium for determining the growth characteristics of streptococci in the identification of isolates to primary physiological groups (see page 271).

Yeast Glucose Lemco Broth + 4 per cent (or 6·5 per cent) sodium chloride

Dissolve 4 g (or 6·5 g) of sodium chloride in *c.* 70 ml of yeast glucose lemco broth and make up volume to 100 ml with more broth. Distribute in 7 ml amounts in 150 × 16 mm test-tubes and sterilise by autoclaving at 115°C for 15 min.

Yeast Glucose Litmus Milk

Litmus milk (*q.v.*)	1 litre
Yeast extract	3·0 g
D-Glucose	10·0 g

Dissolve the yeast extract and glucose in previously bulk-sterilised litmus milk by steaming for ½–1 h. Distribute in test-tubes or as required and sterilise by steaming for 30 min on each of three successive days. Incubate at 37°C for 2 days and at 30°C for 2 days (and at 55°C if required for work with thermophiles) prior to use to check the sterility of the medium.

(Wheater, 1955)

Yeast Extract Agar

Yeast extract	3·0 g
Peptone	5·0 g
Agar	15·0 g
Distilled water	1 litre

pH 7·2

Dissolve the yeast extract, peptone and agar in the water by steaming. Adjust to pH 7·4 and filter while hot through hot paper pulp (*q.v.*). Adjust to pH 7·2, distribute as required and sterilise at 121°C for 15 min.

(Report, 1969)

Yeast Extract Lactate Broth

(For detection and isolation of propionibacteria)

Yeast extract	5·0 g
Sodium lactate	20·0 g
Distilled water	1 litre

pH 7·0

Dissolve the ingredients and adjust to pH 7·0. Distribute as required and sterilise by autoclaving at 121°C for 20 min.

If sodium lactate is not available, it may be prepared by adding 6·2 g of sodium hydroxide® pellets to 14 g of lactic acid® or, alternatively, neutralising the lactic acid with 155 ml of N sodium hydroxide and making up the volume to 1 litre with distilled water.

This medium may be converted into the *semi-solid medium* by the inclusion of 3·0 g of agar per litre, or into the *solid agar medium* by the inclusion of 15·0 g of agar per litre.

(van Neil, 1928)

Yeast Extract Milk Agar*

(General non-selective medium for detecting bacteria in milk and milk products)

Yeast extract	3·0 g
Peptone	5·0 g
Agar	15·0 g
Fresh whole or skim milk	10·0 ml
Distilled water	1 litre

Dissolve the yeast extract and peptone in distilled water by steaming, then cool and adjust to pH 7·4. Add the agar and milk to the broth and autoclave at 121°C for 20 min. Filter the medium while hot through hot paper pulp (*q.v.*). Determine the pH of the filtrate at 50°C and adjust if necessary to pH 7·0. Distribute the medium as required and sterilise by autoclaving at 121°C for 15 min. The final reaction of the medium at room temperature should be pH 7·2.

This medium is available dehydrated as Milk Agar (Oxoid). Its composition differs from that of Wilson (1935) and the similar formulation of the British Standard (1968) in the substitution of yeastel by yeast extract.

(Wilson, 1935)

APPENDIX 2

PROBABILITY TABLES FOR THE ESTIMATION OF MICROBIAL NUMBERS BY THE MULTIPLE TUBE TECHNIQUE

PROBABILITY TABLES FOR THE ESTIMATION OF MICROBIAL NUMBERS BY THE MULTIPLE TUBE TECHNIQUE

In the case of samples for which it is impossible to make an informed guess about the approximate concentration of organisms likely to be found, it is sometimes helpful to set up a 2-tube series (Table 1) over a very wide range of dilutions. Such tests may enable the correct range of dilutions to be chosen for future samples which can be used in the preparation of 3- or 5-tube series in order to obtain better estimates of the microbial load (Tables 2 and 3).

It should be noted that these tables assume that adequate mixing of samples and dilutions has taken place, that the micro-organisms are randomly distributed, that one or more organisms in a tube always results in growth, and that there are no disturbing influences. These assumptions may sometimes be incorrect. If, for example, results are obtained in which no growth occurs in the lower dilutions and growth occurs in the higher dilutions, this *could* be indicative of an antimicrobial inhibitory substance present in the food which becomes diluted below a threshold minimum inhibitory concentration before the micro-organisms have been diluted out to undetectably low numbers.

Determination of MPNs from Series of More than Three Dilutions

If more than 3 dilutions have been used, choose the 3 dilutions for determining the MPN from the Tables according to the following rules.

1. If more than one dilution has all its tubes positive, choose the set of three to include only the most dilute of these.
2. If more than one dilution contains no positive tubes, choose the set of three dilutions so that only the most concentrated of these is included.
3. If only three dilutions have been left by steps 1 and 2, the MPN may be determined from the Tables as indicated. If four dilutions remain, take both sets of three consecutive dilutions, determine the $\log_{10}$ (MPN) of each and take the average of the logarithms to determine a final $\log_{10}$ (MPN) and hence the MPN. In this case the result should however be taken as only a rough indication of the microbial load, and should not be used as the basis for quality control decisions.
4. If more than four sets of tubes remain, the Tables cannot be used. Stevens (1958) described a procedure for determining the probability that any particular range of results may occur (the range being the number of dilution

levels from the lowest dilution at which at least one tube shows no growth to the highest dilution at which at least one tube shows growth). Where the range of results left by steps 1–3 exceeds the number that can be dealt with by these Tables, an estimate may be obtained using the procedure described in Table 8,2 of Fisher and Yates (1963), together with an indication of the significance of the estimate so obtained. Usually, however, it is preferable to review the experimental procedure in order to identify possible deficiencies or faults in technique, and to repeat the test on a further sample.

TABLE 1

VALUES OF THE MPN FOR 2 TUBES INOCULATED FROM EACH OF THREE SUCCESSIVE 10-FOLD DILUTIONS

Number of positive tubes observed at each dilution			MPN of micro-organisms per ml of the first dilution
1st dilution	2nd dilution	3rd dilution	
0	0	0	0
0	0	1	0·45
0	1	0	0·46
1	0	0	0·6
1	0	1	1·2
1	1	0	1·3
1	1	1	2·0
1	2	0	2·1
2	0	0	2·3
2	0	1	5·0
2	1	0	6·2
2	1	1	13
2	1	2	21
2	2	0	24
2	2	1	70
2	2	2	100+

Approximate 95 per cent confidence limits may be calculated (Cochran, 1950) as

$$\frac{\text{MPN}}{6{\cdot}61} \text{ to MPN} \times 6{\cdot}61$$

TABLE 2

VALUES OF THE MPN FOR 3 TUBES INOCULATED FROM EACH OF THREE SUCCESSIVE 10-FOLD DILUTIONS

(After Demeter, Sauer and Miller, 1933)

Number of positive tubes observed at each dilution				
1st dilution	2nd dilution	3rd dilution	MPN of micro-organisms per ml of the first dilution	Category*
0	0	0	0	–
0	0	1	0·3	3
0	1	0	0·3	2
0	1	1	0·6	4
0	2	0	0·6	4
1	0	0	0·4	1
1	0	1	0·7	3
1	0	2	1·1	4
1	1	0	0·7	2
1	1	1	1·1	4
1	2	0	1·1	3
1	2	1	1·5	4
1	3	0	1·6	4
2	0	0	0·9	1
2	0	1	1·4	3
2	0	2	2·0	4
2	1	0	1·5	2
2	1	1	2·0	4
2	1	2	3·0	4
2	2	0	2·0	3
2	2	1	3·0	4
2	2	2	3·5	4
2	2	3	4·0	4
2	3	0	3·0	4
2	3	1	3·5	4
2	3	2	4·0	4
3	0	0	2·5	1
3	0	1	4·0	2
3	0	2	6·5	4
3	1	0	4·5	1
3	1	1	7·5	2
3	1	2	11·5	3
3	1	3	16·0	4
3	2	0	9·5	1
3	2	1	15·0	2

TABLE 2 – *continued*

Number of positive tubes observed at each dilution				
1st dilution	2nd dilution 10^{-1}	3rd dilution 10^{-2}	MPN of micro-organisms per ml of the first dilution	Category*
3	2	2	20·0	3
3	2	3	30·0	4
3	3	0	25·0	1
3	3	1	45·0	1
3	3	2	110·0	1
3	3	3	140·0+	–

Approximate 95 per cent confidence limits may be calculated (Cochran, 1950) as

$$\frac{\text{MPN}}{4{\cdot}68} \text{ to MPN} \times 4{\cdot}68$$

* In the long run, Category 1 combinations may be expected to constitute 67·5 per cent of test results containing both positive and negative tubes; Categories (1 + 2) 91 per cent; and Categories (1 + 2 + 3) 99 per cent of such test results. Category 4 and unlisted combinations are highly unlikely (Woodward, 1957). Combinations from Categories 3 and 4 should not be used as the basis for quality control decisions involving the rejection and/or reprocessing of batches of food—samples should be retested. In the event of large numbers of improbable combinations being obtained, experimental procedures should be examined closely, as this is indicative of the presence of disturbing influences—e.g. improper mixing of samples and/or dilutions, presence of antagonistic substances in the foodstuff, etc.

TABLE 3

VALUES OF THE MPN FOR 5 TUBES INOCULATED FROM EACH OF THREE SUCCESSIVE 10-FOLD DILUTIONS

(after Report, 1969; Woodward, 1957; American Public Health Association, 1970)

Number of positive tubes observed at each dilution				
1st dilution	2nd dilution	3rd dilution	MPN of micro-organisms per ml of the first dilution	Category*
0	0	0	0	–
0	0	1	0·2	3
0	0	2	0·4	4
0	1	0	0·2	3
0	1	1	0·4	4

TABLE 3 – *continued*

Number of positive tubes observed at each dilution				
1st dilution	2nd dilution	3rd dilution	MPN of micro-organisms per ml of the first dilution	Category*
0	1	2	0·6	4
0	2	0	0·4	3
0	2	1	0·6	4
0	3	0	0·6	4
1	0	0	0·2	1
1	0	1	0·4	3
1	0	2	0·6	4
1	1	0	0·4	2
1	1	1	0·6	4
1	1	2	0·8	4
1	2	0	0·6	3
1	2	1	0·8	4
1	3	0	0·8	4
1	3	1	1·0	4
2	0	0	0·5	1
2	0	1	0·7	3
2	0	2	0·9	4
2	0	3	1·2	4
2	1	0	0·7	2
2	1	1	0·9	3
2	1	2	1·2	4
2	2	0	0·9	3
2	2	1	1·2	4
2	2	2	1·4	4
2	3	0	1·2	4
2	3	1	1·4	4
3	0	0	0·8	1
3	0	1	1·1	3
3	0	2	1·3	4
3	1	0	1·1	2
3	1	1	1·4	3
3	1	2	1·7	4
3	1	3	2·0	4
3	2	0	1·4	3
3	2	1	1·7	4

TABLE 3 – *continued*

Number of positive tubes observed at each dilution				
1st dilution	2nd dilution	3rd dilution	MPN of micro-organisms per ml of the first dilution	Category*
3	2	2	2·0	4
3	3	0	1·7	3
3	3	1	2·1	4
3	4	0	2·1	4
3	4	1	2·4	4
4	0	0	1·3	1
4	0	1	1·7	3
4	0	2	2·1	4
4	0	3	2·5	4
4	1	0	1·7	2
4	1	1	2·1	3
4	1	2	2·6	4
4	2	0	2·2	2
4	2	1	2·6	3
4	2	2	3·2	4
4	3	0	2·7	3
4	3	1	3·3	3
4	3	2	3·9	4
4	4	0	3·4	3
4	4	1	4·0	4
4	4	2	4·7	4
4	5	0	4·1	4
4	5	1	4·8	4
4	5	2	5·6	4
5	0	0	2·3	1
5	0	1	3·1	3
5	0	2	4·3	3
5	0	3	5·8	4
5	1	0	3·3	1
5	1	1	4·6	2
5	1	2	6·4	4
5	1	3	8·4	4
5	1	4	11·5	4
5	2	0	4·9	1
5	2	1	7·0	2

TABLE 3 – *continued*

Number of positive tubes observed at each dilution				
1st dilution	2nd dilution	3rd dilution	MPN of micro-organisms per ml of the first dilution	Category*
5	2	2	9·5	3
5	2	3	12·0	4
5	2	4	15·0	4
5	3	0	7·9	1
5	3	1	11·0	2
5	3	2	14·0	3
5	3	3	17·5	3
5	3	4	20·0	4
5	3	5	25·0	4
5	4	0	13·0	1
5	4	1	17·0	2
5	4	2	22·0	2
5	4	3	28·0	3
5	4	4	35·0	3
5	4	5	42·5	4
5	5	0	24·0	1
5	5	1	35·0	1
5	5	2	54·0	1
5	5	3	92·0	1
5	5	4	160·0	1
5	5	5	180·0+	–

Approximate 95 per cent confidence limits may be calculated (Cochran, 1950) as

$$\frac{\text{MPN}}{3{\cdot}30} \text{ to MPN} \times 3.30$$

* See footnote to Table 2.

APPENDIX 3

MANUFACTURERS AND SUPPLIERS

Astell-Hearson. (Astell roll tube apparatus, Astell tube seals, stainless steel dropping pipettes, Miller-Prickett tubes)
172 Brownhill Road, Catford, London, SE6 2DL, England.

Atlas Chemical Industries Inc. (Tweens)
Wilmington, Delaware, U.S.A.
U.K. distribution: Honeywill-Atlas Ltd. (*q.v.*)
Tween 80 "TB culture grade" is available from:
Laboratory Services Division, Matheson Scientific, 12101 Cantron Place, Cincinnatti, Ohio, U.S.A.

Baird and Tatlock (London) Ltd. (Anaerobic jar)
Freshwater Road, Chadwell Heath, Essex, England.

Baltimore Biological Laboratories. (Dehydrated media and constituents, GasPak anaerobic system, fluorescent antibody reagents)
Cockeysville, Maryland 21030, U.S.A.
U.K. distribution: Becton, Dickinson UK Ltd., York House, Empire Way, Wembley, Middlesex HA9 0PS.

BDH Chemicals Ltd. (Oxidation-potential indicators, methylene blue tablets, resazurin tablets, pH indicators, Merck indicator strips, 01/129 (2,4-diamino 6,7-di*iso*propyl pteridine phosphate))
Broom Road, Poole, Dorset, BH12 4NN, England.
U.S. distribution: Gallard-Schlesinger Chemicals Manfg. Corp., 584 Mineola Avenue, Carle Place, New York 11514.

A. Browne Ltd. (Browne's steriliser control tubes)
Chancery Street, Leicester, England.

Calbiochem (O/129 (2,4-diamino 6,7-di*iso*propyl pteridine phosphate), penicillinase)
P.O. Box 12087, San Diego, California 92112, U.S.A.
U.K.: Calbiochem Ltd., 10 Wyndham Place, London W1H 1AS.

Campbell's Soup Company (V-8 vegetable juice)
Camden, New Jersey, U.S.A.
U.K.: King's Lynn, Norfolk.

C. F. Casella & Co. Ltd. (Air samplers for bacteria and fungal spores)
Brittania Walk, London N1, England.
U.S. distribution: B. G. I. Inc., 58 Guinan Street, Waltham, Mass. 02154.

Clinical Sciences Inc. ("Fluoro-Kit" fluorescent antibody detection system for *Salmonella*)
30 Troy Road, Whippany, New Jersey, 07981, U.S.A.

Corning-EEL Ltd. (Nephelometer)
Halstead, Essex, England.

Decon Laboratories Ltd. ("Decon 90" laboratory detergent)
Ellen Street, Portslade, Brighton BN4 1EQ, England.
U.S. distribution (under the name "Contrad 90"): Harleco, Democrat Road, Gibbstown, New Jersey.

Difco Laboratories Inc. (Dehydrated media and constituents, antibiotic sensitivity discs and dispensers, discs for rapid test systems, antisera, fluorescent antibody reagents)
Detroit, Michigan, 48232, U.S.A.
U.K.: P.O. Box 14B, Central Avenue, West Molesey, Surrey, KT8 0SE.

Diversey (U.K.) Ltd. ("Pyroneg" laboratory detergent)
Cockfosters Road, Barnet, Herts, England.
U.S.: Diversey U.S.A., 1855 South Mount Road, Des Plaines, Illinois.

Ellab A/S (Automatic F_0-value computer)
9 Krondalvej, Copenhagen, 2610 Rødovre, Denmark.
U.K. distribution: Siemens Ltd., 15–18 Clipstone Street, London, W1P 8E.

EEL Ltd. *see* Corning-EEL Ltd.

Fisher Scientific Co. (Safety apparatus, safety screens, "Fisher Manual of Laboratory Safety")
711 Forbes Avenue, Pittsburgh, Pennsylvania 15219, U.S.A.

A. Gallenkamp & Co. Ltd. (Bypass filter attachment for screw-capped bottles, safety screens, etc.)
Technico House, Christopher Street, London EC2P 2ER, England.

Gelman Instrument Co. (Membrane filters and accessories)
P.O. Box 1448, Ann Arbor, Michigan 48106, U.S.A.
U.K.: Gelman Hawksley Ltd., 12 Peter Road, Lancing, Sussex.

Honeywill-Atlas Ltd. (Cirrasol ALN-WF, Tweens)
Mill Lane, Carshalton, Surrey, England.
U.S.: Atlas Chemical Industries Inc., Wilmington, Delaware.

Intervet Laboratories Ltd. (Intertest antibiotic detection kit)
Viking Way, Bar Hill, Cambs., England.

Janke & Kunkel KG ("Ultra-Turrax" homogeniser)
IKA-Werk, 7813 Staufen i. Br., Neumagenstrasse 16, West Germany.
U.K. distribution: Astell-Hearson (*q.v.*), Sartorius Instruments Ltd. 18 Avenue Road, Belmont, Surrey.

J. A. Jobling & Co. Ltd., Laboratory Division ("Pyrex" test-tubes etc.)
Stone, Staffs., ST15 0BG, England.

LKB Produkter AB ("BioTec" fermenters)
S161 25, Bromma 1, Sweden.
U.K.: LKB Instruments Ltd., 232 Addington Road, Selsdon, South Croydon, Surrey CR2 8YD.
U.S.: LKB Instruments Inc., 12221 Parklawn Drive, Rockville, Maryland 20852.

Medical Wire and Equipment Co. (Bath) Ltd. (alginate wool swabs, sterile containers, calibrated wire loops)
Potley, Corsham, Wiltshire, England.
U.S. distribution: K.C. Laboratories, P.O. Box 5441, Lenexa, Kansas 66215, U.S.A.

Miles Laboratories ("Clinistix" glucose indicator, "Clintest" indicator)
U.K.: Stoke Court, Stoke Poges, Bucks. SL2 4LY.
U.S.: Ames Co., Miles Laboratories Division, Elkhart, Indiana.

Millipore Filter Corporation (Membrane filters and accessories)
Bedford, Mass., U.S.A.
U.K.: Millipore (U.K.) Ltd., Abbey Road, Park Royal, Middlesex.

Mirfield Agricultural Chemicals Ltd. ("Wescodyne" iodophor)
P.O. Box 1, Steanard Lane, Mirfield, Yorkshire, England.

MSE Scientific Instruments ("Atomix" homogeniser)
Manor Royal, Crawley, Sussex, RH10 2QQ, England.
U.S. distribution: O. H. Johns (Scientific), 219 Broadview Avenue, Toronto, M4M 2G4, Canada.

Oxoid Ltd. (Dehydrated media and constituents, sera)
Wade Road, Basingstoke, RG24 0PW, England.
U.S. distribution: Flow Laboratories Inc., P.O. Box 2226, 1710 Chapman Avenue, Rockville, Maryland 20852, U.S.A.

Ozalid Co. Ltd. ("Safetykling" self-adhesive polyester film)
Langston Road, Laughton, Essex.

John Poulten Ltd. ("Volac" safety pipette controllers)
Barking, Essex, England.
U.S. distribution: Bel-Art Products, Pequannock, New Jersey 07440; and Cole Parmer Instruments Co., 7425 North Oak, Park Avenue, Niles, Illinois 60648.

Reddish Chemical Co. ("Marinol D" and "Reddishquat" detergent-sanitisers)
Stanley Road, Cheadle Hume, Cheshire SK8 6RB, England.

Roche Products Ltd. ("Enterotube" and "ENCISE-System" rapid diagnostic testing media for Enterobacteriaceae)
Roche Diagnostics Division, 15 Manchester Square, London W1M 6AP, England.
U.S.: Roche Diagnostics Division of Hoffman La Roche Inc., Nutley, New Jersey 07110.

Rohm & Haas (Triton X-100 surface active agent)
U.K.: Lennig House, Mason Avenue, Croydon, Surrey.
U.S.: Independence Mall West, Philadelphia, Pennsylvania 19105.

Roussel Laboratories ("Calgitex" calcium alginate wool, and alginate swabs)
Wembley Park, Middlesex, England.

A. J. Seward (Colworth Stomacher, "Droplette")
Atlas Works, Cullum Road, Bury St. Edmunds, Suffolk, IP33 3PD, England.
U.S. distribution: *via* UAC House, Blackfriars Road, London SE1 9UG, England.

Sartorius-Membranfilter GmbH (Membrane filters and accessories)
Göttingen, W. Germany.
U.K. distribution: V. A. Howe & Co. Ltd., 88 Peterborough Road, London SW6.
U.S. distribution: Beckmann (Science Essentials Division), P.O. Box 6100, Anaheim 92806, California.

Sterilin Ltd. (Impression plates, plastic Petri dishes, polypropylene screw caps, autoclavable disposal bags)
43–45 Broad Street, Teddington, Middlesex, England.
U.S. distribution: York Scientific, Bridge and Port Authority Building, Ogdensburg, New York 13669.

Tintometer Sales Ltd. (Lovibond comparators etc.)
Salisbury, Wiltshire, England.

Upjohn International Inc. ("Acti-dione" cycloheximide)
320 Portage Street, Kalamazoo, Michigan 40091, U.S.A.
U.K.: Upjohn Ltd., Fleming Way, Crawley, Sussex.

Wellcome Reagents Ltd. (Sterile sera, dehydrated rabbit plasma, blood, antisera, penicillinase, "Aerosporin" polymyxin B sulphate)
Wellcome Research Laboratories, Beckenham, Kent, England.
U.S.: Burroughs Wellcome Co., Wellcome Reagents Division, 3030 Cornwallis Road, Research Triangle Park, North Carolina 27709.

Whatman (filter paper)
Whatman Reeve Angel Ltd., Springfield Mill, Maidstone, Kent, England.
U.S.: Whatman Inc., 9 Bridewell Place, Clifton, New Jersey 07014.

Wild Heerbrugg (microscopes)
CH 9435 Heerbrugg, Switzerland.
U.K.: Wild Heerbrugg U.K. Ltd., Revenge Road, Lordswood, Chatham, Kent.
U.S.: Wild Heerbrugg Instruments Inc., 465 Smith Street, Farningdale, Long Island, New York 11735.

APPENDIX 4

SELECTED BIBLIOGRAPHY FOR FOOD MICROBIOLOGY

The following list includes some of the books which the authors have found particularly useful, and/or which will provide a good basis for a personal library in food microbiology. The list does not include such items as British Standards, International Dairy Federation Standard Methods, etc., which have been referred to in the text—these will be found detailed in the References section. More extensive bibliographies, which also include books on other aspects of food science and food technology, are provided by:

Baker, E. A., and Foskett, D. J. (1958). "Bibliography of Food: A Select International Bibliography of Nutrition, Food and Beverage Technology and Distribution". London: Butterworths.

Apling, E. C. (1972). "A Select Bibliography and Library Guide to the Literature of Food Science", 2nd Ed. University of Reading: Dept. of Food Science. (A selected list of the books published from 1781–1971, held in the University library.)

International Food Information Service (1975). "Food Annotated Bibliography No. 47: Books 1967–1974". (Monthly or annual updating service available.) Shinfield, Reading, England: International Food Information Service.

General Microbiology (Theory)

Davis, B. D., Dulbecco, R., Eisen, H. N., Ginsberg, H. S., and Wood, W. B. (1974). "Microbiology", 2nd. Ed. London and New York: Harper and Row.

Stanier, R. Y., Doudoroff, M., and Adelberg, E. A. (1971). "General Microbiology", 3rd Ed. London: Macmillan; New Jersey: Prentice Hall.

Safety in the Laboratory

Muir, G. D. (ed.) (1971). "Hazards in the Chemical Laboratory". London: The Royal Institute of Chemistry.

Shapton, D. A., and Board, R. G. (eds.) (1972). "Safety in Microbiology". Society for Applied Bacteriology Technical Series No. 6. London: Academic Press.

"Fisher Manual of Laboratory Safety" (1972). (Lists safety equipment available from Fisher). Pittsburgh: Fisher Scientific Co.

"First Aid Manual" (1972). Authorised manual of the St. John Ambulance Association, the British Red Cross Society, and the St. Andrew's Ambulance Association, 3rd Ed.

General Microbiological Methods

Meynell, G. G., and Meynell, E. (1970). "Theory and Practice in Experimental Bacteriology", 2nd Ed. Cambridge: The University Press.

Norris, J. R., and Ribbons, D. W. (eds.) (1969–1973). "Methods in Microbiology", 12 volumes. London: Academic Press.

Society for Applied Bacteriology Technical Series: published annually from 1966 (individual volumes are listed separately as appropriate, and also listed in the references, under editor). London: Academic Press.

Microscopy

Barer, R. (1956). "Lecture Notes on the Use of the Microscope", 2nd Ed. Oxford: Blackwell.

Casartelli, J. D. (1965). "Microscopy for Students". London: McGraw-Hill.

Stains and Reagents

Silverton, R. E., and Anderson, M. J. (1961). "Handbook of Medical Laboratory Formulae". London: Butterworths.

Media

Manuals for use of dehydrated media

"BBL Manual of Products and Laboratory Procedures" (1968), 5th Ed. Cockeysville, Maryland: Baltimore Biological Laboratory (see page 393).

"Difco Manual" (1953), 9th Ed., and "Difco Supplementary Literature" (1968). Detroit: Difco Laboratories (see page 394).

"Oxoid Manual" (1973), 3rd Ed. revised. London: Oxoid Ltd. (see page 395).

Disinfection and Sterilisation

Sykes, G. (1965). "Disinfection and Sterilization", 2nd Ed. London: Spon.

Richards, J. W. (1968). "Introduction to Industrial Sterilization". London: Academic Press.

With particular reference to laboratory methods

Rubbo, S. D., and Gardner, J. F. (1965). "A Review of Sterilization and Disinfection, as Applied to Medical, Industrial and Laboratory Practice". London: Lloyd-Luke (Medical Books).

Food Microbiology

Theory

Nickerson, J. T. R., and Sinskey, A. J. (1972). "Microbiology of Foods and Food Processing". Amsterdam: Elsevier-North Holland.

Borgstrom, G. (1968). "Principles of Food Science", 2 vols. New York: Macmillan Co.

Methods

Thatcher, F. S., and Clark, D. S. (1968). "Micro-organisms in Foods: Their Significance and Methods of Enumeration". Recommendations of the International Committee on Microbiological Specifications for Foods, a standing committee of the International Association of Microbiological Societies. Toronto: University of Toronto Press.

ICMSF (1974). "Micro-organisms in Foods 2. Sampling for Microbiological Analysis: Principles and Specific Applications". Recommendations of the International Commission on Microbiological Specifications for Foods of the International Association of Microbiological Societies. Toronto: University of Toronto Press.

Society for Applied Bacteriology Technical Series: published annually from 1966. London: Academic Press. Particularly:

No. 3 "Isolation Methods for Microbiologists" (ed. by D. A. Shapton and G. W. Gould)

No. 5 "Isolation of Anaerobes" (ed. by D. A. Shapton and R. G. Board)

No. 7 "Sampling—Microbiological Monitoring of Environments" (ed. by R. G. Board and D. W. Lovelock)

No. 8 "Some Methods for Microbiological Assay" (ed. by R. G. Board and D. W. Lovelock)

American Public Health Association (1966). "Recommended Methods for the Microbiological Examination of Foods", 2nd ed. New York: American Public Health Association.

U.S. Food and Drug Administration (1969). "Bacteriological Analytical Manual". Washington: U. S. Food and Drug Administration, Division of Microbiology.

Quality control (including microbiological)

Herschdoerfer, S. M. (ed.) (1967–1972). "Quality Control in the Food Industry", 3 vols. London: Academic Press.

Food borne diseases

Riemann, H. (1969). "Food-borne Infections and Intoxications". New York: Academic Press.

Hobbs, B. C., and Christian, J. H. B. (eds.) (1973). "The Microbiological Safety of Food". London: Academic Press.

Parry, W. H. (1973). "Communicable Diseases—an Epidemiological Approach". English Universities Press.

Graham-Jones, O. (ed.) (1968). "Some Diseases of Animals Communicable to Man in Britain". Oxford: Pergamon Press.

Goldblatt, L. A. (ed.) (1969). "Aflatoxin: Scientific Background, Control, and Implications". New York: Academic Press. (*See also* I. F. H. Purchase (1975) "Mycotoxins"; Amsterdam: Elsevier, North Holland)

National Academy of Sciences (1969). "An Evaluation of the *Salmonella Problem*". Publication No. 1683. Washington: National Academy of Sciences.

Marine products

American Public Health Association (1970). "Recommended Procedures for the Examination of Seawater and Shellfish", 4th Ed. New York: American Public Health Association.

Kriss, A. E. (1963). "Marine Microbiology (Deep Sea)", translated by J. M. Shewan and Z. Kabata. Edinburgh: Oliver and Boyd.

Dairy products

American Public Health Association (1972). "Standard Methods for the Examination of Dairy Products", 13th Ed. New York: American Public Health Association.

World Health Organisation (1962). "Milk Hygiene", Monograph No. 48. Geneva: World Health Organisation.

Fermented foods

Pederson, C. S. (1971). "Microbiology of Food Fermentations". Westport: Avi Publishing Co.

Carr, J. G., Cutting, C. V., and Whiting, G. C. (eds.) (1975). "Lactic Acid Bacteria in Beverages and Food". *Proc. Symp., Long Ashton, 1973.* London: Academic Press.

Alcoholic beverages

Findlay, W. P. K. (ed.) (1971). "Modern Brewing Technology". London: Macmillan.

Reed, G., and Peppler, H. J. (1973). "Yeast Technology". Westport: Avi Publishing Co.

Carr, J. G., Cutting, C. V., and Whiting, G. C. (eds.) (1975). "Lactic Acid Bacteria in Beverages and Food". *Proc. Symp., Long Ashton, 1973.* London: Academic Press.

Cereals and cereal products

American Association of Cereal Chemists (1962 *et seq.*). "Approved Methods of the AACC", 7th ed. and supplements. St. Paul, Minnesota: American Association of Cereal Chemists Inc.

Kent Jones, D. W., and Amos, A. J. (1967). "Modern Cereal Chemistry", 6th Ed. London: Food Trade Press.

Canned foods and heat-treated foods

Hersom, A. C., and Hulland. E. D. (1969). "Canned Foods (Baumgartner)", 6th Ed. London: Churchill.

Stumbo, C. R. (1973). "Thermobacteriology in Food Processing", 2nd Ed. New York: Academic Press.

National Canners' Association (1968). "Laboratory Manual for Food Canners and Processers", 3rd Ed. 2 vols. Westport: Avi Publishing Co.

Food legislation (British)

O'Keefe, J. A. (1968 *et seq.*). "Bell and O'Keefe's Sale of Food and Drugs", 14th Ed. (with annual service volume). London: Butterworths.

Microbiology of Water

Aaronson, S. (1970). "Experimental Microbial Ecology". New York: Academic Press.

Heukelekian, H., and Dondero, N. C. (eds.) (1964). "Principles and Applications in Aquatic Microbiology". New York: Wiley.

Water supplies

Report (1969). "The Bacteriological Examination of Water Supplies", 4th Ed. Report of the Public Health Laboratory Service Standing Committee on the Bacteriological Examination of Water Supplies. *Reports on Public Health and Medical Subjects*, No. 71. London: H.M.S.O.

American Public Health Association (1971). "Standard Methods for the Examination of Water and Wastewater", 13th Ed. New York: American Public Health Association.

Sykes, G., and Skinner, F. A. (eds.) (1971). "Microbial Aspects of Pollution". *Society for Applied Bacteriology Symposium Series*, No. 1. London: Academic Press.

Analysis of industrial effluents (biological and chemical oxygen demands)

Department of Environment (1972). "Analysis of Raw, Potable and Waste Waters". London: H.M.S.O.

Identification of Bacteria

Buchanan, R. E. *et al.* (1974). "Bergey's Manual of Determinative Bacteriology", 8th Ed. Baltimore: Williams and Wilkins.

Gibbs, B. M., and Skinner, F. A. (eds.) (1966). "Identification Methods for Microbiologists, Part A". *Society for Applied Bacteriology Technical Series*, No. 1. London: Academic Press.

Gibbs, B. M., and Shapton, D. A. (eds.) (1968). "Identification Methods for Microbiologists, Part B". *Society for Applied Bacteriology Technical Series*, No. 2. London: Academic Press.

Cowan, S. T. (1974). "Cowan and Steel's Manual for the Identification of Medical Bacteria", 2nd Ed. Cambridge: The University Press.

Asai, T. (1968). "Acetic Acid Bacteria: Classification and Biochemical Activity". Tokyo: The University of Tokyo Press.

Identification of Yeasts and Moulds

Yeasts

Lodder, J. (1970). "The Yeasts—a Taxonomic Study". Amsterdam: North-Holland Publishing Co.

Barnett, J. A., and Pankhurst, R. J. (1974). "A New Key to the Yeasts". Amsterdam: North-Holland Publishing Co.

Moulds

Ainsworth, G. C., Sparrow, F. K., and Sussman, A. S. (eds.) (1973). "The Fungi", Vols IVA and IVB. New York: Academic Press.

Arx, J. A. von (1970). "The Genera of Fungi Sporulating in Pure Culture". Lehre: J. Cramer.

Barnett, H. L. and Hunter, B. B. (1972). "Illustrated Genera of Imperfect Fungi", 3rd Ed. Minneapolis: Burgess Publishing Co.

Barron, G. L. (1968). "The Genera of Hyphomycetes from Soil". Baltimore: Williams and Wilkins.

Gilman, J. C. (1957). "A Manual of Soil Fungi", 2nd Ed. Iowa: The Iowa State College Press.

Smith, G. (1969). "An Introduction to Industrial Mycology", 6th Ed. London: Edward Arnold.

Aspergillus

Raper, K. B., and Fennell, D. I. (1965). "The Genus *Aspergillus*". Baltimore: Williams and Wilkins.

Penicillium

Raper, K. B., and Thom, C. (1949), "A Manual of the Penicillia". Baltimore: Williams and Wilkins.

REFERENCES

Abd-el-Malek, Y., and Gibson, T. (1948). Studies in the bacteriology of milk. I. The streptococci of milk. *J. Dairy Res.*, **15**, 233.

Adams, M. H. (1959). "Bacteriophages". Reprinted 1966. New York: Interscience.

Ainsworth, G. C., Sparrow, F. K., and Sussman, A. S. (1973). "The Fungi". Vols. IVa and IVb. "A Taxonomic Review with Keys". New York: Academic Press.

Akman, M., and Park, R. W. A. (1974). The growth of salmonellas on cooked cured pork. *J. Hyg., Camb.*, **72**, 369.

Albert, J. O., Long, H. F., and Hammer, B. W. (1944). Classification of the organisms important in dairy products. *Bull. Iowa agric. Exp. Stn.*, 328.

Aldred, J. B., Evans, A. F., and Husbands, V. (1971). Aspects of the Howard mould count. *J. Assoc. Public Analysts*, **9**, 47.

Alton, G. G., and Jones, L. (1963). "Laboratory Techniques in Brucellosis". F.A.O. Animal Health Branch Monograph No. 7. Rome: Food and Agriculture Organisation of the United Nations.

American Association of Cereal Chemists. (1962 *et seq.*). "Approved Methods of the AACC", 7th Ed. and supplements. St. Paul, Minnesota: American Association of Cereal Chemists Inc.

American Public Health Association (1960). "Standard Methods for the Examination of Dairy Products", 11th Ed., New York: American Public Health Association.

American Public Health Association (1966). "Recommended Methods for the Microbiological Examination of Foods", 2nd Ed. New York: American Public Health Association.

American Public Health Association (1970). "Recommended Procedures for the Examination of Sea Water and Shellfish", 4th Ed. New York: American Public Health Association.

American Public Health Association (1971). "Standard Methods for the Examination of Water and Wastewater, Including Bottom Sediments and Sludges", 13th Ed. New York: American Public Health Association.

American Public Health Association (1972). "Standard Methods for the Examination of Dairy Products", 13th Ed. New York: American Public Health Association.

Amos, A. J. (1968). Flour and bread. *In* "Quality Control in the Food Industry", Vol. 2 (ed. S. M. Herschdoerfer), p. 195ff. London: Academic Press.

Anderson, E. C., and Hobbs, B. (1973). Studies of the strain of *Salmonella typhi* responsible for the Aberdeen typhoid outbreak. *Israel J. Medical Sciences*, **9**, 162.

Angelotti, R., and Foter, M. J. (1958). A direct surface agar plate laboratory method for quantitatively detecting bacterial contamination on nonporous surfaces. *Fd Res.*, **23**, 170.

Angelotti, R., Hall, H. E., Foter, M. J., and Lewis, K. H. (1962). Quantitation of *Clostridium perfringens* in foods. *Appl. Microbiol.*, **10**, 193.

Anon. (1975). (*Vibrio parahaemolyticus* food poisoning) *Brit. Food J.*, **77** (869), 175.

Anthony, C., and Zatman, L. J. (1964). The microbial oxidation of methanol. I. Isolation and properties of *Pseudomonas* sp. M27. *Biochem. J.*, **92**, 609.

Asai, T. (1968). "Acetic Acid Bacteria: Classification and Biochemical Activity". Tokyo: The University of Tokyo Press.

Association of Official Analytical Chemists (1975 *et seq.*). "Official Methods of Analysis", 12th Ed. With supplements. Washington: Association of Official Analytical Chemists.

Baird-Parker, A. C. (1962). An improved diagnostic and selective medium for isolating coagulase positive staphylococci. *J. appl. Bact.*, **25**, 12.

Baird-Parker, A. C. (1969). The use of Baird-Parker's medium for the isolation and enumeration of *Staphylococcus aureus*. *In* "Isolation Methods for Microbiologists" (ed. D. A. Shapton and G. W. Gould). *Soc. appl. Bact. Technical Series*, No. 3, pp. 1ff. London: Academic Press.

Baker, D. A., and Park, R. W. A. (1975). Changes in morphology and cell wall structure that occur during growth of *Vibrio sp.* NCTC 4716 in batch culture. *J. gen. Microbiol.*, **86**, 12.

Barber, M., and Kuper, S. W. A. (1951). Identification of *Staphylococcus pyogenes* by the phosphatase reaction. *J. Path. Bact.*, **63**, 65.

Barnes, E. M. (1956). Methods for the isolation of faecal streptococci (Lancefield Group D) from bacon factories. *J. appl. Bact.*, **19**, 193.

Barnes, E. M., and Impey, C. S. (1968). Psychrophilic spoilage bacteria of poultry. *J. appl. Bact.*, **31**, 97.

Barnes, E. M., Impey, C. S., and Parry, R. T. (1973). The sampling of chickens, turkeys, ducks and game birds. *In* "Sampling—Microbiological Monitoring of Environments" (ed. R. G. Board and D. W. Lovelock). *Soc. appl. Bact. Technical Series*, No. 7, pp. 63ff. London: Academic Press.

Barnett, J. A., and Pankhurst, R. J. (1974). "A New Key to the Yeasts". Amsterdam: North-Holland Publishing Co.

Barritt, M. M. (1936). The intensification of the Voges-Proskauer reaction by the addition of α-naphthol. *J. Path. Bact.*, **42**, 441.

Barron, G. L. (1968). "The Genera of Hyphomycetes from the Soil". Baltimore: Williams and Wilkins.

Barrow, G. I., and Miller, D. C. (1972). *Vibrio parahaemolyticus*: a potential pathogen from marine sources in Britain. *Lancet*, i, 485.

Bartholomew, J. W., and Mittwer, T. (1950). A simplified bacterial spore stain. *Stain Technol.*, **25**, 153.

Barton-Wright, E. C. (1963). "Practical Methods for the Microbiological Assay of the Vitamin B Complex". London: United Trade Press Ltd.

Baumann, L., Baumann, P., Mandel, M., and Allen, R. D. (1972). Taxonomy of aerobic marine eubacteria. *J. Bact.*, **110**, 402.

Baumann, P., Baumann, L., and Mandel, M. (1971). Taxonomy of marine bacteria: the genus *Beneckea*. *J. Bact.*, **107**, 268.

Baumann, P., Baumann, L., and Reichelt, J. L. (1973). Taxonomy of marine bacteria: *Beneckea parahaemolytica* and *Beneckea alginolytica*. *J. Bact.*, **113**, 1144.

Baumann, P., Doudoroff, M., and Stanier, R. Y. (1968). Study of the *Moraxella* group. I. Genus *Moraxella* and the *Neisseria catarrhalis* group. *J. Bact.*, **95**, 58.

Baumann, P., Doudoroff, M., and Stanier, R. Y. (1968). A study of the Moraxella Group. II. Oxidative-negative species (Genus *Acinetobacter*). *J. Bact.*, **95**, 1520.

Bean, P. G., and Everton, J. R. (1969). Observations on the taxonomy of chromogenic bacteria isolated from cannery environments. *J. appl. Bact.*, **32**, 51.

Beech, F. W., and Davenport, R. R. (1969). The isolation of non-pathogenic yeasts. *In* "Isolation Methods for Microbiologists" (ed. D. A. Shapton and G. W. Gould). *Soc. appl. Bact. Technical Series* No. 3. p. 71ff. London: Academic Press.

Beerens, H., Sugama, S., and Tahon-Castel, M. (1965). Psychrotrophic clostridia. *J. appl. Bact.*, **28**, 36.

Benedict, S. R. (1908–1909). A reagent for the detection of reducing sugars. *J. biol. Chem.*, **5**, 485.

Bergan, T., and Lystad, A. (1971). Disinfectant evaluation by a capacity use-dilution test. *J. appl. Bact.*, **34**, 741.

Bernaerts, M. J., and De Ley, J. (1963). A biochemical test for crown gall bacteria. *Nature, Lond.*, **197**, 406.

Berry, J. A. (1933). Detection of microbial lipase by copper soap formation. *J. Bact.*, **25**, 433.

Bettes, D. C. (1965). Canning in the dairy industry. *J. Soc. Dairy Technol.*, **18**, 224.

Billing, E. (1969). Isolation, growth and preservation of bacteriophages. *In* "Methods in Microbiology", Vol. 3B (ed. J. R. Norris and D. W. Ribbons) p. 315ff. London: Academic Press.

Billing, E., and Cuthbert, W. A. (1958). Bitty cream: the occurrence and significance of *Bacillus cereus* spores in raw milk supplies. *J. appl. Bact.*, **21**, 65.

Billing, E., and Luckhurst, E. R. (1957). A simplified method for the preparation of egg yolk media. *J. appl. Bact.*, **20**, 90

Blackburn, P. S. (1965). A solution for use in assessing the cell count of cow's milk. *Br. vet. J.*, **121**, 154.

Blackburn, P. S., Laing, C. M., and Malcolm, J. F. (1955). A comparison of the diagnostic value of the total and differential cell counts of bovine milk. *J. Dairy Res.*, **22**, 37.

Blackburn, P. S., and Macadam, I. (1954). The cells in bovine milk. *J. Dairy Res.*, **21**, 31.

Board, R. G. (1966). Review article: the course of microbial infection of the hen's egg. *J. appl. Bact.*, **29**, 319.

Board, R. G., and Holding, A. J. (1960). The utilization of glucose by aerobic Gram-negative bacteria. *J. appl. Bact.*, **23**, xi.

Booth, C. (1971). Fungal culture media. *In* "Methods in Microbiology", Vol.4 (ed. by C. Booth), p. 49ff. London: Academic Press.

Bradstreet, C. M. P. *et al.* (1961). Memorandum No. 4 of the Public Health Laboratory Service: serological reagents for bacteriological diagnosis. *Mon. Bull. Minist. Hlth*, **20**, 134.

Breckinridge, J. C., and Bergdoll, M. S. (1971). Outbreak of food-borne gastroenteritis due to a coagulase-negative enterotoxin producing staphylococcus. *New Engl. J. Med.*, **284**, 541.

Breed, R. S., Murray, E. G. D., and Smith, N. R. (1957). "Bergey's Manual of Determinative Bacteriology", 7th Ed. Baltimore: Williams and Wilkins.

Brewer, J. H., and Allgeier, D. L. (1966). Safe self-contained carbon dioxide-hydrogen anaerobic system. *Appl. Microbiol.*, **14**, 985.

Bridson, E. Y., and Brecker, A. (1970). Design and formulation of microbial culture media. *In* "Methods in Microbiology", Vol. 3A (ed. J. R. Norris and D. W. Ribbons), p. 229ff. London: Academic Press.

Briggs, M. (1953). The classification of lactobacilli by means of physiological tests. *J. gen. Microbiol.*, **9**, 234.

British Standard (1968). B.S. 4285: 1968. "Methods of Microbiological Examination for Dairy Purposes". London: British Standards Institution.

British Standard (1970). Supplement No. 1 (1970) to British Standard 4285: 1968. "Methods of Microbiological Examination of Milk Products". London: British Standards Institution.

British Standard (1974). B.S. 809: 1974. "Sampling Milk and Milk Products". London: British Standards Institution.

Bryan-Jones, G. (1975). Lactic acid bacteria in distillery fermentations. *In* "Lactic Acid Bacteria in Beverages and Food" (ed. J. G. Carr, C. V. Cutting and G. C. Whiting). *Proc. 4th Long Aston Symp., 1973*. p. 165ff. London: Academic Press.

Buchanan, R. E. *et al.* (1974). "Bergey's Manual of Determinative Bacteriology", 8th Ed. Baltimore: Williams and Wilkins.

Buckley, H. R., Campbell, C. K., and Thompson, J. C. (1969). Techniques for the isolation of pathogenic fungi. *In* "Isolation Methods for Microbiologists" (ed. by D. A. Shapton and G. W. Gould). *Soc. appl. Bact. Technical Series*, No. 3. p. 113ff. London: Academic Press.

Bulletin (1968). "Bacteriological Techniques for Dairy Purposes". Ministry of Agriculture, Fisheries and Food, Technical Bulletin, No. 17. London: H.M.S.O.

Calam, C. T. (1969). The culture of micro-organisms in liquid medium. *In* "Methods in Microbiology", Vol. 1 (ed. J. R. Norris and D. W. Ribbons) p. 255ff. London: Academic Press.

Carr, J. G. (1968). Methods for identifying acetic acid bacteria. *In* "Identification Methods for Microbiologists, Part B" (ed. B. M. Gibbs and D. A. Shapton). *Soc. appl. Bact. Technical Series*, No. 2. London: Academic Press.

Chalmers, C. H. (1962). "Bacteria in Relation to the Milk Supply", 4th Ed. London: Edward Arnold.

Charlett, S. M. (1954). An improved staining method for the direct microscopical counting of bacteria in milk. *Dairy Inds.*, **19**, 652.

Charney, J., Fisher, W. P., and Hegarty, C. P. (1951), Manganese as an essential element for sporulation in the genus *Bacillus*. *J. Bact.*, **62**, 145.

Christensen, W. B. (1946). Urea decomposition as a means of differentiating *Proteus* and paracolon cultures from each other and from *Salmonella* and *Shigella* types. *J. Bact.*, **52**, 461.

Clarke, P. H. and Steel, K. J. (1966). Rapid and simple biochemical tests for bacterial identification. *In* "Identification Methods for Microbiologists, Part A" (Ed. B. M. Gibbs and F. A. Skinner) p. 111ff. London: Academic Press.

Cobb, R. W. (1963). Cultural characteristics of some corynebacteria of animal origin, with special reference to *C. bovis* and *C. pyrogenes*. *J. med. Lab. Technol.*, **20**, 199.

Cobb, R. W. (1966). *Corynebacterium bovis*: fermentation of sugars in the presence of serum or Tween 80. *Vet. Rec.*, **78**, 33.

Cochran, W. G. (1950). Estimation of bacterial densities by means of the "most probable number". *Biometrics*, **6**, 105.

Collins, V. G. (1964). The fresh water environment and its significance in industry. *J. appl. Bact.*, **27**, 143.

Collins, V. G. (1970). Recent studies of bacterial pathogens of freshwater fish. *Water Treatment and Examination*, **19**, 3.

Collins, V. G. *et al.* (1973). Sampling and estimation of bacterial populations in the aquatic environment. *In* "Sampling–Microbiological Monitoring of Environments" (ed. R. G. Board and D. W. Lovelock) *Soc. appl. Bact. Technical Series*, No. 7. p. 77ff. London: Academic Press.

Committee on Salmonella. (1969). "An Evaluation of the *Salmonella* problem". A report of the U.S. Dept. of Agriculture and the U.S. Food and Drug Administration (Publication No. 1683). Washington: National Academy of Sciences.

Cooper, K. E. (1955). Theory of antibiotic inhibition zones in agar media. *Nature, Lond.*, **176**, 510.

Corry, J. E. L., Kitchell, A. G., and Roberts, T. A. (1969). Interactions in the recovery of *Salmonella typhimurium* damaged by heat or gamma radiation. *J. appl. Bact.*, **32**, 415.

Cowan, S. T. (1974). "Cowan and Steel's Manual for the Identification of Medical Bacteria", 2nd Ed. Cambridge: The University Press.

Cowan, S. T., and Steel, K. J. (1965). "Manual for the Identification of Medical Bacteria". Cambridge: The University Press.

Cowell, N. D. (1968). Methods of thermal process evaluation. *J. Fd. Technol.*, **3**, 303.

Crawford, R. J. M. (1962). Citrate utilizing activity of certain starter bacteria. *16th Int. Dairy Congr. B*, 322.

Crawford, R. J. M., and Galloway, J. H. (1964). Testing milk for antibiotic residues. *Dairy Inds*, **29**, 256.

Croft, C. C., and Miller, M. J. (1956). Isolation of Shigella from rectal swabs with Hajna "GN" broth. *Amer. J. clin. Path.*, **26**, 411.

Crombach, W. H. J. (1974). Morphology and physiology of coryneform bacteria. *Antonie van Leeuwenhoek*, **40**, 361.

Cross, T. (1968). Thermophilic actinomycetes. *J. appl. Bact.*, **31**, 36.

Crossley, E. L. (1941). The routine detection of certain spore-forming anaerobic bacteria in canned foods. *J. Soc. chem. Ind., Lond.*, **60**, 131.

Crossley, E. L. (1948). Studies on the bacteriological flora and keeping quality of pasteurized liquid cream. *J. Dairy Res.*, **15**, 261.

Cruickshank, R., *et al.* (1975). "Medical Microbiology. Vol. 2: The Practice of Medical Microbiology", 12th Ed. Edinburgh: Churchill Livingstone.

Cure, G. L., and Keddie, R. M. (1973). Methods for the morphological examination of aerobic coryneform bacteria. *In* "Sampling—Microbiological Monitoring of Environments" (ed. R. G. Board and D. W. Lovelock). *Soc. appl. Bact. Technical Series*, No. 7. p. 123ff. London: Academic Press.

Cuthbert, W. A. (1964). The significance of thermoduric organisms in milk. *Annual Bull. I. D. F. IVB*, 10.

Darlow, H. M. (1969). Safety in the microbiological laboratory. *In* "Methods in Microbiology", Vol. 1 (ed. J. R. Norris and D. W. Ribbons), p. 169ff. London: Academic Press.

Davidson, C. M., Mobbs, P., and Stubbs, J. M. (1968). Some morphological and physiological properties of *Microbacterium thermosphactum*. *J. appl. Bact.*, **31**, 551.

Davies, F. L., and Wilkinson, G. (1973). *Bacillus cereus* in milk and dairy products. *In* "The Microbiological Safety of Food" (ed. B. C. Hobbs and J. H. B. Christian) p. 57ff. London: Academic Press.

Davies, R. R. (1971). Air sampling for fungi, pollens and bacteria. *In* "Methods in Microbiology", Vol. 4 (ed. C. Booth), p. 367ff. London: Academic Press.

Davis, J. G. (1958). A convenient semi-synthetic medium for yeast and mould counts. *Lab. Pract.*, **7**, 30.

Davis, J. G. (1963). Microbiological standards for dairy products. *J. Soc. Dairy Technol.*, 16, 150 and 224.

Davis, J. G. (1968). Dairy products. *In* "Quality Control in the Food Industry", Vol. 2, (ed. S. M. Herschdoerfer), p. 29ff. London: Academic Press.

Davis, J. G. (1970). Laboratory control of yoghurt. *Dairy Inds.*, **35**, 139.

Demeter, K. J., Sauer, F., and Miller, M. (1933). Vergleichende Untersuchungen uber verschiedene Methoden zur Coli-aerogenes-Titerbestimmung in Milch. *Milchwirtschaftliche Forschungen*, **15**, 265.

Dewhurst, F., and Cassells, W. (1976). Some pollutants of laboratory atmospheres. *Laboratory News*, No. 110, 14.

Difco Manual (1953) 9th Ed., and *Difco Supplementary Literature* (1968). Detroit: Difco Laboratories Inc.

Dixon, J. M. S., and Wilson, F. N. (1960). Salmonellae in fertilizers containing superphosphate. *Mon. Bull. Minist. Hlth.*, **19**, 79.

Dowdell, M. J., and Board, R. G. (1968). A microbiological survey of British fresh sausage. *J. appl. Bact.*, **31**, 378.

Dowdell, M. J., and Board, R. G. (1971). The microbial associations in British fresh sausages. *J. appl. Bact.*, **34**, 317.

Druce, R. G., and Thomas, S. B. (1959). The microbiological examination of butter. *J. appl. Bact.*, **22**, 52.

Duguid, J. P., and Wilkinson, J. F. (1961). Environmentally induced changes in bacterial morphology. *In* "Microbial Reaction to Environment" *Symp. Soc. gen. Microbiol.*, **11**, 69.

Duncan, C. L. (1973). Time of enterotoxin formation and release during sporulation of *Clostridium perfringens* type A. *J. Bact.*, **113**, 932.

Eddy, B. P. (1960*a*). The use and meaning of the term "psychrophilic". *J. appl. Bact.*, **23**, 189.

Eddy, B. P. (1960*b*). Cephalotrichous, fermentative Gram-negative bacteria: the genus *Aeromonas*. *J. appl. Bact.*, **23**, 216.

Eddy, B. P., and Carpenter, K. P. (1964). Further studies on *Aeromonas*. II. Taxonomy of *Aeromonas* and C.27 strains. *J. appl. Bact.*, **27**, 96.

Edwards, P. R., and Ewing, W. H. (1972). "Identification of Enterobacteriaceae", 3rd Ed. Minneapolis: Burgess Publishing Co.

Edwards, S. J. (1933). Studies on bovine mastitis. IX. A selective medium for the diagnosis of streptococcus mastitis. *J. comp. Path. Ther.*, **46**, 211.

Egdell, J. W., Thomas, S. B., Clegg, L. F. L., and Cuthbert, W. A. (1950). Thermoduric organisms in milk. *Proc. Soc. appl. Bact.*, **13**, 132.

Ellner, P. D. (1956). A medium promoting rapid quantitative sporulation in *Clostridium perfringens*. *J. Bact.*, **71**, 495.

Evans, H. J., Kwantes, W., Jenkins, D. C., and Phillips, J. I. (1956). Sucrose loss from ice-cream on storage. *Analyst*, *Lond.*, **81**, 204.

Ewing, W. H., and Edwards, P. R. (1960). The principal divisions and groups of *Enterobacteriaceae* and their differentiation. *Int. Bull. bact. Nomencl. Taxon.*, **10**, 1.

Farrell, J., and Rose, A. H. (1967). Temperature effects in micro-organisms. *In* "Thermobiology" (ed. A. H. Rose). London: Academic Press.

Favero, M. S. *et al.* (1968). Microbiological sampling of surfaces. *J. appl. Bact.*, **31**, 336.

Fidler, J. C., Wilkinson, B. G., Edney, K. L., and Sharples, R. O. (1973). "The Biology of Apple and Pear Storage". *Commonwealth Bureau of Horticulture and Plantation Crops, Research Review No. 3.* Farnham Royal: Commonwealth Agricultural Bureaux.

Fifield, C. W., and Hoff, J. E. (1957). Dilute malachite green: a background stain for the Millipore filter. *Stain Technol.*, **32**, 95.

Findlay, W. P. K. (ed.) (1971). "Modern Brewing Technology". London: Macmillan.

Finley, N., and Fields, M. L. (1962). Heat activation and heat-induced dormancy of *Bacillus stearothermophilus* spores. *Appl. Microbiol.*, **10**, 231.

Fishbein, M., and Wentz, B. (1975). Enumeration, laboratory identification, and serotypic analyses of *Vibrio parahaemolyticus*. *In* "Microbiology—1974" (ed. by D. Schlessinger). Washington: American Society of Microbiology.

Fisher, P. J. (1963). The effect of freeze drying on the viability of *Chromobacterium lividum*. *J. appl. Bact.*, **26**, 502.

Fisher, R. A., and Yates, F. (1963). "Statistical Tables for Biological, Agricultural and Medical Research", 6th Ed. Edinburgh: Oliver and Boyd.

Ford, J. E. (1962). A microbiological method for assessing the nutritional value of proteins. 2. The measurement of 'available' methionine, leucine, isoleucine, arginine, histidine, tryptophan and valine. *Brit. J. Nutr.*, **16**, 409.

Ford, J. E. (1964). A microbiological method for assessing the nutritional value of proteins. 3. Further studies on the measurement of available amino acids. *Brit. J. Nutr.*, **18**, 449.

Franklin, J. G., Williams, D. J., and Clegg, L. F. L. (1956). A survey of the number and types of aerobic mesophilic spores in milk before and after commercial sterilization. *J. appl. Bact.*, **19**, 46.

Freame, B., and Fitzpatrick, B. W. F. (1971). The use of differential reinforced clostridial medium for the isolation and enumeration of clostridia from food. *In* "Isolation of Anaerobes" (ed. D. A. Shapton and R. G. Board), *Soc. appl. Bact. Technical Series* No. 5, p. 49ff. London: Academic Press.

Futter, B. V. (1967). "The detection and viability of anaerobic spores surviving bactericidal influences". Ph.D. thesis (C.N.A.A.), Portsmouth College of Technology.

Futter, B. V., and Richardson, G. (1970). Viability of clostridial spores and the requirements of damaged organisms. II. Gaseous environment and redox potentials. *J. appl. Bact.*, **33**, 331.

Futter, B. V., and Richardson, G. (1971). Anaerobic jars in the quantitative recovery of clostridia. *In* "Isolation of Anaerobes" (ed. D. A. Shapton and R. G. Board), *Soc. appl. Bact. Technical Series* No. 5, p. 81ff. London: Academic Press.

Galesloot, T. E., and Hassing, F. (1962). A rapid and sensitive paper disc method for the detection of penicillin in milk. *Ned. Melk-en Zuiveltijdschr.*, **16**, 89.

Galesloot, T. E., Hassing, F., and Stadhouders, J. (1961). Agar media for the isolation and enumeration of aroma bacteria in starter cultures. *Ned. Melk-en Zuiveltijdschr.*, **15**, 127.

Galesloot, T. E., and Stadhouders, J. (1968). The microbiology of spray-dried milk products with special reference to *Staphylococcus aureus* and salmonellae. *Ned. Melk-en Zuiveltijdschr.*, **22**, 158.

Gardner, G. A. (1966). A selective medium for the enumeration of *Microbacterium thermosphactum* in meat and meat products. *J. appl. Bact.*, **29**, 455.

Gardner, G. A. (1969). Physiological and morphological characteristics of *Kurthia zopfii* isolated from meat products. *J. appl. Bact.*, **32**, 371.

Garvie, E. I. (1953). Some group N streptococci isolated from raw milk. *J. Dairy Res.*, **20**, 41.

Garvie, E. I. (1960). The genus *Leuconostoc* and its nomenclature. *J. Dairy Res.*, **27**, 283.

Garvie, E. I., and Rowlands, A. (1952). The role of micro-organisms in dye reduction and keeping quality tests. II. The effect of micro-organisms when added to milk in pure and mixed culture. *J. Dairy Res.*, **19**, 263.

Geldreich, E. E. (1972). Water-borne pathogens. *In* "Water Pollution Microbiology" (ed. R. Mitchell), p. 207ff. New York: Wiley-Interscience.

Georgala, D. L., and Boothroyd, M. (1964). A rapid immunofluorescence technique for detecting salmonellae in raw meat. *J. Hyg., Camb.*, **62**, 319.

Georgala, D. L., and Boothroyd, M. (1968). Immunofluorescence—a useful technique for microbial identification. *In* "Identification Methods for Microbiologists, Part B" (ed. B. M. Gibbs and D. A. Shapton), *Soc. appl. Bact. Technical Series, No. 2*, p. 187ff. London: Academic Press.

Georgala, D. L., and Boothroyd, M. (1969). Methods for the detection of salmonellae in meat and poultry. *In* "Isolation Methods for Microbiologists" (ed. D. A. Shapton and G. W. Gould), *Soc. appl. Bact. Technical Series, No. 3*, p. 29ff. London: Academic Press.

Georgala, D. L., Boothroyd, M., and Hayes, P. A. (1965). Further evaluation of a rapid immunofluorescence technique for detecting salmonellae in meat and poultry. *J. appl. Bact.*, **28**, 421.

Gibbs, B. M., and Freame, B. (1965). Methods for the recovery of clostridia in foods. *J. appl. Bact.*, **28**, 95.

Gibbs, B. M., and Hirsch, A. (1956). Spore formation by *Clostridium* species in an artificial medium. *J. appl. Bact.*, **19**, 129.

Gibson, T. (1965). Clostridia in silage. *J. appl. Bact.*, **28**, 56.

Gibson, T., and Abd-el-Malek, Y. (1945). The formation of carbon dioxide by lactic acid bacteria and *Bacillus licheniformis* and a cultural method of detecting the process. *J. Dairy Res.*, **14**, 35.

Gilbert, R. J., Kendall, M., and Hobbs, B. C. (1969). Media for the isolation and enumeration of coagulase-positive staphylococci from foods. *In* "Isolation Methods for Microbiologists" (ed. D. A. Shapton and G. W. Gould), *Soc. appl. Bact. Technical Series*, *No. 3*, p. 9ff. London: Academic Press.

Gilbert, R. J., Stringer, M. F., and Peace, T. C. (1974). The survival and growth of *Bacillus cereus* in boiled and fried rice in relation to outbreaks of food poisoning. *J. Hyg.*, *Camb.*, **73**, 433.

Gillies, R. R. (1956). An evaluation of two composite media for preliminary identification of *Shigella* and *Salmonella*. *J. clin. Path.*, **9**, 368.

Gillespie, T. G. (1951). Estimation of sterilizing values of processes as applied to canned foods. I. Packs heating by conduction. *J. Sci. Fd. Agric.*, **2**, 107.

Gilman, J. C. (1957). "A Manual of Soil Fungi", 2nd Ed. Iowa: The Iowa State College Press.

Giolitti, G., and Cantoni, C. (1966). A medium for the isolation of staphylococci from foodstuffs. *J. appl. Bact.*, **29**, 395.

Goepfert, J. M., Spira, W. M., Glatz, B. A., and Kim, H. U. (1973). Pathogenicity of *Bacillus cereus*. *In* "The Microbiological Safety of Food" (ed. B. C. Hobbs and J. H. B. Christian), p. 69ff. London: Academic Press.

Goldenberg, N. (1964). Food hygiene: standards in manufacture, retailing and catering. *R. Soc. Hlth. J.*, **84**, 195.

Goldenberg, N., and Edmonds, G. (1973). Education in microbiological safety standards. *In* "The Microbiological Safety of Food" (ed. B. C. Hobbs and J. H. B. Christian), p. 435ff. London: Academic Press.

Goldenberg, N., and Elliott, D. W. (1973). The value of agreed non-legal specifications. *In* "The Microbiological Safety of Food" (ed. by B. C. Hobbs and J. H. B. Christian), p. 359ff. London: Academic Press.

Gomez, R. F., Sinskey, A. J., Davies, R., and Labuza, T. P. (1973). Minimal medium recovery of heated *Salmonella typhimurium* LT 2. *J. gen. Microbiol.*, **74**, 267.

Gordon, R. E., and Mihm, J. M. (1957). A comparative study of some strains received as Nocardiae. *J. Bact.*, **73**, 15.

Gordon, R. E., and Mihm, J. M. (1959). A comparison of four species of mycobacteria. *J. gen. Microbiol.*, **21**, 736.

Gordon, R. E., and Mihm, J. M. (1962). The type species of the genus *Nocardia*. *J. gen. Microbiol.*, **27**, 1.

Gordon, R. E., Haynes, W. C., and Pang, C. H.-N. (1973). "The Genus *Bacillus*". *Agriculture Handbook No. 427*. Washington: U.S. Dept. of Agriculture.

Gorrill, R. H., and McNeil, E. M. (1960). The effect of cold diluent on the viable count of *Pseudomonas pyocyanea*. *J. gen. Microbiol.*, **22**, 437.

Goudkov, A. V., and Sharpe, M. E. (1965). Clostridia in dairying. *J. appl. Bact.*, **28**, 63.

Gould, G. W. (1971). Methods for studying bacterial spores. *In* "Methods in Microbiology", Vol. 6A, (ed. J. R. Norris and D. W. Ribbons), p. 327ff. London: Academic Press.

Graham-Jones, O. (ed.) (1968). "Some Diseases of Animals Communicable to Man in Britain". Oxford: Pergamon Press.

Greene, R. A., and Larks, G. G. (1955). A quick method for the detection of gelatin-liquefying bacteria. *J. Bact.*, **69**, 224.

Gregory, P. H., Lacey, M. E., Festenstein, G. N., and Skinner, F. A. (1963). Microbial and biochemical changes during the moulding of hay. *J. gen. Microbiol.*, **33**, 147.

Grinsted, E., and Clegg, L. F. L. (1955). Spore-forming organisms in commercial sterilized milk. *J. Dairy Res.*, **22**, 178.

Gunther, H. L. (1959). Mode of division of pediococci. *Nature, Lond.*, **183**, 903.

Gunther, H. L., and White, H. R. (1961). The cultural and physiological characters of the pediococci. *J. gen. Microbiol.*, **26**, 185.

Hajna, A. A. (1955). A new enrichment broth medium for Gram-negative organisms of the intestinal group. *Public Health Laboratory*, **13**, 83.

Hajna, A. A., and Perry, C. A. (1943). Comparative study of presumptive and confirmative media for bacteria of the coliform group and for fecal streptococci. *Amer. J. Public Hlth.*, **33**, 550.

Hanks, J. H., and James, D. F. (1940). The enumeration of bacteria by the microscopic method. *J. Bact.*, **39**, 297.

Hannay, C. L., and Norton, I. L. (1947). Enumeration, isolation and study of faecal streptococci from river water. *Proc. Soc. appl. Bact.*, No. 1, p. 69.

Harrigan, W. F. (1966). The nutritional requirements and biochemical reactions of *Corynebacterium bovis*. *J. appl. Bact.*, **29**, 380.

Harrigan, W. F. (1967). "A study of coryneform bacteria from milk and other sources with particular reference to *Corynebacterium bovis*". Ph.D. thesis, University of Glasgow.

Harris, R. F., and Somers, L. E. (1968). Plate-dilution frequency technique for assay of microbial ecology. *Appl. Microbiol.*, **16**, 330.

Harrison, J. (1938). Numbers and types of bacteria in cheese. *Proc. Soc. agric. Bact.*, p. 12.

Harvey, R. W. S. (1965). "A study of the factors governing the isolation of salmonellae from infected materials and the application of improved techniques to epidemiological problems". M.D. thesis, University of Edinburgh. *Cited by* Harvey and Price (1974).

Harvey, R. W. S., and Phillips, W. P. (1961). An environmental survey of bakehouses and abattoirs for salmonellae. *J. Hyg., Camb.*, **59**, 93.

Harvey, R. W. S. and Price, T. H. (1961). An economical and rapid method for H antigen phase change in the Salmonella group. *Mon. Bull. Minist. Hlth.*, **20**, 11.

Harvey, R. W. S., and Price, T. H. (1967). The isolation of salmonellas from animal feeding stuffs. *J. Hyg., Camb.*, **65**, 237.

Harvey, R. W. S., and Price, T. H. (1968). Elevated temperature incubation of enrichment media for the isolation of salmonellas from heavily contaminated materials. *J. Hyg., Camb.*, **66**, 377.

Harvey, R. W. S., and Price, T. H. (1970). Sewer and drain swabbing as a means of investigating salmonellosis. *J. Hyg., Camb.*, **68**, 611.

Harvey, R. W. S., and Price, T. H. (1974). "Isolation of Salmonellas". *Public Health Laboratory Service Monograph Series*, No. 8. London: H.M.S.O.

Hastings, E. G. (1904). The action of various classes of bacteria on casein as shown by milk agar plates. *Zbl. Bakt. 2 Abt.*, **12**, 590.

Hauge, S. (1955). Food poisoning caused by aerobic spore-forming bacilli. *J. appl. Bact.*, **18**, 591.

Hayes, G. L., and Reister, D. W. (1952). The control of "off-odor" spoilage in frozen concentrated orange juice. *Fd Technol.*, **6**, 386.

Hayes, P. R. (1963). Studies on marine flavobacteria. *J. gen. Microbiol.*, **30**, 1.

Hayward, A. C. (1966). Methods of identification in the genus *Xanthomonas*. *In* "Identification Methods for Microbiologists, Part A" (ed. B. M. Gibbs and F. A. Skinner), *Soc. appl. Bact. Technical Series*, No. 1, p. 9ff. London: Academic Press.

Hendrie, M. S., Hodgkiss, W., and Shewan, J. M. (1970). The identification, taxonomy and classification of luminous bacteria. *J. gen. Microbiol.*, **64**, 151.

Henriksen, S. D. (1973). *Moraxella*, *Acinetobacter*, and the *Mimae*. *Bact. Revs.*, **37**, 522.

Hermann, J. E., and Cliver, D. O. (1968). Methods for detecting food-borne enteroviruses. *Appl. Microbiol.*, **16**, 1564.

Hersom, A. C., and Hulland, E. D. (1969). "Canned Foods (Baumgartner)", 6th Ed. London: J. A. Churchill Ltd.

Higginbottom, C. (1945). The technique of the bacteriological examination of dried milks. *J. Dairy Res.*, **14**, 184.

Higgins, M. (1950). A comparison of the recovery rate of organisms from cotton-wool and calcium alginate wool swabs. *Mon. Bull. Minist. Hlth.*, **9**, 50.

Hill, E. C., and Wenzel, F. W. (1957). The diacetyl test as an aid for quality control of citrus products. I. Detection of bacterial growth in orange juice during concentration. *Fd. Technol.*, **11**, 240.

Hirsch, A., and Grinsted, E. (1954). Methods for the growth and enumeration of anaerobic spore-formers from cheese, with observations on the effect of nisin. *J. Dairy Res.*, **21**, 101.

Holbrook, R., Anderson, J. M., and Baird-Parker, A. C. (1969). The performance of a stable version of Baird-Parker's medium for isolating *Staphylococcus aureus*. *J. appl. Bact.*, **32**, 187.

Holding, A. J. (1960). The properties and classification of the predominant Gram-negative bacteria occurring in soil. *J. appl. Bact.*, **23**, 515.

Howie, J. W. (1968). Typhoid in Aberdeen, 1964. *J. appl. Bact.*, **31**, 171.

Hugh, R., and Leifson, E. (1953). The taxonomic significance of fermentative versus oxidative metabolism of carbohydrate by various Gram negative bacteria. *J. Bact.*, **66**, 24.

Hungate, R. E. (1969). A roll-tube method for cultivation of strict anaerobes. *In* "Methods in Microbiology", Vol. 3B (ed. J. R. Norris and D. W. Ribbons), p. 117ff. London: Academic Press.

ICMSF (1974). "Micro-organisms in Foods. 2. Sampling for Microbiological Analysis: Principles and Specific Applications". Recommendations of the International Commission on Microbiological Specifications for Foods of the International Association of Microbiological Societies. Toronto: University of Toronto Press.

Ingram, M., and Kitchell, A. G. (1970). Introductory paper to a symposium on microbiological standards for foods. *Chemy Ind.*, p. 186.

Insalata, N. F., Dunlap, W. G., and Mahnke, C. W. (1973). Evaluation of the Salmonella Fluoro-kit for fluorescent antibody staining. *Appl. Microbiol.*, **25**, 202.

International Dairy Federation (1962a). FIL-IDF18:1962. "Standard Capacity Test for the Evaluation of the Disinfectant Activity of Dairy Disinfectants". Brussels, Belgium: International Dairy Federation.

International Dairy Federation (1962b). FIL-IDF19:1962. "Standard Suspension Test for the Evaluation of the Disinfectant Activity of Dairy Disinfectants". Brussels: International Dairy Federation.

International Dairy Federation (1964). FIL-IDF31:1964. "Count of Yeasts and Moulds in Butter". Brussels: International Dairy Federation.

International Dairy Federation (1966a). FIL-IDF39:1966. "Standard Routine Method for the Count of Coliform Bacteria in Raw Milk". Brussels: International Dairy Federation.

International Dairy Federation (1966b). FIL-IDF40:1966. "Standard Routine Method for the Count of Coliform Bacteria in Pasteurized Milk". Brussels: International Dairy Federation.

International Dairy Federation (1966c). FIL-IDF41:1966. "Standard Method for the Count of Lipolytic Organisms". Brussels: International Dairy Federation.

International Dairy Federation (1967). FIL-IDF44:1967. "Tube Test for the Evaluation of Detergent-Disinfectants for Dairy Equipment". Brussels: International Dairy Federation.

International Dairy Federation (1969). FIL-IDF50:1969. "Standard Methods for Sampling Milk and Milk Products". Brussels: International Dairy Federation.

International Dairy Federation (1970a). FIL-IDF49:1970. "Standard Method for Determining the Colony Count of Dried Milk and Whey Powder". Brussels: International Dairy Federation.

International Dairy Federation (1970b). FIL-IDF57:1970. "Detection of Penicillin in Milk by a Disk Assay Technique". Brussels: International Dairy Federation.

International Dairy Federation (1971a). FIL-IDF60:1971. "Detection of Coagulase Positive Staphylococci in Dried Milk". Brussels: International Dairy Federation.

International Dairy Federation (1971b). FIL-IDF62:1971. "Ice Cream and Milk Ices: Count of Coliform Bacteria". Brussels: International Dairy Federation.

International Dairy Federation (1971c). FIL-IDF64:1971. "Dried Milk and Dried Whey: Count of Coliforms." Brussels: International Dairy Federation.

International Dairy Federation (1971d). FIL-IDF65:1971. "Fermented Milks: Count of Coliforms." Brussels: International Dairy Federation.

International Dairy Federation (1971e). FIL-IDF66:1971. "Fermented Milks: Count of Microbial Contaminants". Brussels: International Dairy Federation.

International Dairy Federation (1971f). FIL-IDF67:1971. "Fermented Milks: Count of Yeasts and Moulds". Brussels: International Dairy Federation.

Jarvis, B. (1973). Comparison of an improved rose bengal-chlortetracycline agar with other media for the selective isolation and enumeration of moulds and yeasts in foods. *J. appl. Bact.*, **36**, 723,

Jarvis, B., and Moss, M. O. (1973). Bioassay methods for mycotoxins. In "Microbiological Safety of Food" (ed. B. C. Hobbs and J. H. B. Christian), p. 293ff. London: Academic Press.

Jayne-Williams, D. J., and Skerman, T. M. (1966). Comparative studies on coryneform bacteria from milk and dairy sources. *J. appl. Bact.*, **29**, 72.

Jeffries, L. (1969). Menaquinones in the classification of *Micrococcaceae*, with observations on the application of lysozyme and novobiocin sensitivity tests. *Internat. J. Syst. Bact.*, **19**, 183.

Johnson, E. A. (1946). An improved slide culture technique for the study and identification of pathogenic fungi. *J. Bact.*, **51**, 689.

Joint Committee (1972 *et seq.*). "Code of Practice for Assessment of Milk Quality", 2nd Ed. Joint Committee of the Milk Marketing Board and the Dairy Trade Federation: Thames Ditton, Surrey, England.

Jones, A., and Richards, T. (1952). Night blue and victoria blue as indicators in lipolysis media. *Proc. Soc. appl. Bact.*, **15**, 82.

Jones, B. D. (1972). "Methods of Aflatoxin Analysis". TPI Report No. G13. London: Tropical Products Institute.

Jones, D. (1975). A numerical taxonomic study of coryneform and related bacteria. *J. gen. Microbiol.*, **87**, 52.

Kaplan, M. M., Abdussalam, M. and Bijlenga, G. (1962). Diseases transmitted through milk. *In* "Milk Hygiene", *W.H.O. Monograph Series No. 48*, p. 11ff. Geneva: World Health Organization.

Keddie, R. M. (1951). The enumeration of lactobacilli on grass and in silage. *Proc. Soc. appl. Bact.*, **14**, 157.

Kelsey, J. C., Beeby, M. M., and Whitehouse, C. W. (1965). A capacity use-dilution test for disinfectants. *Mon. Bull. Minist. Hlth.*, **24**, 152.

Kelsey, J. C., and Sykes, G. (1969). A new test for the assessment of disinfectants with particular reference to their use in hospitals. *Pharmaceutical J.*, **202**, 607.

Kenner, B. A., Clark, H. F., and Kabler, P. W. (1961). Fecal streptococci. I. Cultivation and enumeration of streptococci in surface water. *Appl. Microbiol.*, **9**, 15.

King, E. O., Ward, M. K., and Raney, D. E. (1954). Two simple media for the demonstration of pyocyanin and fluorescin. *J. Lab. clin. Med.*, **44**, 301.

Kirsop, B. H., and Dolezil, L. (1975). Detection of lactobacilli in brewing. *In* "Lactic Acid Bacteria in Beverages and Food" (Ed. J. G. Carr, C. V. Cutting, and G. C. Whiting), Proc. 4th Long Ashton Symp., 1973. p. 159ff. London: Academic Press.

Kissinger, J. C. (1969). Modified resazurin test for estimating bacterial population in maple sap. *J. Assoc. offic. Analyt. Chem.*, **52**, 714.

Knight, B. C. J. G., and Proom, H. (1950). A comparative survey of the nutrition and physiology of mesophilic species in the genus *Bacillus. J. gen. Microbiol.*, **4**, 508.

Knisely, R. F. (1965). Differential media for the identification of *Bacillus anthracis. J. Bact.*, **90**, 1778.

Knott, F. A. (1951). "Memorandum on the principles and standards employed by the Worshipful Company of Fishmongers in the bacteriological control of shellfish in the London markets". London: Fishmongers' Company.

Knox, R., Gell, P. G. H., and Pollock, M. R. (1942). Selective media for organisms of the *Salmonella* group. *J. Path. Bact.*, **54**, 469.

Kohn, J. (1953). A preliminary report of a new gelatin liquefaction method. *J. clin. Path.*, **6**, 249.

Kohn, J. (1954). A two-tube technique for the identification of organisms of the Enterobacteriaceae group. *J. Path. Bact.*, **67**, 286.

Kovacs, N. (1956). Identification of *Pseudomonas pyocyanea* by the oxidase reaction. *Nature, Lond.*, **178**, 703.

Krogh, P. (ed.) (1973). "Control of Mycotoxins", special lectures of Symposium on the Control of Mycotoxins, Göteborg, Sweden, 1972. London: Butterworths.

Kusay, R. G. P. (1972). Preservation of micro-organisms. *Process Biochem.*, **7**, 24.

Laing, C. M., and Malcolm, J. F. (1956). The incidence of bovine mastitis with special reference to the non-specific condition. *Vet. Rec.*, **68**, 447.

Lapage, S. P., Shelton, J. E., and Mitchell, T. G. (1970). Media for the maintenance and preservation of bacteria. *In* "Methods in Microbiology", Vol. 3A (ed. J. R. Norris and D. W. Ribbons), p. 1ff. London: Academic Press.

Lapage, S. P., Shelton, J. E., Mitchell, T. G., and MacKenzie, A. R. (1970). Culture collections and the preservation of bacteria. *In* "Methods in Microbiology", Vol. 3A (ed. J. R. Norris and D. W. Ribbons), p. 135ff. London: Academic Press.

Lederberg, J., and Lederberg, E. M. (1952). Replica plating and indirect selection of bacterial mutants. *J. Bact.*, **63**, 399.

Leifson, E. (1951). Staining, shape, and arrangement of bacterial flagella. *J. Bact.*, **62**, 377.

Leifson, E. (1958). Timing of the Leifson flagella stain. *Stain Technol.*, **33**, 249.

Lepper, E., and Martin, C. J. (1929). The chemical mechanisms exploited in the use of meat media for the cultivation of anaerobes. *Br. J. exp. Path.*, **10**, 327.

Linton, A. H. (1961). Interpreting antibiotic sensitivity tests. *J. Medical Laboratory Technology*, **18**, 1.

Lloyd, T. P. (1969). Bacteriological control of ice-cream manufacture. Part 3. Interpretation of results: plant and personal hygiene. *Dairy Inds.*, **34**, 363.

Lodder, J. (1970). "The Yeasts: a Taxonomic Study". Amsterdam: North-Holland Publishing Co.

Lodder, J., and Kreger-van Rij, N. J. W. (1952). "The Yeasts – a Taxonomic Study". Amsterdam: North-Holland Publishing Co.

Lowe, G. H. (1962). The rapid detection of lactose fermentation in paracolon organisms by the demonstration of β-D-galactosidase. *J. med. Lab. Technol.*, **19**, 21.

Lück, H. (1972). Bacteriological quality tests for bulk-cooled milk. *Dairy Science Abstracts*, **34**, 101.

Mabbitt, L. A., and Zielinska, M. (1956). The use of a selective medium for the enumeration of lactobacilli in Cheddar cheese. *J. appl. Bact.*, **19**, 95.

MacKelvie, R. M., Gronlund, A. F., and Campbell, J. J. R. (1968). Influence of cold shock on the endogenous metabolism of *Pseudomonas aeruginosa*. *Canadian J. Microbiol.*, **14**, 633.

Mackenzie, E. F. W., Taylor, E. W., and Gilbert, W. E. (1948). Recent experiences in the rapid identification of *Bacterium coli* type 1. *J. gen. Microbiol.*, **2**, 197.

MacLeod, R. A. (1965). The question of the existence of specific marine bacteria. *Bact. Revs.*, **29**, 9.

Mallman, W. L., and Darby, C. W. (1941). Uses of lauryl sulphate tryptose broth for the detection of coliform organisms. *Am. J. publ. Hlth.*, **31**, 127.

Man, J. C. de, Rogosa, M., and Sharpe, M. E. (1960). A medium for the cultivation of lactobacilli. *J. appl. Bact.*, **23**, 130.

Marier, R., Wells, J. G., Swanson, R. C., Callahan, W., and Mehlman, I. J. (1973). An outbreak of enteropathogenic *Escherichia coli* food-borne disease traced to imported French cheese *Lancet*, **ii**, 1376.

Martin, J. P. (1950). Use of acid, rose bengal, and streptomycin in the plate method for estimating soil fungi. *Soil Sci.*, **69**, 215.

McCoy, J. H., and Spain, G. E. (1969). Bismuth sulphite media in the isolation of salmonellae. *In* "Isolation Methods for Microbiologists" (ed. D. A. Shapton and G. W. Gould), *Soc. appl. Bact. Technical Series*, No. 3, p. 17ff. London: Academic Press.

McKell, J., and Jones, D (1976). A numerical taxonomic study of *Proteus*—providence bacteria. *J. appl. Bact.*, **41**, 143.

McLean, J., and Henderson, A. (1966). Test for the presence of nitrite not involving carcinogenic reagents. *J. clin. Path.*, **19**, 632.

McMeekin, T. A., Patterson, J. T., and Murray, J. G. (1971). An initial approach to the taxonomy of some Gram negative yellow pigmented rods. *J. appl. Bact.*, **34**, 699.

McMeekin, T. A., Stewart, D. B., and Murray, J. G. (1972). The Adansonian taxonomy and the deoxyribonucleic acid base composition of some Gram negative, yellow pigmented rods. *J. appl. Bact.*, **35**, 129.

Meers, P. D., and Goode, D. (1965). The influence of preservative salts on the anaerobic growth of *Salmonella typhi*. *Mon. Bull. Minist. Hlth.*, **24**, 334.

Memorandum (1948). Ice-cream. Memorandum 43 (Tech.), Dept. of Health for Scotland. Edinburgh: H.M.S.O.

Meynell, G. G. (1958). The effect of sudden chilling on *Escherichia coli*. *J. gen. Microbiol.*, **19**, 380.

Meynell, G. G., and Meynell, E. (1970). "Theory and Practice in Experimental Bacteriology", 2nd Ed. Cambridge: The University Press.

Miles, A. A., and Misra, S. S. (1938). The estimation of the bactericidal power of the blood. *J. Hyg., Camb.*, **38**, 732.

Milk (Special Designation) Regulations (1963). Statutory Instruments No. 1571. London: H.M.S.O.

Mitchell, R. G., and Baird-Parker, A. C. (1967). Novobiocin resistance and the classification of staphylococci and micrococci. *J. appl. Bact.*, **30**, 251.

Moffett, M., Young, J. L., and Stuart, R. D. (1948). Centralized gonococcus culture for dispersed clinics. *Br. med. J*, **2**, 421.

Møller, V. (1955). Simplified tests for some amino acid decarboxylases and for the arginine dihydrolase system. *Acta path. microbiol. scand.*, **36**, 158.

Montie, T. R., Kadis, S., and Ajl, S. J. (eds.) (1970). "Microbial toxins", Vol. III. New York: Academic Press.

Moore, B. (1955). Streptococci and food poisoning. *J. appl. Bact.*, **18**, 606.

Morris, E. O., and Eddy, A. A. (1957). Method for the measurement of wild yeast infection in pitching yeast. *J. Inst. Brew.*, **63**, 34.

Mossel, D. A. A. (1957). The presumptive enumeration of lactose negative as well as lactose positive *Enterobacteriaceae* in foods. *Appl. Microbiol.*, **5**, 379.

Mossel, D. A. A. (1964). Essentials of the assessment of the hygienic condition of food factories and their products. *J. Sci. Fd. Agric.*, **15**, 349.

Mossel, D. A. A. (1967). Ecological principles and methodological aspects of the examination of foods and feeds for indicator micro-organisms. *J. Assoc. offic. Analyt. Chem.*, **50**, 91.

Mossel, D. A. A., van Diepen, H. M. J., and de Bruin, A. S. (1957). The enumeration of faecal streptococci in foods, using Packer's crystal violet sodium azide blood agar. *J. appl. Bact.*, **20**, 265.

Mossel, D. A. A., Koopman, M. J., and Jongerius, E. (1967). Enumeration of *Bacillus cereus* in foods. *Appl. Microbiol.*, **15**, 650.

Mossel, D. A., Mengerink, W. H. J., and Scholts, H. H. (1962). Use of a modified MacConkey agar medium for the selective growth and enumeration of Enterobacteriaceae. *J. Bact.*, **84**, 381.

Moussa, R. S. (1975). Evaluation of the API, the PathoTec, and the Improved Enterotube systems for the identification of Enterobacteriaceae. *In* "New Approaches to the Identification of Micro-organisms" (ed. C.-G. Heden and T. Illeni), p. 407ff. New York: John Wiley.

Muir, G. D. (ed.) (1971). "Hazards In the Chemical Laboratory". London: The Royal Institute of Chemistry.

Mulder, E. G. (1964). Iron bacteria, particularly those of the *Sphaerotilus-Leptothrix* group, and industrial problems. *J. appl. Bact.*, **27**, 151.

Muller, E. G. (1972). The sugar industry. *In* "Quality Control in the Food Industry", Vol. 3 (ed. S. M. Herschdoerfer), p. 229ff. London: Academic Press.

Mulvany, J. G. (1969). Membrane filter techniques in microbiology. *In* "Methods in Microbiology", Vol. 1 (ed. J. R. Norris and D. W. Ribbons), p. 205ff. London: Academic Press.

Murdock, D. I. (1968). Diacetyl test as a quality control tool in processing frozen concentrated orange juice. *Fd. Technol.*, **22**, 90.

Murdock, D. I., Folinazzo, J. F., and Troy, V. S. (1952). Evaluation of plating media for citrus concentrates. *Fd. Technol.*, **6**, 181.

Naylor, J., and Sharpe, M. E. (1958a). Lactobacilli in Cheddar cheese. I. The use of selective media for isolation and of serological typing for identification. *J. Dairy Res.*, **25**, 92.

Naylor, J., and Sharpe, M. E. (1958b). Lactobacilli in Cheddar cheese. III. The source of lactobacilli in cheese. *J. Dairy Res.*, **25**, 431.

Neal, C. E., and Calbert, H. E. (1955). The use of 2,3,5-triphenyltetrazolium chloride as a test for antibiotic substances in milk. *J. Dairy Sci.*, **38**, 629.

Newell, K. W. (1955). Outbreaks of paratyphoid B fever associated with imported frozen egg. I. Epidemiology. *J. appl. Bact.*, **18**, 462.

Niel, C. B. van (1928). "The Propionic Bacteria". Haarlem: N. V. Uitgeverszaak J. W. Boissevain.

Niel, C. B. van (1955). Natural selection in the microbial world. *J. gen. Microbiol.*, **13**, 201.

Norris, J. R., and Ribbons, D. W. (1970). "Methods in Microbiology", Vol. 2. London: Academic Press.

North, W. R. (1961a). Lactose pre-enrichment method for isolation of Salmonella from dried egg albumen. *Appl. Microbiol.*, **9**, 188.

North, W. R. (1961b). Use of crystal violet or brilliant green dyes for the determination of salmonellae in dried food products. *J. Bact.*, **80**, 861.

North, W. R., and Bartram, M. T. (1953). The efficiency of selenite broth of different compositions in the isolation of *Salmonella*. *Appl. Microbiol.*, **1**, 130.

Oakley, C. L. (1956). The classification and biochemical activities of the genus *Clostridium*. *J. appl. Bact.*, **19**, 112.

Oakley, C. L. (1971). Antigen-antibody reactions in microbiology. *In* "Methods in Microbiology", Vol. 5A (ed. J. R. Norris and D. W. Ribbons) p. 173ff. London: Academic Press.

O'Meara, R. A. Q. (1931). A simple and delicate and rapid method of detecting the formation of acetylmethylcarbinol by bacteria fermenting carbohydrates. *J. Path. Bact.*, **34**, 401.

Onishi, H., McCance, M. E., and Gibbons, N. E. (1965). A synthetic medium for extremely halophilic bacteria. *Canad. J. Microbiol.*, **11**, 365.

Orla-Jensen, S. (1943). "The Lactic Acid Bacteria" (Erganzungsband). Copenhagen: Ejnar Munksgaard.

Ormay, L., and Novotny, T. (1969). The significance of *Bacillus cereus* food poisoning in Hungary. *In* "The Microbiology of Dried Foods", *Proceedings of 6th Internat. Symp. on Fd. Microbiol., 1968*. Published by International Association of Microbiological Societies; printed by Grafische Industrie, Haarlem, The Netherlands.

Orr, M. J., McLarty, R. M., McCance, M. E., and Baines, S. (1964). Alternate day collection of bulk milk. *Dairy Inds*, **29**, 169.

Packer, R. A. (1943). The use of sodium azide (NaN_3) and crystal violet in a selective medium for streptococci and *Erysipelothrix rhusiopathiae*. *J. Bact.*, **46**, 343.

Palleroni, N. J., and Doudoroff, M. (1972). Some properties and taxonomic subdivisions of the genus *Pseudomonas*. *Ann. Rev. Phytopath.*, **10**, 73.

Park, R. W. A. (1967). A comparison of two methods for detecting attack on glucose by pseudomonads and achromobacters. *J. gen. Microbiol.*, **46**, 355.

Park, R. W. A., and Holding, A. J. (1966). Identification of some common Gram-negative bacteria. *Lab. Pract.*, **15**, 1124.

Parker, M. T. (1962). Phage-typing and the epidemiology of *Staphylococcus aureus* infection. *J. appl. Bact.*, **25**, 389.

Paton, A. M., and Gibson, T. (1953). The use of hydrogenated fats in tests for the detection of microbiological lipases. *Proc. Soc. appl. Bact.*, **16**, iii.

Patterson, J. T. (1969). Salmonellae in meat and poultry, poultry plant cooling waters and effluents and animal feeding stuffs. *J. appl. Bact.*, **32**, 329.

Patton, J. (1950). Bacteriological testing of ice-cream in Northern Ireland. *Proc. Soc. appl. Bact.*, **13**, 100.

Payne, W. J. (1973). Reduction of nitrogenous oxides by micro-organisms. *Bact. Rev.*, **37**, 409.

Perry, L. B. (1973). Gliding motility in some non-spreading flexibacteria. *J. appl. Bact.*, **36**, 227.

Pike, E. B. (1962). The classification of staphylococci and micrococci from the human mouth. *J. appl. Bact.*, **25**, 448.

Poelma, P. L. (1968). Recommended changes in the method for detection and identification of *Salmonella* in egg products. *J. Assoc. offic. Anal. Chem.*, **51**, 870.

Post, F. J., and Krishnamurty, G. B. (1964). Suggested modifications of the calcium alginate swab technique. *J. Milk Fd. Technol.*, **27**, 62.

Post, F. J., Krishnamurty, G. B., and Flanagan, M. D. (1963). Influence of sodium hexametaphosphate on selected bacteria. *Appl. Microbiol.*, **11**, 430.

Powell, E. O. (1963). Photometric methods in bacteriology. *J. Sci. Fd. Agric.*, **14**, 1.

Proom, H., and Knight, B. C. J. G. (1955). The minimum nutritional requirements of some species in the genus *Bacillus*. *J. gen. Microbiol.*, **13**, 474.

Quesnel, L. B. (1971). Microscopy and micrometry. *In* "Methods in Microbiology", Vol. 5A (ed. J. R. Norris and D. W. Ribbons) p. 1ff. London: Academic Press.

Raper, K. B., and Fennell, D. I. (1965). "The Genus *Aspergillus*". Baltimore: Williams and Wilkins.

Raper, K. B., and Thom, C. (1949). "A Manual of the Penicillia". Baltimore: Williams and Wilkins.

Recommendations (1965). Recommendations of the Subcommittee on taxonomy of staphylococci and micrococci. *Int. Bull. bact. Nomencl. Taxon.*, **15**, 109.

Reed, G., and Peppler, H. J. (1973). "Yeast Technology". Westport, Connecticut: Avi Publishing Co.

Reiter, B., and Møller-Madsen, A. (1963). Reviews of the progress of dairy science. Section B. Cheese and dairy starters. *J. Dairy Res.*, **30**, 419.

Report (1947). The bacteriological examination and grading of ice-cream. *Mon. Bull. Minist. Hlth.*, **6**, 60.
Report (1950). The bacteriological examination and grading of ice-cream. *Mon. Bull. Minist. Hlth.*, **9**, 231.
Report (1956). "The Bacteriological Examination of Water Supplies", 3rd. Ed. Report of the Public Health Laboratory Service Water Committee. *Reports on Public and Medical Subjects*, No. 71. London: H.M.S.O.
Report (1958a). Report of the Enterobacteriaceae Subcommittee of the Nomenclature Committee of the International Association of Microbiological Societies. *Int. Bull. bact. Nomencl. Taxon.*, **8**, 25.
Report (1958b). The bacteriological examination and grading of fresh cream. *Mon. Bull. Minist. Hlth.*, **17**, 77.
Report (1959). Sterilisation by steam under increased pressure. *Lancet*, **i**, 425.
Report (1964). "The Aberdeen Typhoid Outbreak, 1964". Report of the Committee of enquiry. Cmnd. 2542. Edinburgh: H.M.S.O.
Report (1969). "The Bacteriological Examination of Water Supplies", 4th Ed. Report of the Public Health Laboratory Service Standing Committee on the Bacteriological Examination of Water Supplies. *Reports on Public Health and Medical Subjects*, No. 71. London: H.M.S.O.
Report (1971). The hygiene and marketing of fresh cream as assessed by the methylene blue test. A report by a Working Party to the Director of the Public Health Laboratory Service. *J. Hyg., Camb.*, **69**, 155.
Report (1972). A comparative assessment of media for the isolation and enumeration of coagulase staphylococci from foods: a report from a Working Party of the Public Health Laboratory Service. *J. appl. Bact.*, **35**, 673.
Report (1974). "On the state of the public health": the Annual Report of the Chief Medical Officer of the Department of Health and Social Security for the Year 1973. London: H.M.S.O.
Rhodes, M. E. (1958). The cytology of *Pseudomonas* spp. as revealed by a silver-plating method. *J. gen. Microbiol.*, **18**, 639.
Rice, E. B. (1962). Hygienic control of dairy equipment. *In* "W.H.O. Monograph Series No. 48: Milk Hygiene", p. 457ff. Geneva: World Health Organization.
Robertson, D. S. F. (1970). Selenium – a possible teratogen. *Lancet*, **i**, 518.
Rogosa, M., Mitchell, J. A., and Wiseman, R. F. (1951). A selective medium for the isolation and enumeration of oral and fecal streptococci. *J. Bact.*, **62**, 132.
Rosenberger, R. F. (1956). The isolation and cultivation of obligate anaerobes from silage. *J. appl. Bact.*, **19**, 173.
Rosenthal, S. L. (1974). A simplified method for single carbon source tests with *Pseudomonas* species. *J. appl. Bact.*, **37**, 437.
Rubbo, S. D., and Gardner, J. F. (1965). "A Review of Sterilization and Disinfection, as Applied to Medical, Industrial and Laboratory Practice". London: Lloyd-Luke (Medical Books).
Scarr, M. P. (1959). Selective media used in the microbiological examination of sugar products. *J. Sci. Fd. Agric.*, **10**, 678.
Scholefield, J. (1964). "A comparison of the biochemical activities of psychrophilic bacteria". M.Sc. thesis, University of Leeds.

Shapton, D. A., and Board, R. G. (eds) (1972). "Safety in Microbiology". *Society for Appl. Bact. Technical Series*, No. 6. London: Academic Press.

Sharpe, A. N. (1973). Automation and instrumentation developments for the bacteriology laboratory. *In* "Sampling – Microbiological Monitoring of Environments" (ed. R. G. Board and D. W. Lovelock), p. 197ff. *Soc. appl. Bact. Technical Series*, No. 7. London: Academic Press.

Sharpe, A. N., and Jackson, A. K. (1972). Stomaching: a new concept in bacteriological sample preparation. *Appl. Microbiol.*, **24**, 175.

Sharpe, A. N., and Kilsby, D. C. (1971). A rapid, inexpensive bacterial count technique using agar droplets. *J. appl. Bact.*, **34**, 435.

Sharpe, M. E. (1962). Taxonomy of the lactobacilli. *Dairy Sci. Abstr.*, **24**, 109.

Sharpe, M. E., and Fryer, T. F. (1966). Identification of the lactic acid bacteria. *In* "Identification Methods for Microbiologists, Part A" (ed. B. M. Gibbs and F. A. Skinner) p. 65ff. London: Academic Press.

Sharpe, M. E., Neave, F. K., and Reiter, B. (1962). Staphylococci and micrococci associated with dairying. *J. appl. Bact.*, **25**, 403.

Shattock, P. M. F., and Hirsch, A. (1947). A liquid medium buffered at pH 9.6 for the differentiation of *Streptococcus faecalis* from *Streptococcus lactis*. *J. Path. Bact.*, **59**, 495.

Shaw, C., and Clarke, P. H. (1955). Biochemical classification of Proteus and Providence cultures. *J. gen. Microbiol.*, **13**, 155.

Sherman, J. M. (1937). The streptococci. *Bact. Rev.*, **1**, 3.

Shewan, J. M. (1970). Bacteriological standards for fish and fishery products. *Chemy Inds*, (**2**), 193.

Shewan, J. M. (1971). The microbiology of fish and fishery products – a progress report. *J. appl. Bact.*, **34**, 299.

Shewan, J. M., and Hobbs, G. (1967). The bacteriology of fish spoilage and preservation. *Progr. Indust. Microbiol.*, **6**, 169.

Shorrock, C., and Ford, J. E. (1973). An improved procedure for the determination of available lysine and methionine with *Tetrahymena*. *In* "Proteins in Human Nutrition" (ed. J. W. G. Porter and B. A. Rolls), p. 207ff. London: Academic Press.

Shuval, H. I., and Katzenelson, E. (1972). The detection of enteric viruses in the water environment. In "Water Pollution Microbiology" (ed. R. Mitchell), p. 347ff. New York: Wiley-Interscience.

Sierra, G. (1957). A simple method for the detection of lipolytic activity of micro-organisms and some observations on the influence of the contact between cells and fatty substrates. *Antonie van Leeuwenhoek*, **23**, 15.

Sierra, G. (1964). Hydrolysis of triglycerides by a bacterial proteolytic enzyme. *Can. J. Microbiol.*, **10**, 926.

Silva, G. A. N. da, and Holt, J. G. (1965). Numerical taxonomy of certain coryne-form bacteria. *J. Bact.*, **90**, 921.

Silverstolpe, L., Plazikowski, U., Kjellander, J., and Vahlne, G. (1961). An epidemic among infants caused by *Salmonella muenchen*. *J. appl. Bact.*, **24**, 134.

Silverton, R. E., and Anderson, M. J. (1961). "Handbook of Medical Laboratory Formulae". London: Butterworths.

Skean, J. D., and Overcast, W. W. (1962). Another medium for enumerating citrate-fermenting bacteria in lactic cultures. *J. Dairy Sci.*, **45**, 1530.

Skerman, V. B. D. (1959). "A Guide to the Identification of the Genera of Bacteria". Baltimore: Williams and Wilkins.

Slanetz, L. W., Chichester, C. O., Gaufin, A. R., and Ordal, Z. J. (eds) (1963). "Microbiological Quality of Foods". New York: Academic Press.

Smith, B. A., and Baird-Parker, A. C. (1964). The use of sulphamezathine for inhibiting *Proteus* spp. on Baird-Parker's isolation medium for *Staphylococcus aureus*. *J. appl. Bact.*, **27**, 78.

Smith, D. G. (1975). Inhibition of swarming in *Proteus* spp. by tannic acid. *J. appl. Bact.*, **38**, 29.

Smith, G. (1969). "An Introduction to Industrial Mycology", 6th Ed. London: Edward Arnold.

Smith, N. R., Gordon, R. E., and Clark, F. E. (1946). "Aerobic mesophilic spore-forming bacteria". U.S. Dept. Agric. Misc. Publ. No. 559. Washington: U.S. Dept. of Agriculture.

Smith, N. R., Gordon, R. E., and Clark, F. E. (1952). "Aerobic spore-forming bacteria". U.S. Dept. Agric. Monograph No. 16. Washington: U.S. Dept. of Agriculture.

Sneath, P. H. A. (1974). Test reproducibility in relation to identification. *Int. J. Systematic Bacteriol.*, **24**, 508.

Sneath, P. H. A., and Collins, V. G. (eds) (1974). A study in test reproducibility between laboratories: Report of a Pseudomonas Working Party. *Antonie van Leeuwenhoek*, **40**, 481.

Society of American Bacteriologists (1957). "Manual of Microbiological Methods". New York: McGraw-Hill.

Sojka, W. J. (1965). "*Escherichia coli* in animals". Review Series No. 7 of the Commonwealth Bureau of Animal Health, Weybridge. Farnham Royal: Commonwealth Agricultural Bureaux.

Stableforth, A. W., and Galloway, I. A. (eds) (1959). "Infectious Diseases of Animals: Diseases due to Bacteria", 2 vols. London: Butterworths.

Stamer, J. R., Albury, M. N., and Pederson, C. S. (1964). Substitution of manganese for tomato juice in the cultivation of lactic acid bacteria. *Appl. Microbiol.*, **12**, 165.

Stanier, R. Y., Palleroni, N. J., and Doudoroff, M. (1966). The aerobic pseudomonads: a taxonomic study. *J. gen. Microbiol.*, **43**, 159.

Starr, M. P., and Stephens, W. L. (1964). Pigmentation and taxonomy of the genus *Xanthomonas*. *J. Bact.*, **87**, 293.

Steel, K. J. (1961). The oxidase reaction as a taxonomic tool. *J. gen. Microbiol.*, **25**, 297.

Steel, K. J. (1962). The oxidase activity of staphylococci. *J. appl. Bact.*, **25**, 445.

Steiner, E. H. (1967). Statistical methods in quality control. *In* "Quality Control in the Food Industry", Vol. 1 (ed. S. M. Herschdoerfer), London: Academic Press.

Stephens, R. L. (1970). Bacteriological standards for foods: a retailer's point of view. *Chemy Ind.*, p. 220.

Stevens, W. L. (1958). Dilution series: a statistical test of technique. *J. R. statist. Soc.*, **B20**, 205.

Stevenson, I. L. (1963). Some observations on the so-called "cystites" of the genus *Arthrobacter. Can. J. Microbiol.*, **9**, 467.

Stott, J. A., and Smith, H. (1966). Microbiological assay of protein quality with *Tetrahymena pyriformis* W. *Brit. J. Nutr.*, **20**, 663.

Stott, J. A., Smith, H., and Rosen, G. D. (1963). Microbiological evaluation of protein quality with *Tetrahymena pyriformis* W. 3. A simplified assay procedure. *Brit. J. Nutr.*, **17**, 227.

Straka, R. P., and Stokes, J. L. (1957a). A rapid method for the estimation of the bacterial content of precooked frozen foods. *Food Res.*, **22**, 412.

Straka, R. P., and Stokes, J. L. (1957b). Rapid destruction of bacteria in commonly used diluents and its elimination. *Appl. Microbiol.*, **5**, 21.

Straka, R. P., and Stokes, J. L. (1959). Metabolic injury to bacteria at low temperature. *J. Bact.*, **78**, 181.

Strange, R. E., and Dark, F. A. (1962). Effect of chilling on *Aerobacter aerogenes* in aqueous suspension. *J. gen. Microbiol.*, **29**, 719.

Stuart, R. D., Toshach, S. R., and Patsula, T. M. (1954). The problem of transport of specimens for culture of gonococci. *Can. J. publ. Hlth*, **2**, 73.

Stumbo, C. R. (1973). "Thermobacteriology in Food Processing", 2nd Ed. New York: Academic Press.

Sullivan, R., and Read, R. B. (1968). Method for recovery of viruses from milk and milk products. *J. Dairy Sci.*, **51**, 1748.

Sykes, G. (1962). The philosophy of the evaluation of disinfectants and antiseptics. *J. appl. Bact.*, **25**, 1.

Sykes, G. (1965). "Disinfection and Sterilization", 2nd Ed. London: Spon.

Symposium on Myxobacteria and Flavobacteria (1969). *J. appl. Bact.*, **32**, 1–67

Taylor, E. W., Burman, N. P., and Oliver, C. W. (1955). Membrane filtration technique applied to the routine bacteriological examination of water. *J. Instn Wat. Engrs*, **9**, 248.

Taylor, J. (1961). Host specificity and enteropathogenicity of *Escherichia coli. J. appl. Bact.*, **24**, 316.

Taylor, J. (1966). Host-parasite relations of *Escherichia coli* in man. *J. appl. Bact.*, **29**, 1.

Taylor, M. M. (1975). The water agar test: a new test to measure the bacteriological quality of cream. *J. Hyg., Camb.*, **74**, 345.

Taylor, W. I. (1965). Isolation of shigellae. I. Xylose lysine agars; new media for isolation of enteric pathogens. *Amer. J. clin. Path.*, **44**, 471.

Taylor, W. I., and Achanzar, D. (1972). Catalase test as an aid to the identification of Enterobacteriaceae. *Appl. Microbiol.*, **24**, 58.

Taylor, W. I., and Harris, B. (1965). Isolation of shigellae. II. Comparison of plating media and enrichment broths. *Amer. J. clin. Path.*, **44**, 476.

Taylor, W. I., and Schelhart, D. (1967). Isolation of shigellae. IV. Comparison of plating media with stools. *Amer. J. clin. Path.*, **48**, 356.

Ten Cate, L. (1965). A note on a simple and rapid method of bacteriological sampling by means of agar sausages. *J. appl. Bact.*, **28**, 221.

Thatcher, F. S., and Clark, D. S. (eds) (1968). "Micro-organisms in Foods: their Significance and Methods of Enumeration". Recommendations of the International Committee on Microbiological Specifications for Foods, a Standing Committee of the International Association of Microbiological Societies. Toronto: University of Toronto Press.

Thomas, M. (1961). The sticky film method of detecting skin staphylococci. *Mon. Bull. Minist. Hlth*, **20**, 37.

Thomas, M. (1966). Bacterial penetration in raw meats. Comparisons using a new technique. *Mon Bull. Minist. Hlth*, **25**, 42.

Thomas, S. B. (1969). Methods of assessing the psychrotrophic bacterial content of milk. *J. appl. Bact.*, **32**, 269.

Thornley, M. J. (1960). The differentiation of *Pseudomonas* from other Gram-negative bacteria on the basis of arginine metabolism. *J. appl. Bact.*, **23**, 37.

Thornley, M. J. (1967). A taxonomic study of *Acinetobacter* and related genera. *J. gen. Microbiol.*, **9**, 211.

Tomlins, R. I., and Ordal, Z. J. (1976). Thermal injury and inactivation in vegetative bacteria. *In* "Inhibition and Inactivation of Vegetative Microbes". *Soc. appl. Bact. Symp. Series*, No. 5 (in press). London: Academic Press.

Trust, T. J. (1975). Bacteria associated with the gills of salmonid fishes in freshwater. *J. appl. Bact.*, **38**, 225.

Turner, N., Sandine, W. E., Elliker, P. R., and Day, E. A. (1963). Use of tetrazolium dyes in an agar medium for the differentiation of *Streptococcus lactis* and *Streptococcus cremoris*. *J. Dairy Sci.*, **46**, 380.

Tuttlebee, J. W. (1975). The Stomacher – its use for homogenization in food microbiology. *J. Fd Technol.*, **10**, 113.

Varnam, A. H., and Grainger, J. M. (1972). Enumeration of certain lactic acid bacteria from Wiltshire bacon curing brines. *J. Sci. Fd Agric.*, **23**, 546.

Varnam, A. H., and Grainger, J. M. (1973). Methods for the general microbiological examination of Wiltshire bacon curing brines. *In* "Sampling – Microbiological Monitoring of Environments" (ed. R. G. Board and D. W. Lovelock), p. 29ff. *Soc. appl. Bact. Technical Series*, No. 7. London: Academic Press.

Varnam, A. H., and Grainger, J. M. (1975). The nature of the stimulatory effect of pork extract on the growth of bacteria of Wiltshire bacon curing brines. *J. appl. Bact.*, **39**, (3), vii.

Vernon, E., and Tillett, H. E. (1972). Food poisoning and Salmonella infections in England and Wales, 1969–1972. *Public Health, Lond.*, **88**, 225.

Waes, G. (1968). The enumeration ot aromabacteria in B D starters. *Ned. Melk-en Zuiveltijdschr.*, **22**, 29.

Walker, W. (1965). The Aberdeen typhoid outbreak of 1964. *Scot. med. J.*, **10**, 466.

Weibull, C. (1960). Movement. *In* "The Bacteria: A Treatise on Structure and Function", Vol. 1. Structure. (ed. I. C. Gunsalus and R. Y. Stanier), p. 153ff. New York: Academic Press.

Wheater, D. M. (1955). The characteristics of *Lactobacillus plantarum*, *L. helveticus* and *L. casei*. *J. gen. Microbiol.*, **12**, 133.

Whittenbury, R. (1963). The use of soft agar in the study of conditions affecting the utilization of fermentable substrates by lactic acid bacteria. *J. gen. Microbiol.*, **32**, 375.

Whittenbury, R. (1964). Hydrogen peroxide formation and catalase activity in the lactic acid bacteria. *J. gen. Microbiol.*, **35**, 13.

Wickerham, L. J., Flickinger, M. H., and Burton, K. A. (1946). A modification of Henrici's vegetable-juice sporulation medium for yeasts. *J. Bact.*, **52**, 611.

Wilkinson, J. F. (1958). The extracellular polysaccharides of bacteria. *Bact. Rev.*, **22**, 46.

Williams, H. A. (1968). The detection of rot in tomato products. *J. Assoc. Public Analysts*, **6**, 69.

Williams, S. T., Davies, F. L., and Cross, T. (1968). Identification of genera of the Actinomycetales. *In* "Identification Methods for Microbiologists, Part B" (ed. B. M. Gibbs and D. A. Shapton), p. 111ff. *Soc. appl. Bact. Technical Series*, No. 2. London: Academic Press.

Willis, A. T. (1962). Some diagnostic reactions of clostridia. *Lab. Pract.*, **11**, 526.

Willis, A. T. (1965). Media for clostridia. *Lab. Pract.*, **14**, 690.

Willis, A. T. (1969). Techniques for the study of anaerobic spore-forming bacteria. *In* "Methods in Microbiology", Vol. 3B (ed. J. R. Norris and D. W. Ribbons), p. 79ff. London: Academic Press.

Wilson, M. M., and MacKenzie, E. F. (1955). Typhoid fever and salmonellosis due to the consumption of infected desiccated coconut. *J. appl. Bact.*, **18**, 510.

Wilson, G. S. (1922). The proportion of viable bacteria in young cultures with especial reference to the technique employed in counting. *J. Bact.*, **7**, 405.

Wilson, G. S. (1935). "The bacteriological grading of milk". Medical Research Council Special Report Series, No. 206. London: H.M.S.O.

Wilson, G. S., and Miles, A. A. (1975). "Topley and Wilson's Principles of Bacteriology, Virology and Immunity", 6th Ed. London: Edward Arnold.

Wilson, J. M., and Davies, R. (1976). Minimal medium recovery of thermally injured *Salmonella seftenberg* 4969. *J. appl. Bact.*, **40**, 365.

Wilson, W. J. (1938). Isolation of *Bact. typhosum* by means of bismuth sulphite medium in water- and milk-borne epidemics. *J. Hyg., Camb.*, **38**, 507.

Woodward, R. L. (1957). How probable is the most probable number? *J. Amer. Water Works Assoc.*, **49**, 1060.

Wright, R. C., and Tramer, J. (1961). The estimation of penicillin in milk. *J. Soc. Dairy Technol.*, **14**, 85.

SUBJECT INDEX

B

M

N

U

V

W